From Airline Reservations to Sonic the Hedgehog

History of Computing

I. Bernard Cohen and William Aspray, editors

William Aspray, *John von Neumann and the Origins of Modern Computing*

Charles J. Bashe, Lyle R. Johnson, John H. Palmer, and Emerson W. Pugh, *IBM's Early Computers*

Martin Campbell-Kelly, *From Airline Reservations to Sonic the Hedgehog: A History of the Software Industry*

Paul E. Ceruzzi, *A History of Modern Computing*

I. Bernard Cohen, *Howard Aiken: Portrait of a Computer Pioneer*

I. Bernard Cohen and Gregory W. Welch, editors, *Makin' Numbers: Howard Aiken and the Computer*

John Hendry, *Innovating for Failure: Government Policy and the Early British Computer Industry*

Michael Lindgren, *Glory and Failure: The Difference Engines of Johann Müller, Charles Babbage, and Georg and Edvard Scheutz*

David E. Lundstrom, *A Few Good Men from Univac*

René Moreau, *The Computer Comes of Age: The People, the Hardware, and the Software*

Emerson W. Pugh, *Building IBM: Shaping an Industry and Its Technology*

Emerson W. Pugh, *Memories That Shaped an Industry*

Emerson W. Pugh, Lyle R. Johnson, and John H. Palmer, *IBM's 360 and Early 370 Systems*

Kent C. Redmond and Thomas M. Smith, *From Whirlwind to MITRE: The R&D Story of the SAGE Air Defense Computer*

Raúl Rojas and Ulf Hashagen, editors, *The First Computers—History and Architectures*

Dorothy Stein, *Ada: A Life and a Legacy*

John N. Vardalas, *The Computer Revolution in Canada: Building National Technological Competence*

Maurice V. Wilkes, *Memoirs of a Computer Pioneer*

From Airline Reservations to Sonic the Hedgehog
A History of the Software Industry

Martin Campbell-Kelly

The MIT Press
Cambridge, Massachusetts
London, England

© 2003 Massachusetts Institute of Technology

All rights reserved. No part of this book may be reproduced in any form by any electronic or mechanical means (including photocopying, recording, or information storage and retrieval) without permission in writing from the publisher.

Set in New Baskerville by The MIT Press. Printed and bound in the United States of America.

Library of Congress Cataloging-in-Publication Data

Campbell-Kelly, Martin.
From airline reservations to Sonic the Hedgehog : a history of the software industry / Martin Campbell-Kelly.
p. cm. — (History of computing)
Includes bibliographical references and index.
ISBN 0-262-03303-8 (hc. : alk. paper)
1. Computer software industry—History. I. Title. II. Series.
HD9696.63.A2 C35 2003
338.4'70053—dc21   2002075351

10 9 8 7 6 5 4 3 2

Contents

Preface vii
Acknowledgments xiii

1
The Software Industry 1

2
Origins of the Software Contractor, the 1950s 29

3
Programming Services, the 1960s 57

4
Origins of the Software Products Industry, 1965–1970 89

5
The Shaping of the Software Products Industry, the 1970s 121

6
The Maturing of the Corporate Software Products Industry, 1980–1995 165

7
Early Development of the Personal Computer Software Industry, 1975–1983 201

8
Not Only Microsoft: The Maturing of the Personal Computer Software Industry, 1983–1995   *231*

9
Home and Recreational Software   *269*

10
Reflections on the Success of the US Software Industry   *303*

Notes   *313*
Sources of Chapter Frontispieces   *347*
Bibliography   *349*
Index   *361*

Preface

In 1997, when I first began to plan a book on the history of the software industry, software was already a significant factor in my life. However, while I knew in principle that software is an important enabling technology of the modern economy, from moment to moment I was unaware of its presence. This is probably true for most people. As is true of the plumbing beneath the city streets, it takes an effort of will to bring software to conscious attention. Indeed, one of the great fascinations of software is its invisibility and intangibility.

On Monday, July 7, 1997, I made a conscious effort to reflect on my software interactions for a day. That Monday was otherwise a perfectly typical though unusually sunny day. For a British academic, it was one of those pleasant days at the beginning of summer when the demands of undergraduates have abated and working life takes on a new rhythm of scholarly research and writing. After a leisurely breakfast to celebrate the first day of the long vacation, I drove to the campus at 9:30, well after the peak-hour traffic had died down, stopping at the recycling center on the way to deposit a month's accumulation of newspapers and a disturbingly large collection of wine bottles.

My first task at work was to revise an article I was writing for a newspaper, using my aging but trusty office PC. I printed a copy of the article and faxed it to the commissioning editor. Having got that out of the way, I grabbed a coffee and checked first my letter post and then my e-mail. The letter post consisted mainly of publishers' brochures, and none of it required a reply. The e-mail, however, brought a dozen personal communications, about half from inside the university and half from colleagues in Britain and overseas. Several of the messages required detailed answers, which took nearly an hour.

Most of the rest of the day was taken up by meetings and by various social exchanges, such as reviewing the work of research students and gossiping on the turn of the staircase. I spent a full hour in the library (increasingly a vacation-only indulgence) catching up on back numbers of the *Harvard Business Review* and checking out some novels to read on a forthcoming holiday. On the way back from the library, I used an ATM.

At about 4 P.M., before leaving for the day, I decided to book flights for a September conference in Florence. The previous year, I would have had to entrust the entire operation to my travel agent, and I was never too sure how thoroughly she had searched for the lowest fare or the most convenient times. However, for the last several months it had been possible to obtain scheduled fight information on the Internet. I therefore fired up my Netscape browser and logged onto EasySabre, a publicly available flight information system. After I spent a few minutes feeding it information about when and where, EasySabre came up with a list of available flights. I then phoned my travel agent, who was able to give me pricing information for the various flights; together we made a selection, and she made the booking for me. I also took out some travel insurance and paid with my Visa card.

I hit the peak-hour traffic on my return journey home. However, the traffic lights were with me—as I went into the home stretch, red lights flicked to green as I approached each intersection, and I was home in 10 minutes. "Honey, I'm home" I didn't say as I unlocked the front door—partly because my partner was still at work, but mainly because only Americans say that, and probably only in movies. There was a Post-It note stuck to the inside of the front door asking me to ring my son to arrange to collect him from college—he had just acquired a cell phone; I hadn't and felt sure I never would. That's the generation gap. Upstairs I could hear my youngest son in mortal combat—he was playing a computer game called Doom. I don't play videogames, and I dare say I never will.

During this very ordinary day, I used software on many occasions, consciously and unconsciously. To begin at the beginning, I first used software unconsciously when I started my car. Modern automobiles contain tiny computers that look after functions such as engine management and controlling the radio. This type of software usually is produced by the auto or radio manufacturer or under contract by a specialist software firm.

Typically I spend two or more hours a day working directly on a computer, composing articles and lecture notes, answering e-mail, and—increasingly—browsing the World Wide Web. Ninety percent of desktop computers are "IBM compatibles" that run the Microsoft Windows operating system and Microsoft Office software, and most members of the Macintosh minority also use Microsoft software for office tasks. Although

Microsoft dominates the desktop software industry, there are many other players, some of whom will feature in this book. They all have histories that stretch back to the 1970s or the early 1980s. The transformation these firms have wrought on the office scene has been remarkable, although it is easy to overlook. In 20 years, the typewriter has been consigned to the dustbin of history.

In my university's library, the transition to the information age is far less complete. For example, in 1997 it was still necessary to read back numbers of the *Harvard Business Review* in hard-copy form. Some publishers had started to produce electronic versions, but the transformation was (and is) only at its very beginning. However, much of the library's back-office work has been computerized for 15 years. When I checked out my books, bar codes on my readers' ticket and on the books were scanned, and computerized records were adjusted. The market for university library systems has never been sufficiently remunerative to attract major industrial players. This is evident from the rag-bag of non-compatible and somewhat archaic and unreliable systems readers have to deal with. The system my university uses is produced by a nonprofit consortium funded by a group of British universities. Besides employing programmers, the organization makes use of software tools and packages from various suppliers, most of them American. My encounter with an ATM was quite different. Since the early 1970s these electronic money boxes have mushroomed around the world, and most of them use a piece of software plumbing called CICS, made by IBM, to connect with their home bank's information system. Although most people are blissfully unaware of CICS, they probably make use of it several times a week, for almost every commercial electronic transaction they make. In the whole scheme of things, CICS is *much* more important than Microsoft Windows.

Before there were ATMs, there were airline reservation systems. From the very beginning of the commercial use of computers, airline reservations was *the* path-breaking application. When I used the EasySabre system, I was rummaging in the very basement of software history. EasySabre is a user-friendly connection to Sabre, perhaps the most venerable piece of software. It began as a collaboration between IBM and American Airlines in the mid 1950s, but eventually became a business in its own right. Today it is a huge computer services organization with 6,000 workers in 45 countries. But note how even this most sophisticated software system did not do the whole job. To actually make a booking, I had to use a travel agent to negotiate the system. This was not a restrictive trade practice. Dealing with an airline reservation system was a complicated business requiring considerable training. The travel clerk also interacted with two further information systems: those of an insurance broker and a credit card company.

More software encounters were to come that day. I guess we have all had that happy experience of cooperating traffic lights. What we have encountered is a traffic management system, whose function is to maximize traffic throughput. Again, when we use the telephone system, we harness software that took thousands of programmer-years to write. My son's mobile phone contained 300,000 computer instructions etched on a tiny memory chip. And the sounds of Doom coming from upstairs were produced by a hot-in-1997 videogame made by ID Software. It seems that software has a generational component. Most people of my parents' generation, silver surfers apart, have little direct contact with software, while working individuals of my generation use software a great deal of the time in the office environment. Our children, and other people under 30, seem immersed in the stuff for work and play.

Software is commercially produced by thousands of firms. This book is the history of the firms that make this software. It is an account of the interplay of myriad suppliers. Some (such as Microsoft) everyone has heard of; others (such as the high-ranking software firms SAP and Computer Associates) few outside the computer industry have heard of.

This book, however, is not a sales pitch for the software industry, which constantly falls below our expectations. And software, it should be noted, is ubiquitous rather than all-embracing; it is a part of life, but only a small part. Consider the things I did that Monday in which software played no part. It didn't help me unload my waste for recycling (though the recycling agency undoubtedly uses information technology in many of its operations). And though I spent 2 hours in front of a computer screen, most of the day was spent in social exchanges and information transactions in which the computer played very little part at all. Back home, while one son was playing Doom, the other was reading a book. And that Post-It on the front door was a convenience for which software has provided no substitute.

This book will take the story of the software industry up to about 1995. There are very good reasons for this, apart from the historian's professional reluctance to write about very recent events. Software is very much a work in progress, and anything written today is likely to look foolish in 5 years' time. Even since my software day in July 1997, things have moved on. For example, the newspaper I write for has moved into the software age, and now I can submit copy by e-mail instead of by fax. Increasingly I use e-mail at home as well as in the office (and I have some misgivings about the merging of my home and university life). I no longer use a travel agent; now I book and pay directly on the Internet—a miracle of software integration that still takes my breath away. Microsoft has been in and out of favor, and in and out of the antitrust courts. I have more or less given up on Netscape and switched to Microsoft's Internet Explorer. To my chagrin I have acquired a cell phone, but I have not yet succumbed to videogames. And, yes, we still have yellow Post-Its stuck to the front door.

Acknowledgments

Many individuals and organizations helped me in producing this book. It is a particular pleasure to acknowledge the help of my friends at the Charles Babbage Institute, at the University of Minnesota: Director Arthur Norberg and Associate Director Jeff Yost allowed me to participate in the organization of a conference on The Emergence of the Software Product (Palo Alto, September 2000) and to slightly subvert the agenda to my own ends. Archivists Bruce Brummer and Beth Kaplan facilitated my access to the Institute's holdings. Luanne Johnson and Burton Grad of the Software History Center, in Benicia, California, have been very supportive of my work, particularly by allowing me to participate in reunion meetings of software industry pioneers. The Information Technology Association of America, in Washington, granted me access to the records of its forerunner, the Association of Data Processing Service Organizations. Peter Cunningham, founder and president of INPUT, granted me access to statistical data and industry reports and guided me through the minefield of software industry statistics. Ulf Hashagen and his colleagues at the Heinz Nixdorf MuseumsForum, in Paderborn, Germany, allowed me to participate in the organizing of a conference on Mapping the History of Computing—Software Issues (Paderborn, April 2000). Terry Gourvish, director of the Business History Unit of the London School of Economics, encouraged me with a visiting fellowship.

I am greatly indebted to my research fellows at Warwick University, Mary Croarken and Ross Hamilton. Additional help was provided by several senior undergraduates who researched and wrote dissertations on various aspects of the software industry: Lauren Gray, Jonathan Flatt, Tracy Kimberley, Andrew Kwok, Khilan Shah, and Debbie White. The following individuals provided information and/or commented substantially on early drafts: William Aspray of the Computing Research Association; Walter Bauer, former president of Informatics; Paul Cerruzi of the Smithsonian Institution; James Cortada, senior consultant at IBM; Robert Damuth, economist and consultant to the Business Software Alliance; Marty Goetz, former president of ADR; David Mounce of IBM UK's Hursley Laboratory.

The book was largely written using IBM's ViaVoice voice-recognition software. Like all software products, it seems to improve year on year, and by release 6, when this book was started, its accuracy was (just) sufficiently encouraging. I owe a special debt of gratitude to Burton Grad, who checked the manuscript with an industry analyst's eye for detail, eliminating ViaVoice's unintentional interlocutions and homonyms and my somewhat less numerous factual errors. As always, the remaining errors lie at my doorstep.

My thanks to Larry Cohen, acquisitions editor at The MIT Press. Finally, I wish to thank my wife, Jane, and my sons, George, Dave, and Rob, for putting up with the inevitable clutter and occasional bad temper that accompanies a writing project.

The research for the book was funded by the Economic and Social Research Council of Great Britain under award R000237065.

From Airline Reservations to Sonic the Hedgehog

Our Motto: 124 seats, 124 tickets.

Having the number of passengers and the number of seats come out the same has not always been easy.

(Major airlines have ticket offices all over the country. To err was inevitable.)

So we got together with IBM to make sure the seat we reserve for you is reserved for you.

And you can imagine what a collaboration like this would come up with. A computer. (A giant that took 10 years and 30 million dollars to develop. We call it Sabre.)

It not only "memorizes" every seat on every flight we have—it also memorizes the name and address of everybody on a waiting basis.

The moment there's a cancellation, it tells us you're next on the list—and even gives us your number to call.

(And it doesn't wait until we ask. It barges right in and tells us there's an empty seat on Flight 61 and to get hold of Paul Zoellner in Riverdale, New York.)

There it is. 124 seats. 124 tickets. Want one?

American Airlines

The SABRE airline reservation system was a path-breaking application that gave American Airlines a great commercial advantage over competitors.

1
The Software Industry

In January 1952, half a century ago, *Fortune* magazine ran an article titled "Office Robots." It was one of the very first general articles on computers. No mention was made of programming or software—indeed, the latter term had not yet been invented. However, the general-purpose nature of the computer was well understood. Having described the use of the Univac computer at the US Census Bureau the writer explained: "At the flip of a few switches, UNIVAC can be turned from such mass statistical manipulations to solving differential equations for scientists or handling payroll lists, computations, and check writing for businessmen."[1] The slightly inappropriate metaphor "flipping a few switches" suggests that programming was either something *Fortune*'s writer did not understand or something he thought his readers did not need to know about.

It would be nearly 15 years before a major business magazine devoted a feature article to software. In November 1966 *Business Week* carried a report titled "Software Gap—A Growing Crisis for Computers."[2] The article bemoaned the shortage of programmers, but hinted at the glorious opportunities ahead for the nascent software industry. For almost another 15 years, however, the software industry remained a hidden world—known mainly to computer-industry professionals, investors, and analysts. Not until September 1980 did *Business Week* carry a special report on the software business, its first in-depth look at the industry since 1966. That this article appeared at all was a tribute to the Association of Data Processing Service Organizations (ADAPSO), the trade association of the US software industry, whose Image Committee had worked tirelessly behind the scenes to get the industry noticed by the media. Titled "Missing Computer Software," the 1980 report (like its 1966 predecessor) highlighted the shortages of software applications and programmers and trumpeted the recent spectacular growth of the industry.[3] The article is especially interesting now because, like a fly in amber, it is caught in a time when computing still meant "big iron" and the only kind of software most business people knew about was bought for huge sums by corporations. Yet even as this article was going to press the personal computer was changing everyone's perception of information processing.

Business Week's next special report on the software industry appeared in February 1984, and it could not have been more different in tone from the 1980 article.[4] In the intervening 3½ years, the rise of the personal computer had made the world at large aware of software, and brand names such as VisiCalc, WordStar, and Lotus 1-2-3 had entered the lingua franca of many office workers. There was no longer any need for *Business Week* to explain as it had in 1980—like a kindergarten teacher explaining to the class—that software was "the long lists of commands or instructions that tell the computer what to do."[5] Instead, the 1984 article, titled "Software: The New Driving Force," spoke of a $10 billion industry with boundless opportunities.[6] No longer was software only for corporations; now it was in the shopping malls too.

Writing at the very end of the twentieth century, the authors of a book titled *Secrets of Software Success* claimed: "Life without software is hard to imagine. Without software, paper letters would be the fastest form of written correspondence. No fax, no e-mail, and no business voice mail. But that's just the beginning of the impact of software. Across industries, software now enables and fuels economic growth. . . . Software tasks today range from controlling nuclear power plants, recognizing customer purchasing patterns, enabling stock trading, and running banking systems all the way to running cell phone systems and exploring for oil."[7] Warming to their theme, the authors continued: "Software—nothing but pure knowledge in codified form—largely drives and enables today's economy."

If the writers seem somewhat hyperbolic, they should be forgiven. From its first glimmerings in the 1950s, the software industry has evolved to become the fourth largest industrial sector in the US economy.[8] This book is the story of that evolution.

Understanding the Software Industry

Although today most people are aware of the software industry, not many would claim to "understand" it. In contrast, most people have a sense of knowing, say, the automobile industry—they are familiar with its products, they know or can envision the production processes, and they understand the links between producer and consumer. Perhaps this understanding is naive and illusory, for beneath the surface there is a fantastically intricate set of industrial networks. However, in recent times no one has felt the need to "explain" the automobile industry. The same is true of most other producer industries, whether they be chemicals, airplanes, building materials, or food.

Yet when it comes to software, people are much less comfortable. This is due in part to the intangible nature of software, evocatively described by one prominent software scientist as "only slightly removed from pure thought-stuff."[9] But it is also attributable to the fact that traditional industries have been around for so long that we have unconsciously internalized a great deal of knowledge about them. The software industry is relatively new. Twenty-five years ago it was invisible and unacknowledged; today it is ubiquitous.

The aim of this book is to explain the software industry by a historical account of its evolution. Because no simple one-dimensional framework is adequate for this purpose, I use three main vectors of explanation. The first vector is that of time—the historical development and periodization of the industry. This vector informs the whole structure of the book, which traces the evolution of the industry from its first glimmerings in the mid 1950s to the mid 1990s in a series of partially overlapping but chronologically progressing narratives. The second vector of explanation is the sectorization of the industry, which can be divided into three main types of firm: software contractors, producers of corporate software products, and makers of mass-market software products. The third vector is that of products and markets. Software comes in many prices, sizes, and genres; sometimes one copy is sold, sometimes 100, sometimes 10 million. Clearly this range of possibilities leads to a significant variety that suggests an explanation though classification or taxonomy.[10]

Periodization, Sectorization, and Capabilities

The software industry can be divided into three sectors: software contracting, corporate software products, and mass-market software products.[11] Each of these three sectors emerged at a moment when contemporary computer technology created a business opportunity for a new mode of software delivery. Rather neatly (though purely coincidentally), the three sectors arrived at intervals of a decade.

Software contracting developed alongside the corporate mainframe computer in the mid 1950s. A software contractor wrote a one-of-a-kind program for a corporate or government customer. Custom-written programs were hugely expensive, $1 million being not untypical.

Corporate software products emerged after the launch of the IBM System/360 computer family in the mid 1960s. The new IBM computer was relatively inexpensive, sold in large numbers, and thus created a much broader market for lower-cost software than could ever have been satisfied by software contractors. A software product was a program that could be used without modification by a large number of corporate users. Software products typically automated common business functions, such as payroll or inventory management, or ran an entire medium-size business, such as a manufacturing operation or a savings bank. They were typically priced between $5,000 and $100,000. The more successful ones sold in the hundreds, and a few in the thousands.

The arrival of the personal computer in the mid 1970s created an opportunity for mass-market software. The most characteristic form of distribution was a shrink-wrapped box of software sold in a retail store or by mail order. Software for personal computers was relatively cheap (typically between $100 and $500) and sold in large volumes, often several hundred thousand copies. In parallel with the personal computer revolution, there was a revolution in software-based home entertainment. Entertainment software was a major subsector of the mass-market software industry.

The terminology used to describe each of the three sectors is somewhat problematic, or at least ahistorical, because all three sectors have continued to flourish since their inception and have adopted the preferred terminology of the day. For example, "software contracting," which began in 1955, pre-dated the invention of the word "software"; it originally went by such names as "custom programming" and "programming services." Similarly, the first pre-packaged programs were simply called "software products," no further distinction being necessary. With the rise of the personal computer software industry, it became necessary to distinguish between the markets for corporate software and personal computer software by introducing terms such as "enterprise software" and "shrink-wrapped software."

The division of the software industry into three sectors is natural both in market terms and in terms of the distinctive business models that firms evolved. The software firms' competencies and their knowledge of their specialized markets enabled the more successful firms to maintain dominant positions in their own sector but made it difficult for them to cross over into either of the other sectors. Thus, the very strengths that enabled a firm to succeed in one market segment became institutional rigidities in another. This is the main reason why few firms have successfully escaped the confines of their particular sector.[12]

Software Contractors

The defining event for the software contracting industry came in 1956, when the US-government-owned RAND Corporation created the Systems Development Corporation (SDC) to develop the computer programs for the huge SAGE air defense project. This was the first of several multi-billion-dollar defense projects in the 1950s and the 1960s, known as the L-Systems, that provided an important market for early software contractors. At the same time, computer manufacturers and private corporations were also creating a demand for software, albeit on a smaller scale. In response to the latter demand, small startup firms such as the Computer Usage Company (CUC) and the Computer Sciences Corporation (CSC) came into existence. These firms ultimately developed into major corporations that competed successfully for the largest software contracts.

The business model consciously or unconsciously adopted by custom programming firms was that of an engineering or construction contractor. They existed by bidding for and winning contracts executed on a time-and-materials basis or a fixed-price basis. The critical capabilities for a software contracting firm were exploitation of scope, cost estimation, and project management. A successful software contractor exploited the economies of scope by specializing in particular sub-markets. For example, SDC specialized in real-time defense projects, while CSC focused on systems software for computer manufacturers. By concentrating on these narrow markets, firms could reduce costs by reusing software from one project in the next and could develop specialized human resources by working in a consistent application domain. Specialized domestic knowledge enabled non-American firms to survive against multi-national competitors. The profits on software contracting were surprisingly low, typically less than 15 percent of sales, so cost-estimation and project-management skills were essential. Accurate cost estimation was needed to prepare a price-competitive bid, and project-management skills were needed to ensure completion within time and cost constraints. In contrast, marketing was a relatively unimportant competence, since most of the selling was done through the personal contacts of senior staff members or by responding to openly published requests for quotation.

Corporate Software Products

Two packaged programs, Applied Data Research's Autoflow and Informatics' Mark IV (announced in 1965 and 1967, respectively), are generally agreed to be, if not the first, certainly the most influential of the early software products. These products and a few others had already proved viable in the market in January 1970, when IBM implemented its "unbundling" decision. Previously, IBM had provided programs free of charge to customers on request, as had the other computer manufacturers. This made it difficult for software entrepreneurs to establish a market. Therefore, the software products that succeeded were ones that satisfied needs not yet anticipated by the computer manufacturers. Under antitrust pressure (perhaps assisted by an independent lawsuit from Applied Data Research), IBM decided to charge separately for software and other services. Unbundling had the effect of establishing a vibrant market for software products, which previously had been merely embryonic. It was a turning point for the industry.

At first, because of the analogy between the low incremental costs of reproducing programs and recorded music, the software products business was likened to the recorded-music industry. This turned out to be an illusion. Because of their high marketing costs and the need for sales support, corporate software products were classic capital goods. Thus, the business model adopted by the software products firms, often quite consciously, was that of a producer of capital goods—and the firms often looked to computer manufacturers, particularly IBM, for role models. The critical capabilities that the firms developed were exploitation of scale, corporate marketing, quality assurance, and pre- and after-sale support. Exploitation of scale was the most important of these capabilities, because selling in volume was the only way to recover the high initial development costs of a generalized software product, which were much higher than for custom software. Because sales volume was so important, it was necessary to develop quota-based sales operations, typically on the IBM model, and for this reason firms often recruited former IBM salespeople. Software products, such as database programs or industrial applications, were usually "mission critical," and for this reason product reliability was paramount. The software firms developed skills in quality assurance, using such techniques as beta testing to ensure that programs were ruggedly "productized" and reliable in use. Finally, as with all capital goods, pre- and after-sale support was needed to establish a long-term relationship with the customer. In the case of software products, this took the forms of product customization, user training, and regular upgrades.

These services turned out to be unexpected sources of income for which the pioneers of the industry had not initially planned.

Mass-Market Software Products

The personal computer software industry began in the late 1970s with the establishment of hundreds of very small software firms, almost none of which had any connection with the existing software industry. Microsoft is one of the few firms from this early period to have survived. The industry really took off in 1979–80, with the arrival of mass-market software such as Software Arts' VisiCalc spreadsheet and MicroPro's WordStar word processor. In many popular histories, VisiCalc is credited as the "killer app" that kicked off the personal computer revolution. The concept of the "killer app" is attractive and superficially plausible, but it has been neither taken up nor refuted by academic economists. The view taken in this book is that the "killer app" hypothesis probably confuses cause and effect. Thus, one could argue that the personal computer established a platform on which many software products could exist, and that VisiCalc was simply a prominent example. Had VisiCalc not existed, the personal computer revolution would still have happened, and perhaps another software product would have earned the epithet "killer app."

Along with VisiCalc and WordStar came a slew of popular products that became, if not exactly household names, certainly well-recognized brands: Supercalc, Lotus 1-2-3, dBase II, WordPerfect, and many others. Closely related to the personal computer software industry (whose products were used in corporations and homes) was the recreational software industry (whose products were used exclusively in domestic and learning environments). Recreational software products tended to be cheaper (typically $50) and more ephemeral. Early firms active in the recreational software industry included Activision and Broderbund.

The personal computer software products industry was completely disjoint from the corporate software products industry. The essential difference between the two was that their markets differed by two orders of magnitude in numbers of units sold. For example, in 1984 the world's best-selling corporate software product was Informatics' Mark IV, with 3,000 installations; the best-selling personal computer software product was WordStar, with sales of 700,000.

The analogy to the recorded-music business held, and the business model adopted by the industry was that of a producer of information goods. Another parallel was drawn with the pharmaceutical industry, which had a similar cost structure based on high R&D inputs, low production costs, and high marketing expenses.

The critical capabilities developed by the personal computer software firms included exploitation of scale, mass marketing, and ease of use. Exploitation of scale through high volumes was the defining characteristic of the mass-market software industry, whose product cost structure was entirely different from that of the corporate software industry. For example, the cost of a Mark IV installation was about $100,000, while WordStar cost $495. Personal computer software firms targeted their products at the end user rather than at the corporate information systems manager, making use of low-cost distribution channels such as retail outlets and mail order. This required the development of a set of marketing competencies much different from that of corporate software firms, with their IBM-type sales forces. In time, personal computer software firms with strong corporate sales, such as Lotus and Microsoft, developed conventional sales forces. The sales messages, as expressed in advertising, continued to be largely directed toward end users, however.

So that personal computer software products could be used by many thousands of customers without any after-sale support, programs had to have intuitive interfaces and had to require no customization. This again required the development of a set of skills different from that of corporate software makers, who could rely on training courses and third parties to install and customize software.

Figure 1.1 illustrates the three-sector structure of the software industry, populated with a few of the firms that will feature in this book. Though this view of the industry is widely accepted by academics and industry analysts, it is an artificial construction designed to bring some coherence to what would otherwise be a star field of random firms. Though the great majority of firms can be unequivocally placed in a particular sector, this is not true of every firm. Most of the software contractors established in the 1950s and the 1960s subsequently sold software products in the 1960s and the 1970s as a subsidiary activity. For example, Informatics began as software contractor in 1962, and programming services remained its primary business, even though its Mark IV became the biggest-selling independent software product in the industry. Hence, Informatics can be properly located in both software contracting and corporate software products. Such multiple activities were usually reflected in a firm's organizational structure. In Informatics, for example, there was a separate "Software Products Division" that ran the Mark IV operation. Later in the development

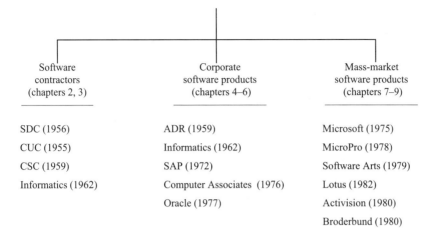

Figure 1.1
A taxonomy of the software industry.

of the industry, producers of corporate software, such as Computer Associates and Oracle, attempted to establish personal computer software operations.

An inestimable advantage of structuring this book around the three-sector model is that it ensures an allocation of space that bears some relation to the importance of any given subject. Thus, Microsoft, the leading player in personal computer software, gets a rightful and prominent place in the chapters devoted to mass-market software, but nowhere else in the book. Microsoft deservedly dominates about 10 percent of the book, just as it dominates 10 percent of the software industry. If this book serves no other purpose, I hope it will at least provide a corrective to the common misconception that Microsoft is the center of the software universe around which all else revolves.

Scope: Exclusions and Limitations

Packaged software often has a list of exclusions and limitations that, when read carefully, explain to a disappointed user why the software does not do the very thing he or she bought it for. In the spirit of exclusion and limitation, I should declare up front what this book is not about.

The most severe limitation is that the book has a strong US focus. To some extent this is justified in a single-volume work. The United States dominates the world's software industry, especially in software products.

However, other countries do have indigenous software industries, and readers may feel that these countries have been slighted. In fact, I am British, I wrote this book in Britain with only occasional visits to the United States, and I probably have a better knowledge of the British software industry than of the American. However, I felt unable to incorporate much material on the British software industry in this book because it would have been disproportionate, would have appeared chauvinistic, and would have raised this question: What about the software industries of Germany, France, Japan, Israel, Ireland, the former Eastern Bloc countries, and so on? All these countries certainly deserve attention, but I think this would be best done by a set of monographic studies written from the viewpoint of the individual software nation. Indeed, there is a fine study of the Indian software industry by my colleague Richard Heeks at Manchester University.[13] In 1996, when Heeks published *India's Software Industry*, the exports of the entire Indian software industry were estimated at about $700 million. At the time, the total revenues of Microsoft were $9 billion and those of Computer Associates were $3.5 billion, about half from overseas sales. Thus, India's entire software exports were less than the software exports of any of the top five US producers, and only a fraction of 1 percent of world output. Purely on a space argument, it is hard to make a case for a long discussion of Indian software here. Many industry observers foresee a glittering future for the Indian software industry, but it is a complex industry with its own rich history, its own economic and political shaping, and its own structure (for example, it has a distinctive "offshore" software writing sector). Indeed, it is so different from the American software industry that it might almost be on another planet. The Japanese software industry is completely different again, with a strong software contracting sector but a weak software products sector (with the exception of videogames).[14] I believe that the same is true of the other nations, and one could not begin to incorporate these different histories into a single volume except in the most superficial way. These other software nations all deserve to be studied on their merits, not as a set of potted plants set out for the inspection of the hurried Western reader.

A second limitation of this book is the cutoff date of 1995. Naturally, historians have a professional reluctance to write about very recent events on which they lack a proper perspective, so I have no fear of criticism from other historians on that score. However, any self-respecting industry analyst or software journalist would bring the story up to date and would, for good measure, project a few years into the future. This involves a set of skills different from that of the historian. It is not mere pusillanimity that makes me reluctant to attempt to do the same, but the fact that such projections are often wrong and therefore that contemporary obsessions often miss the real drama and turning points. For example, in the last 5 years there has been an enormous amount of press coverage of the Java programming language, the Linux operating system, and open-source software. I have no idea whether these will turn out to be turning points in the industry or not, and my opinion is certainly no better than the average pundit's. On the other hand, I find it quite fascinating that in the business press of the early 1990s the Internet was one of the least-written-about subjects, getting perhaps one-tenth the column inches devoted to Microsoft Windows or the tribulations of WordPerfect. I don't know what it is, but I bet there is something much more important going on right now than Java, Linux, or open-source software, and that it will be 2010 before it becomes fully apparent.

The year 1995 also seems a good cutoff point because there is a sense that in the mid 1990s, with the rise of the Internet, the software industry entered a new phase of its development. For example, the electronic delivery of software has made the metaphor of "shrink wrapped" software inappropriate. More significantly, software firms appear to be extending their reach as the boundaries between corporate and consumer software become more diffuse. Thus, since 1995, Microsoft has increasingly strayed from its traditional desktop software into corporate networks at one end of the spectrum and into videogame consoles at the other. Likewise, Oracle, once an archetypal producer of corporate software, has been making forays into desktop software and video entertainment. In *Secrets of Software Success,* Detlev Hoch and his co-authors project the "Internet Era," a new period in the development of the software industry, and suggest boundary dates of 1994 and 2008.[15] My expectation is that we will indeed, as historians, move toward such a periodization in due time. However, my aim in this book is to set out a history that is robust enough to make sense 10 or 15 years from now.

Numbers: Software Industry Statistics and Trade Associations

Recently I read an article in the business press that referred to "the $300 billion software industry." Three years of researching the software industry had made me cautious about interpreting such statements. Before I became immersed in the subject, I had naively assumed that making software was not much different from making photocopiers, refrigerators, or automobiles. I assumed it would be possible to determine the number sold, the total revenues of the industry, and the value of the market. And, allowing for distribution costs, it would be possible to reconcile these numbers. Unfortunately, this is just not possible for the software industry. Indeed, it is not even possible to find out the total revenues of the industry from any public-domain source of which I am aware. Here are some of the reasons.

According to *Software Magazine*, the sales of the top 500 software firms in the year 2000 amounted to $259 billion. However, the authors of *Secrets of Software Success* cite two estimates for the total number of firms in the software industry worldwide, one stating that there are 35,000 firms with more than five employees and the other stating that there are 150,000 firms "regardless of size."[16] Take your pick. The US Census Bureau stated that in 1997 there were 12,000 software publishing "establishments" and a further 31,000 establishments engaged in programming services in the United States. Thus, the figure given by *Software Magazine*—a well-regarded proxy for industry size—speaks only for the biggest firms; it leaves out many thousands of small firms. This has always been the case. In the late 1970s, analysts computed revenues for the top 50 firms, then the top 100, then the top 200, then the top 500.

Another problem with measuring the software industry is that it is in one respect (though only in one respect) like the chemical industry: its products are sometimes consumed within the industry and sometimes by end users. (You might occasionally buy a couple of pounds of washing soda, but when did you last buy a retail gallon of sulfuric acid?) For example, a major activity in the software industry is the installation of Enterprise Resource Planning (ERP) software made by firms such as SAP and Oracle. When SAP or Oracle uses its in-house personnel to install the software for an end user, the combined cost of software and consulting services shows up in its year-end revenues. However, when a computer services firm such as EDS or CSC undertakes the same activity, the cost of the ERP software bought in from SAP or Oracle is passed on to the end user together with the installation charges, and the original ERP software appears in the accounting books of both firms and adds to their year-end revenues. This is not an isolated example. Today most software products are constructed using at least some tools or components bought from other software vendors, or some development activity may be subcontracted to a specialist software house. Because of this "double counting," total sales to end-users is arguably a better measure of the software industry than total industry revenues.[17]

Table 1.1 presents data on the US software market in the period 1970–2000. I believe this is the only 30-year time series for the software industry in the public domain, and it appears here courtesy of the industry analyst INPUT. INPUT has a rigorous, proprietary methodology—which I have seen but cannot disclose—based on questionnaires and interviews with end users and subsequent reconciliation with vendors.[18] Table 1.1 covers only the US market, because INPUT did not begin to capture worldwide data until much later. However, the value of table 1.1 is the trend. The time series gives a real insight into the growth of the industry. This is demonstrated visually in figure 1.2, which shows the classic "hockey stick" growth of the industry—growth that has not yet begun to flatten. Figure 1.3 reveals some interesting subplots. It shows that software products became the dominant mode of US of software consumption around 1980, and that the growth of programming services then began to slow, relatively. Figure 1.3 also shows that software products did not really take off until the early 1980s, long after IBM liberated the industry by unbundling.

There are two main time series for US software industry revenues in the public domain. One series (which lapsed in 1990) was published by ADAPSO; statistics supplied by INPUT were used in its later years. The other series is produced at taxpayers' expense by the US Census Bureau.

Table 1.2 gives the ADAPSO-INPUT industry statistics for the years 1966–1990.[19] This is the only long-term time series for US software industry revenues in the public domain. ADAPSO, founded in 1961, became the leading trade association for the US computer services and software industries.[20] In its first 25 years of existence, the number of member firms grew from about 40 to 850; these tended to be the larger firms, and between them they represented approximately half the sales of the industry. ADAPSO was like other trade associations in that its role was to promote the industry's interests through lobbying, public relations activities, education, standards setting, and so on. In 1991 it was renamed the Information Technology Association of America (ITAA).

ADAPSO's many activities were complemented by a program to gather statistics on the industry. Annual industry surveys were published from 1966 to 1990. Early on, before the emergence of professional software industry analysts, ADAPSO made use of an academic consultant. Beginning in 1970, it employed the International Data Corporation (IDC) and Quantum Sciences, two early providers of information on the computer industry. At first, ADAPSO simply tracked the two main classes of industry participant: processing services firms and software houses.

Table 1.1
User expenditures, US software market.

	Systems software products	Applications software products	Total software products	Programming services	Total software
1970	$150,000,000	$100,000,000	$250,000,000	$744,000,000	$994,000,000
1971	$210,000,000	$140,000,000	$350,000,000	$856,000,000	$1,206,000,000
1972	$270,000,000	$170,000,000	$440,000,000	$952,000,000	$1,392,000,000
1973	$330,000,000	$210,000,000	$540,000,000	$1,072,000,000	$1,612,000,000
1974	$390,000,000	$270,000,000	$660,000,000	$1,200,000,000	$1,860,000,000
1975	$490,000,000	$320,000,000	$810,000,000	$1,352,000,000	$2,162,000,000
1976	$590,000,000	$390,000,000	$980,000,000	$1,528,000,000	$2,508,000,000
1977	$720,000,000	$480,000,000	$1,200,000,000	$1,728,000,000	$2,928,000,000
1978	$890,000,000	$590,000,000	$1,480,000,000	$1,984,000,000	$3,464,000,000
1979	$1,150,000,000	$720,000,000	$1,870,000,000	$2,346,000,000	$4,216,000,000
1980	$1,401,000,000	$1,325,000,000	$2,726,000,000	$2,985,000,000	$5,711,000,000
1981	$1,974,000,000	$2,191,000,000	$4,165,000,000	$3,994,000,000	$8,159,000,000
1982	$2,685,000,000	$3,080,000,000	$5,765,000,000	$4,335,000,000	$10,100,000,000
1983	$3,534,000,000	$4,168,000,000	$7,702,000,000	$5,023,000,000	$12,725,000,000
1984	$5,333,000,000	$5,741,000,000	$11,074,000,000	$5,387,000,000	$16,461,000,000
1985	$6,322,000,000	$6,964,000,000	$13,286,000,000	$6,233,000,000	$19,519,000,000
1986	$8,022,000,000	$9,015,000,000	$17,037,000,000	$6,833,000,000	$23,870,000,000

Year				
1987	$9,880,000,000	$10,670,000,000	$20,550,000,000	$7,540,000,000	$28,090,000,000
1988	$12,095,000,000	$12,970,000,000	$25,065,000,000	$8,805,000,000	$33,870,000,000
1989	$14,512,000,000	$16,208,000,000	$30,720,000,000	$10,185,000,000	$40,905,000,000
1990	$16,390,000,000	$17,676,000,000	$34,066,000,000	$10,402,000,000	$44,468,000,000
1991	$18,370,000,000	$18,923,000,000	$37,293,000,000	$10,872,000,000	$48,165,000,000
1992	$19,825,000,000	$21,582,000,000	$41,407,000,000	$11,657,000,000	$53,064,000,000
1993	$21,702,000,000	$24,176,000,000	$45,878,000,000	$12,663,000,000	$58,541,000,000
1994	$23,845,000,000	$28,003,000,000	$51,848,000,000	$13,627,000,000	$65,475,000,000
1995	$26,200,000,000	$32,111,000,000	$58,311,000,000	$15,319,000,000	$73,630,000,000
1996	$28,753,000,000	$36,998,000,000	$65,751,000,000	$19,774,000,000	$85,525,000,000
1997	$31,618,000,000	$42,724,000,000	$74,342,000,000	$23,100,000,000	$97,442,000,000
1998	$33,426,000,000	$50,365,000,000	$83,790,000,000	$27,250,000,000	$111,040,000,000
1999	$37,217,000,000	$61,227,000,000	$98,444,000,000	$31,910,000,000	$130,354,000,000
2000	$41,689,000,000	$63,000,000,000	$104,689,000,000	$33,400,000,000	$138,089,000,000

Courtesy of INPUT.

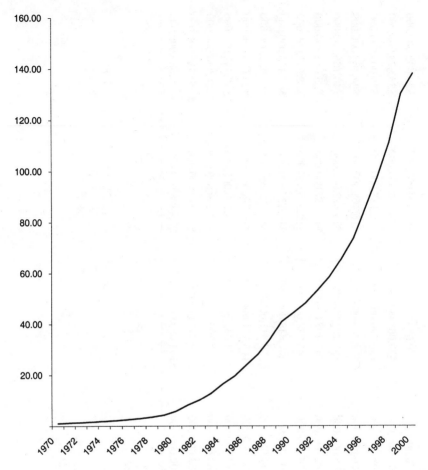

Figure 1.2
The total US software market (user expenditures in billions of dollars), 1970–2000. Courtesy of INPUT.

Processing services firms included organizations (such as ADP and Computers & Software Inc.) that undertook data processing activity for client firms and whose programming activities, if any, were incidental. Software houses were organizations that undertook custom programming for client firms or, beginning in the late 1960s, sold software products.

In the absence of an official census of the software industry, information providers used ad hoc collection methods based on surveys of member firms and on sampling. For example, ADAPSO's fourth annual survey (1970) estimated the number of firms in the industry "from examining the firms listed under 'Data Processing Services' in the yellow pages for

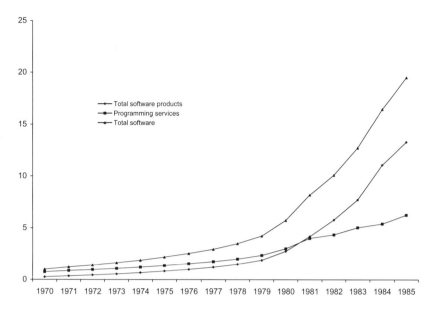

Figure 1.3
The US software market (user expenditures in billions of dollars), 1970–1985. Courtesy of INPUT.

31 cities."[21] From 1976 to 1990, the computer industry analyst INPUT produced all ADAPSO's statistics. Beginning in 1977, INPUT disaggregated software houses into programming services firms and software products firms. Beginning in 1980, integrated systems (also known as turnkey systems) were included. INPUT's figures are estimates of the total domestic and international revenues of all US-owned software producers, excluding recreational software firms.

The statistics produced by INPUT for ADAPSO were the best-regarded in the industry, and they were adopted by the US Department of Commerce and by the Organization for Economic Cooperation and Development for their official reports on the software industry in the mid 1980s.[22] Since 1991, ADAPSO's successor, ITAA, has no longer published annual surveys of the industry, primarily because the ITAA's board decided to suspend funding for the statistics program; once terminated, it was never resumed.

Two other trade associations also published statistics on the industry. One of these was the Software Publishers Association (SPA), established in 1984 to exclusively represent the mass-market software industry, including producers of business, educational, domestic, and recreational

Table 1.2
Revenues, US computer services and software industry. ("Software" column includes software products and programming services.)

	Total		Processing services		Software		Software products		Programming services		Turnkey		Systems integrators	
	Firms	$ million	Firms	$ million	Firms	$ million	Firms	$ million	Firms	$ million	Firms	$ million	Firms	$ million
1966	700	540		440		100								
1967	840	740		560		180								
1968	1,300	1,040		770		270								
1969	1,300	1,550		1,100		450								
1970	1,400	1,900		1,460		440								
1971	1,500	2,350		1,900		450								
1972	1,600	2,760		2,300		460								
1973	1,700	3,230		2,750		480								
1974	1,900	4,410		3,760		650								
1975	2,550	4,580	1,557	3,290	993	1,290								
1976	2,584	5,325	1,556	3,605	1,028	1,720								
1977	2,977	6,300	1,942	4,700	1,035	1,600	618	600	417	1,000				
1978	3,391	7,750	2,089	5,580	1,302	2,170	752	940	550	1,230				
1979	4,055	9,446	2,140	6,706	1,915	2,760	1,095	1,210	820	1,550				
1980	4,336	14,903	2,132	8,800	2,203	6,103	1,225	2,631	978	3,472				
1981	6,178	22,311	2,259	11,220	2,889	8,335	1,605	3,765	1,284	4,570	1,030	2756		

1982	6,470	26,430	2,130	12,484	3,227	10,624	1,879	5,295	1,348	5,329	1,113	3,322	
1983	7,000	32,600	2,150	14,600	3,650	13,900	2,250	7,500	1,400	6,400	1,200	4,100	
1984		41,200		14,800		20,000		11,100		8,900		6,400	
1985	7,313	47,000	2,121	16,300	3,963	23,600	2,488	13,000	1,475	10,600	1,229	7,100	
1986	7,532	54,000	2,110	21,300	4,260	25,800	2,705	14,800	1,555	11,000	1,162	6,900	
1987		67,400		21,800		33,300		20,600		12,700		8,500	5,500
1988	8,070	79,400	2,225	25,000	4,510	40,200	2,970	25,100	1,540	15,100	1,255	9,500	6,900
1989		89,800		22,700		45,900		30,700		15,200		9,500	80 7,500
1990		100,400		25,100		51,300		34,500		16,800		10,300	8,900

Source: ADAPSO annual reports and internal sources. Data attributed to unnamed consultants, IDC, Quantum Sciences (various years), and INPUT (1976–1990).

software. Statistics were published from 1984 to 1997, when publication was discontinued owing to funding problems. The SPA's activities were primarily directed at reducing software piracy.[23] This activity was always politically difficult, not least because the Association's strong-arm tactics risked alienating the industry's customers.[24] For this reason, another trade association, the Business Software Alliance (BSA), was formed by a group of the largest personal computer software firms in 1987. Its overt mission was to promote anti-piracy legislation and enforcement, particularly overseas, in a more politically adept way. In its lobbying activities, the BSA has used census data on the US software industry to demonstrate that the software products industry is a large sector of the US economy, is a major net exporter, and deserves anti-piracy legislation.

Table 1.3 shows the US Census Bureau statistics for the computer and software industries, which it began to publish in the mid 1980s. Standard Industry Classification 737 consists of "computer programming, data processing, and other computer related services." In 1987, the SIC 737 classification for the computer services and software industries was subdivided into nine subclasses, numbered 7371 through 7379. The first three of these constituted the "core" software industry—programming services (SIC 7371), packaged software (SIC 7372), and integrated systems (SIC 7373). Statistics on these three classes have been published since 1990 (table 1.4). Statistics on the software industry are collected as follows: Every five years (those ending in 2 and 7), all significant establishments are required by law to complete a schedule detailing their sources of income, while smaller establishments are sampled or their contributions estimated from other sources. In other years, firms are sampled in lieu of a full census. Respondents are required to complete a detailed questionnaire allocating their revenues to programming services, packaged software (including recreational software), and integrated systems. The resulting statistics eventually enter the public domain through publication in the *Statistical Abstract of the United States*. Because the Census Bureau carries out much more exhaustive data collection than commercial industry analysts (who do not have the force of law to compel responses or the machinery to analyze them), the census statistics are probably more accurate, but they are not necessarily more useful in understanding the industry.

It is important to appreciate that the ADAPSO-INPUT and Census Bureau statistics on the software industry are not really comparable. The ADAPSO statistics, which will mostly be used in this book, measure the total software revenues of US software firms, wherever located; the Census

Table 1.3
Annual receipts (in billions) of US computer and data processing services industries (Standard Industrial Classification 737), 1985–1998.

1985	1986	1987	1988	1989	1990	1991	1992	1993	1994	1995	1996	1997	1998
$45.1	$50.6	$56.0	$67.7	$78.7	$88.3	$94.4	$104.7	$117.9	$135.6	$156.4	$184.4	$215.3	$258.8

Source: US Census Bureau, Statistical Abstract of the United States, various years.

Table 1.4
Annual receipts (in billions) of US core software industry.

SIC		1990	1991	1992	1993	1994	1995	1996	1997	1998
7371	Programming services	$21.3	$23.4	$25.0	$27.4	$31.1	$35.1	$42.1	$50.1	$64.2
7372	Packaged software	$16.5	$18.3	$21.2	$25.2	$28.9	$33.2	$39.3	$43.1	$50.5
7373	Integrated systems	$12.9	$13.8	$15.2	$16.2	$17.0	$17.5	$20.2	$26.1	$31.8
7371–7373	Total software	$50.8	$55.6	$61.6	$68.6	$77.0	$85.8	$101.6	$119.5	$160.9

Source: US Census Bureau, Statistical Abstract of the United States, various years.

Bureau statistics report the worldwide revenues of software producers arising from their operations located in the United States, regardless of ownership (though in fact there are few non-American software firms with operations in the United States). Thus, the ADAPSO data would capture all the software revenues of Microsoft (say), while the Census Bureau data would capture only the part derived from domestic operations. Similarly, CSC's programming services and consulting operations in many non-US countries are not captured by the census data.

Although statistics for the software industries of individual nations exist, there are few time series, there is no universal basis for defining the sectors of the industry, and there are similar problems in interpreting industrial census statistics. Table 1.5 shows two of the more plausible estimates for the worldwide software industry in 1982 and 1990. The 1982 statistics were provided to the US Department of Commerce by INPUT and IDC. There is no estimate for Germany—then a major software producer, comparable to France or Britain—because "information was not available from government or private-sector sources [although] we presume, that based on the size of West Germany's computer equipment market, that the size of the country's software industry would place it at a level near the four shown."[25] This speaks eloquently for the poverty of the available statistics. More reliance can, perhaps, be placed on the 1990 data, attributed to the IDC.[26]

As table 1.5 shows dramatically, the United States dominates the world software industry, with a 70 percent market share in 1982 and 57 percent in 1990. What the table does not convey is the relative importance of the three sectors of the software industry in different countries. In 1990, an authoritative source estimated that Japan had a $7.9 billion market for programming services but only a $1.4 billion market for packaged software, and that most of the latter was imported from the United States.[27] Western Europe was much closer to the American model. In Britain, for example, the market for programming services was slightly smaller than that for package software, but the latter was heavily penetrated by US imports.[28] Without doubt, the packaged software industry was and remains an American phenomenon.

Sources

This book is perhaps the first attempt at writing a full-length history of the software industry broader than the study of an individual firm. To date, most histories of software firms, particularly the numerous books

Table 1.5
Revenues (in US dollars) of world software industry, 1982 and 1990.

	1982	1990
US	$10.3 billion (70%)	$62.7 billion (57%)
Japan	$1.2 billion (8%)	$14.3 billion (13%)
France	$1.3 billion (9%)	$8.8 billion (8%)
Germany	—	$7.7 billion (7%)
UK	$0.7 billion (5%)	$6.6 billion (6%)
Other		$9.9 billion (9%)
Total	$14.7 billion (100%)	$110.0 billion (100%)

Sources: 1982 data from US Department of Commerce, *A Competitive Assessment of the United States Software Industry*, p. 32-6 (data attributed to IDC and INPUT); 1990 data from Richard Brandt, "Can the US Stay Ahead in Software?" *Business Week*, March 11, 1991: 62–67 (data attributed to IDC).

about Microsoft, have been journalistic endeavors based on interviews and press clippings. These books are usually lively and full of incident, but historically flawed because they lack wider context and because they rely on hearsay not supported by documentation. Much more satisfactory have been the two major corporate histories written to date: Claude Baum's history of SDC, *The System Builders*, and Richard Foreman's history of Informatics, *Fulfilling the Computer's Promise*.[29] The authors of these two fine histories made extensive use of a major corporate archive in addition to oral testimony and the general business literature.

Since the present book does not focus on any one firm, heavy use of corporate archives was not appropriate, nor indeed was it possible. Instead, it is based largely on monographic studies of the software industry, the periodical literature, and reports by industry analysts. These sources are discussed in detail below. It is hoped, first, that this analysis of the literature will prove of value to future historians of the software industry, whether of individual firms or of broader sectors and themes. A second purpose of this analysis is to discuss the shortcomings of the literature, which have inevitably impinged on this book. Readers who have no interest in either concern may safely advance to chapter 2.

In discussing the monographic literature, one should perhaps begin with the bad news: There are more books written about Microsoft and Bill Gates than about the rest of the industry put together. Most of this literature has been produced by journalists seeking to satisfy the immense curiosity about how Bill Gates came to be the richest man in the world. The result has been a gross distortion of the public perception of the structure of the software industry and Microsoft's place in it. That Microsoft generates at least half of the literature on the software industry but accounts for no more than 10 percent of traded software perhaps says everything.

In contrast with the torrent of Microsoft histories, there are barely a dozen worthwhile accounts of other software companies. Apart from the excellent histories of SDC and Informatics mentioned above, the majority of the rest are autobiographical accounts by movers and shakers in the software industry. They include Sandra Kurtzig's *CEO*, W. E. Peterson's *AlmostPerfect*, John Walker's *The Autodesk File*, Douglas Carlston's *Software People*, John Imlay's *Jungle Rules*, and Ben Voth's *A Piece of the Computer Pie*.[30] All these books contain valuable firsthand accounts of individual firms. Some of the better books by journalists (and not about Microsoft) are Mike Wilson's account of Oracle, *The Difference between God and Larry Ellison*, Gerd Meissner's *SAP: Inside the Software Power*, and Tristan Gaston-Breton's *La Saga Cap Gemini*.[31] There are also a few article-length reminiscences by industry pioneers, including Elmer Kubie's recollections of the early years of the Computer Usage Company and J. Lesourne and R. Armand's description of the first decade of the French software house SEMA.[32] Most histories of individual firms focus on the largest companies—those with hundreds or thousands of employees and with revenues of $1 billion or more. Since the average software company has fewer than 30 employees, this literature is highly unrepresentative. Hence, for me one of the hidden gems of software history is the little-known privately published volume *The MacNeal-Schwendler Corporation: The First Twenty Years*, written by that corporation's founding president, Richard MacNeal, in 1988.[33]

The biographies of individual firms, however many they number, are not truly representative of the software industry, any more than a random collection of biographies—of say Mozart, Marconi, and Kissinger—are representative of the human race. The best sources for the broader industrial scene are the reports resulting from government-sponsored software policy studies, a few academic monographs, and the publications of market-research organizations. Three major national policy reports related to software were written in the mid 1980s, when the software industry experienced its most dramatic growth spurt. These reports were published by the US Department of Commerce, by the Organization for Economic Cooperation and Development, and by the UK government's Advisory Council for Applied Research and Development.[34]

These reports, all in the public domain, give an excellent 1980s view of the software industry. During the 1990s, the software industry has had some attention from mainstream economists and business analysts. Publications in this category include David Mowery's edited volume *The International Computer Software Industry*, Salvatore Torrisi's *Industrial Organisation and Innovation: An International Study of the Software Industry*, Stephen Siwek and Harold Furchtgott-Roth's *International Trade in Computer Software*, and *Secrets of Software Success* by Detlev Hoch et al.[35] Although these works are not primarily historical, they are all informed by history, and in time they will become important historical sources in their own right.

The periodical literature of the software industry comprises, in order of usefulness, the trade press, general business periodicals, and newspapers. The trade press includes several long-running and well-respected general computer periodicals, such as *Datamation*, *Computer World*, and *Byte*, and many narrower titles. Among the latter are two that focused on the software industry: *Software News* and *Business Software Review*. *Software News* commenced publication in 1984 and was renamed *Software Magazine* in 1988; its profiles of individual companies and its industry rankings were particularly useful.[36] In the 1980s, *Business Software Review* was published by International Computer Programs as a spinoff of its *ICP Quarterly* software catalog. Its most interesting features were early rankings of the industry and reports of ICP's annual "million dollar awards" celebrating cumulative sales achievements of individual software products.[37] Among the general computer periodicals, by far the most useful is *Datamation*, which was published in print from 1958 until 1997 (when it became web based). *Datamation* is the only periodical to span almost the entire history of the software industry. For its first 30 years, when it was aimed squarely at senior data processing managers, it was the best source on the computer industry; for example, the annual Datamation 100 was perhaps the best survey of the computer industry ever produced. *Datamation* became less useful in the early 1990s, when it began to focus more on low-level technical issues and dropped costly features such as the Datamation 100. It is a loss that future computer historians will mourn.

The trade press has published dozens of titles over the years, journals flourishing for a few years before vanishing into oblivion. There are few complete holdings of any serials other than *Datamation* and *Byte*, and none of them are indexed. However, in the age of the press release, the trade papers often carried the same story, at the same time, in almost the same words. For the historian, there are few gems in this ephemeral literature that justify the difficulty of excavation.

In the general business press, the best sources on the history of the software industry are *Business Week* and *Fortune*. Alas, neither is indexed, but there are less pleasant occupations than leafing through the pages of these journals in a warm library on a cold afternoon. The long and authoritative articles *Fortune* ran until about 1980 are important sources for historians of the computer industry; since then, *Fortune* has published shorter articles that have proved less useful for the historian. *Business Week* tracked the software industry only intermittently from 1966 through the 1970s. After 1980, however, as software became a significant sector of the economy, articles on it appeared with increasing frequency,. Today it is a rare issue of *Business Week* that does not contain at least one reference to the industry. Newspapers and the financial press have covered the software industry since the software stock boom of 1966–1969. The articles generally are brief and contain little more than financial details; however, newspapers are well indexed, so at least the articles are easy to find.

Business historians traditionally make much use of company archives, which typically contain annual reports, minutes of board meetings, planning documents, product literature, the correspondence of senior officers, and ephemera of various kinds. Usually a company first thinks about creating an archive on the occasion of commissioning a history for a 25th or a 50th anniversary. Few software firms have reached this stage of maturity. Indeed, the only major corporate archive available to scholars is that of SDC, now housed by the Charles Babbage Institute at the University of Minnesota. The SDC Collection covers the period from 1956 to 1981, when SDC was acquired by the Burroughs Corporation. The impressive volume of information in the SDC collection hints at the riches that will become available when software companies organize their archives. For the moment, we can only be patient.[38]

The software industry came into being at about the same time as the industry analyst. The software industry attracted a number of general industry analysts, such as Frost & Sullivan (established in 1961), Creative Strategies International (formed in 1969), and Business Communications (formed in 1971). Several analysts were established in the 1960s and the 1970s to track, exclusively, the computer software and service industries. These include the International Data Corporation (formed in 1964), International Computer Programs (formed in 1967), the Yankee Group (formed in 1970), and INPUT (formed in 1976). And there have been numerous other industry watchers reporting on software, including

Auerbach Publications, Communications Trends, Electronic Trend Publications, the Gartner Group, International Resource Development, and Knowledge Industry Publications.[39]

Industry and market-research reports were produced for subscribers only (primarily software firms and large users), and few have migrated to the public domain. In the last 30 years, thousands of such reports have been published. For example, a catalog of INPUT reports published between 1976 and 1993 lists about 1,500 titles. And INPUT was just one of more than a dozen industry analysts. Of this huge volume of material, no more than 50 reports survive in the Library of Congress. Although so few reports survive, they were indispensable to the writing of this book.[40] It is hard to believe that all this information has simply vanished, but it truly has—few of the analysts who responded to my inquiry reported holding materials more than 10 years old. There are, however, a few bright spots. In the summer of 2000, in time for the final draft of this book, ICP transferred its archives to the Charles Babbage Institute. INPUT is establishing a corporate archive of its 25 years of industry reports. In February 2000, the Software History Center was incorporated (in Benicia, California) with the support of seed money from Computer Associates International. The Software History Center is dedicated to preserving the history of the software industry by ensuring that records of the companies, the individuals, and the events that shaped the industry's growth are preserved and made accessible to anyone seeking to understand how the industry evolved.[41]

"Computer Programming

at SDC is a fundamental discipline rather than a service. This approach to programming reflects the special nature of SDC's work—developing large-scale computer-centered systems.

"Our computing facility is the largest in the world. Our work includes programming for real time systems, studies of automatic programming, machine translation, pattern recognition, information retrieval, simulation, and a variety of other data processing problems. SDC is one of the few organizations that carries on such broad research and development in programming.

"When we consider a complex system that involves a high speed computer, we look on the computer program as a system component—one requiring the same attention as the hardware, and designed to mesh with other components. We feel that the program must not simply be patched in later. This point of view means that SDC programmers are participants in the development of a system and that they influence the design of components such as computers and communication links, in much the same way as hardware design influences computer programs:

"Major expansion in our work at both our New Jersey and California divisions has created a number of new positions for those who wish to accept new challenges in programming. Senior positions are open. I suggest you write or phone Mr. Rodman A. Frank at SDC's New York office. He may be reached by phoning ELdorado 5-0776 or 5-0777, or, by writing him at Box 2651, Grand Central Station, New York 17."

T. B. Steel
Senior Computer Systems Specialist

SYSTEM DEVELOPMENT CORPORATION Santa Monica, California / Lodi, New Jersey

In 1956, SDC, the biggest US software contractor, began a nationwide advertising search for programming talent.

2
Origins of the Software Contractor, the 1950s

All the pioneers of computers, and all the firms that built the first commercial models, greatly underestimated how difficult it would be to get programs to work and how much code would be needed.

Only weeks after the first prototype laboratory computers sprang to life, it became clear that programs had a life of their own—they would take weeks or months to shake down, and they would forever need improvement and modification in response to the demands of a changing environment. At first, a program containing 10,000 instructions was considered an awesome achievement. When it turned out that major defense and space projects needed programs containing hundreds of thousands, or even a million, lines of code, it was perceived as being the kind of undertaking for which only a government agency could assume the risk.

However, it soon became clear that even the most apparently mundane of programs—say, for payroll or insurance billing—exhibited similar characteristics. Programs were very hard to "debug," they were logically complex artifacts containing thousands of lines of code, and they needed constant modification. The software industry owed its beginnings to the desire among computer users for commercial enterprises that would take on this burden.

Early Sources of Software

In the first half of the 1950s, there were three main sources of software: users could write it for themselves, could obtain programs from a computer manufacturer, or could share programs among themselves.

At first, most software was written by the users, for the simple reason that virtually no software was provided by computer manufacturers. For example, IBM's first production computer, the 701, came with little more than a user's manual: "The vendor delivered . . . a number of copies of

Principles of Operation (IBM Form 24-6042-1) of 103 pages (including four pages of octal-decimal conversion tables), a primitive assembler, and some utility programs (such as a one-card bootstrap loader, a one-card bootstrap clear memory, and the like). That was it."[1]

Maintaining a large programming staff was a necessary and not disproportionate part of the cost of running a computer. The IBM 701, for example, rented at $15,000 a month, while a programmer earned $350 a month "at most."[2] A cohort of 30 or more programmers for a single mainframe was typical, and what users expected from the manufacturer was not so much programs as customer support and training.

IBM's Technical Computing Bureau

Of all the computer manufacturers, IBM provided by far the most comprehensive programmer support and training, although even here the activity was "a small marketing support function" with a staff no larger than that of the average customer.[3] Other manufacturers, such as Univac, provided far less support, and users paid the price. For example, after General Electric's Appliances Division acquired a Univac computer, in January 1954, it took nearly 2 years to get a set of basic accounting applications working satisfactorily, one entire programming group having been fired in the process. As a result, "the whole area of automated business computing, and Univac especially, had become very questionable in the eyes of many businessmen."[4]

IBM's awareness of the need to provide programming support was in large part a legacy of its punch-card accounting machine tradition. Since the 1930s, IBM had regularly published *Pointers*, pamphlets describing accounting applications developed by or for customers in various industries; these "machine set-ups" were analogous to programs for the new computers.[5] IBM viewed the coding of programs as simply an additional step in its "application development" process. That process, which long predated IBM's entry into computers, consisted of the systematic analysis of the data processing task, the design of machine records and files, and the division of the task into machine runs.[6] IBM was accustomed to working with customers to develop accounting machine applications as a part of the marketing process. To a degree this was true of the other suppliers of office machines that had now begun to sell computers (such as NCR, Burroughs, and Remington Rand), but engineering firms that had also moved into computers (such as RCA and Honeywell) had little knowledge of applications, and even less of how to go about helping their customers to implement them.

The task of providing programming support for the science-oriented 701 fell to IBM's Applied Science Division, which was headed by Cuthbert C. Hurd. Hurd assigned to the job the Technical Computing Bureau, headed by John Sheldon.[7] (Sheldon was soon to be a co-founder of the Computer Usage Company, probably the world's first software contracting firm, and a few years later Hurd would become its chairman.) The Technical Computing Bureau was allocated the first production 701, which was located in IBM's world headquarters on Madison Avenue in New York. The machine was inaugurated in April 1953 with enormous publicity to herald IBM's arrival on the computer scene. Among the 200 guests were J. Robert Oppenheimer (the former director of the Manhattan Project), John von Neumann (one of the inventors of the stored-program computer), and William Shockley (co-inventor of the transistor).

The Technical Computing Bureau's machine was in place about 6 months before the first deliveries of 701s to customers, so customers' programmers were able to become familiar with the 701 before their machines arrived. Most programmers in 701 installations took a basic course in programming organized by IBM (a course from which hundreds eventually graduated). The more seasoned or capable programmers were encouraged to develop application programs on the Technical Computing Bureau's machine so that their own 701's would be immediately productive upon arrival. Bureau staffers worked with customers to help them develop and debug their programs. In the first year of operation, about 100 applications were developed for 71 customers.[8] In addition to customers' application programs, the bureau developed about three dozen "utilities." These included "bootstrap" routines for loading programs into memory from punched cards and routines for transferring data from cards and magnetic tapes and for printing.

IBM's Technical Computing Bureau was a significant force in the tiny new world of programming in United States, and its staff members were well placed to shape the nascent world of software. Two of them, John Backus and Edgar Codd, did pioneering work that later surfaced as the FORTRAN programming language and the relational database, respectively.

SHARE

Early 701 users, having acquired machines with only the most rudimentary programming aids, had to develop their own programming systems. "No large collection of programs was available from IBM or anywhere else. Numerous assemblers were used—each one tailored to a particular installation. Each installation had its own subroutines, and each library routine was written in assembly language; most of these programs were written entirely at the installation. A set of binary-to-decimal and decimal-to-binary conversion programs were needed at every installation, and most people wrote their own."[9] According to a joke of the day, "there were 17 customers with 701s and 18 different assembly programs."[10] This duplication of effort, unavoidable during the early learning period, clearly was untenable in the long run. Some form of cooperative association, it was felt, might alleviate the problem.

The idea of a cooperative association was first proposed by R. Blair Smith, a 701 sales manager in IBM's Santa Monica sales office.[11] Smith had sold 701s to the RAND Corporation and to the Douglas Aircraft Company, and their early experiences had left him "afraid that the cost of programming would rise to the point where users would have difficulty in justifying the total cost of computing."[12] Before joining IBM, Smith had been an accounting machine manager and a founder of the Los Angeles chapter of the National Machine Accountants Association, which seemed to him an appropriate model for a computer user group. IBM's president, Thomas J. Watson Sr., was initially hostile to the idea, but his resistance was overcome by the possibility that a user group might alleviate the programming bottleneck. Smith named the new user group the Digital Computer Association, and in November 1952 he hosted its inaugural meeting at the Santa Ynez Inn in Santa Monica. Smith picked up the tab, which he later recalled as "the largest expense account I had ever submitted." "Some of the attendees," he explained, "were angry with each other for pirating programmer employees. Therefore, I had to buy three rounds of drinks before they became somewhat congenial."[13] This was an early hint of a shortage of programming manpower that would become endemic.

The Digital Computer Association met regularly as an informal dining club. This fostered a cooperative spirit, one outcome of which was PACT, a new programming system developed jointly by a consortium of users. PACT was considerably better than anything IBM was then offering. Two of the movers and shakers behind it were Jack Strong and Frank Wagner of North American Aviation, both later important figures in the early software contracting industry.

In May 1954, IBM announced its model 704, a technologically improved successor to the 701. Machines were ordered by most of the existing 701 users, and also by some first-time users. The Digital Computer

Association was now replaced by a more formal structure, ownership of a 704 being "the single qualification for membership."[14] The new group's name, SHARE, though usually capitalized, was not an acronym; rather, it represented the objective of sharing information and programs. Another objective was to serve as a conduit between users and IBM's future developments in hardware and programming. In August 1955, SHARE's "secretary pro tem," Fletcher Jones of the RAND Corporation, sent out invitations to all seventeen organizations that owned 704s to the inaugural meeting of SHARE.

At a time when the cost of programming ran as high as $10 an instruction, dramatic savings could be achieved through cooperation. It was estimated that each of the 704's early users spent the equivalent of the first year's rental (at least $150,000) establishing a basic programming regime. This cost had been borne by all the early users, but thereafter the fruits of their labors were available free to all through SHARE. For example, the most popular 704 assembly program—one developed by the United Aircraft Corporation under the technical leadership of Roy Nutt—consisted of 2,500 lines of code that would have cost another user $25,000 to replicate. By the first anniversary of SHARE, this program alone was estimated to have saved members "on the order of $1.5 million." In all, approximately 300 programs had been distributed to the membership. Besides organizing a library of programs, SHARE established a terminology and a classificatory framework for software, vestiges of which persist today: "The discussions surrounding the initial coding effort helped to establish what a basic suite of programs . . . ought to look like. Share also created a formal system of classification, which it used to organize its program library. While categories such as compiler, utility, and mathematical subroutine were based on distinctions already found within member installations, these terms gained further stability and definition though Share."[15]

On its first anniversary, SHARE's membership stood at 62 organizations (not only in the United States; some were in Canada, some in France, and some in England) owning a total of 76 IBM 704s. Another 88 organizations had signed up to receive the distribution list for programming literature. Soon other manufacturers followed IBM's example and established SHARE-like user groups, the best known of which was Remington Rand's Univac Scientific Exchange (USE), established in December 1955. The following year, IBM established GUIDE for its business-oriented 702, 705, and 650 computers. GUIDE would eventually become the world's largest user group, with chapters in virtually every country where IBM sold computers. The user group remains the preeminent way for computer users to interact with one another and with computer manufacturers.

Programming Languages—FORTRAN and COBOL

The provision of utilities and assembly programs by manufacturers and user groups went some way toward reducing the programming problem. Utilities made it possible to eliminate the programming of some entire functions (e.g., controlling printers and magnetic-tape units and sorting data). However, applications programming still had to be done in machine code, instruction by instruction, perhaps with the aid of an assembly program. The assembly program eliminated some the housekeeping chores of machine-code programming, but programs still had to be painstakingly constructed.

The crucial breakthrough in programmer productivity was the development of programming languages in the second half of the 1950s.[16] In a programming language, a single line of code could generate several machine instructions, improving productivity by a factor of 5 or 10. Unfortunately, early programming languages were so inefficient that gains in programmer productivity were squandered by inordinate translation times and long, inefficient programs. For example, at Univac, Grace Hopper, the doyenne of automatic programming, had developed a system called the "A-0 Compiler" in 1952, but programs could take as long as an hour to translate and were chronically inefficient. There were many similar academic and research laboratory developments.[17]

The most important early development in programming languages was FORTRAN for the IBM 704.[18] The FORTRAN project took place in IBM's Technical Computing Bureau under the leadership of John Backus, a 29-year-old mathematician and programmer. Backus had learned from meetings with users of the IBM 701 that they typically employed between 30 and 40 programmers, and that as much as half of the available machine time was spent on debugging programs. He estimated that three-fourths of the cost of running a computer (either in staff or in machine time) was spent on developing programs.[19] In November 1954, Backus produced a specification for a Formula Translator (FORTRAN for short), which was then distributed to interested users. Backus promised a system that would generate code 90 percent as good as the code that could be produced by a human programmer. Most programmers and their managers were highly skeptical. However, there was an encouraging response from Roy Nutt of United Aircraft, who was loaned by his employer to IBM to work on the project.

Backus estimated that the FORTRAN system would take 6 months to complete. Like most early software projects, it took much longer. FORTRAN was finally released to users in April 1957, after 2½ years of work by a dozen programmers. The system contained 18,000 lines of code. The responses from users was immediate and ecstatic. General Motors estimated that the productivity of programmers was increased by a factor of between 5 and 10, and that, even with the machine time consumed by the compiler taken into account, the overall cost of programming was reduced by a factor of 2.5.[20] Within 18 months of its release, 60 installations were using the system. Backus and his team refined the system in response to feedback from users and released a new version in 1959. FORTRAN II contained 50,000 lines of code and took 50 programmer-years to develop.

FORTRAN rapidly became the industry standard for scientific and engineering applications programming, and other manufacturers were forced to adopt it in order to compete with IBM. There was no conspiracy: FORTRAN was simply the first efficient and reliable programming language. What economists later called "network effects" made FORTRAN a standard language with little or no help from IBM. Users made software investments of tens of millions of lines of FORTRAN code, and when selecting a new computer—whether from IBM or another manufacturer—they needed a compatible FORTRAN system to protect their software investment, and also to share programs with others. (Though FORTRAN was the de facto standard for scientific programming, the data processing language COBOL later became an officially sanctioned standard.)

Data processing compilers for business applications came onto the scene a year or two after scientific programming systems. In 1956, Grace Hopper's Univac programming group established an early lead with its B-0 compiler (subsequently branded as FLOW-MATIC).[21] The Univac compiler was a strong influence on the two other major business languages of the 1950s, IBM's Commercial Translator, COMTRAN, and Honeywell's FACT, begun in 1957 and 1959 respectively. These were impressive developments, but both were eclipsed by COBOL.[22]

In April 1959, the US Department of Defense convened a meeting of computer manufacturers and major computer users with the aim of agreeing on a standard business language. The upshot of this meeting was the creation of the Committee on Data Systems and Languages (CODASYL), which comprised a short-range committee and an intermediate-range committee. The former was to produce a pragmatic solution that would last "at least two or three years"; the latter was to take a more considered view. The short-range committee produced its proposal for COBOL in June 1960. Like all business programming languages of the period, COBOL was strongly influenced by Univac's FLOW-MATIC.

The user community was generally enthusiastic over COBOL. For manufacturers that had not yet committed to a business language, COBOL provided a clear direction—RCA, for example, contracted with the newly established software contractor Applied Data Research to develop a compiler for its model 601 computer. IBM and Honeywell, however, were unwilling to abandon the large-scale business language development efforts they already had underway. In late 1962, the Department of Defense resolved the problem for them by declaring that in the future "COBOL would be the *preferred language* for problems classed as 'business data processing' unless reason could be given for not using this language."[23] With that, all resistance was gone. During 1962 and 1963, COBOL become the standard language for business applications, which it remained for the next quarter-century.

The "twin peaks" of FORTRAN and COBOL were to account for two-thirds of the applications programming activity of the 1960s and the 1970s.[24] With each passing year, the legacy of software written in these languages, and the ever-increasing human capital of programmers trained to use them, caused them to become ever more securely entrenched. This was software's first example of "lock-in."

While user groups and manufacturers' programming departments were ameliorating some of the problems of programming productivity in the mid 1950s, this did nothing to address the supply of programmers. Largely through an accident of history, the System Development Corporation (SDC) was to play a major role in supplying programming manpower.

SDC, the "University of Programmers"

To understand the context in which SDC was created, we must go back to August 1949, when the United States received intelligence that the Soviet Union had exploded its first atomic bomb. Suddenly the US appeared dangerously exposed, for the Soviet Union also possessed bombers capable of flying over the North Pole and delivering a bomb to the nation's heartland. America's best defense against such an attack was early detection of incoming bombers by radar and immediate deployment of interceptor aircraft. The US air defense system of 1949 was not up to the task, as it relied on the slow manual processing of data gathered from radar stations, which were themselves a legacy from World War II. By the time interceptor aircraft could be put in the air, the bomb would have landed.

In December 1949, the Air Force Scientific Advisory Board established a committee, under the chairmanship of Professor George E. Valley of the Massachusetts Institute of Technology, to review America's air-defense system. The Valley Committee's report recommended that the entire air-defense system be upgraded, with improved interceptor aircraft, ground-to-air missiles, anti-aircraft artillery, comprehensive radar coverage, and automatic processing of data in computer-based command-and-control centers. The last of these requirements was to result in America's largest computer and software project.[25]

The Air Force endorsed the Valley Committee's report, and in December 1950 MIT was contracted to undertake a study of the air-defense problem and then to develop a prototype computer-based system. Known as Project Lincoln, the latter development took place at MIT's Lincoln Laboratory. The computer selected for the system was Whirlwind, a prototype machine then under development at MIT. The selection of Whirlwind owed less to its origins at MIT (which was admittedly convenient) than to the requirement that the computer be able to process data in real time. The Whirlwind was the only computer in the world that had this capability; it was at least 10 times faster than any comparable machine then under development. During 1952 a prototype defense system was developed that used the XD-1 computer—an engineered version of Whirlwind—to process data from radars based on Cape Cod. The system supported 30 Air Force operators working at consoles equipped with large CRT display screens on which digitized radar data could be selected for tactical analysis by means of a light pen.

In the summer of 1954, on the basis of the Cape Cod system, the decision was made to go ahead with a full-scale nationwide defense network. The new system was called SAGE, for Semi-Automatic Ground Environment. The term "semi-automatic" emphasized the interaction of man and machine: the computers would perform high-speed data processing, while humans would be responsible for high-level information processing. Multiple defense contractors were brought in to develop the system. Western Electric was the prime contractor, while IBM, RCA, Bendix, General Electric, Bell Labs, and Burroughs were subcontractors for various aspects of radar, computers, communications, and technical analysis. IBM was given the contract to manufacture the XD-1 computer in quantity. Redesignated AN/FSQ-7 and often referred to as Q-7, the machine weighed 250 tons, contained 49,000 electronic tubes, and consumed 3 megawatts of power. The SAGE project would earn IBM revenues of $500 million during the 1950s, and would employ between 7,000 and 8,000 people at its peak—25 percent of IBM's work force. Thomas Watson Jr., his father's successor as president of IBM, later recalled that the Cold War "helped IBM make itself the king of the computer business."[26] However, IBM contributed only hardware. Although the company had been invited to take on the programming task, it declined. A manager later explained: "We estimated that it could grow to several thousand people before we were through . . . we couldn't imagine where we could absorb 2,000 programmers at IBM when this job would be over."[27] Bell Labs and MIT were also approached, but they too declined.

A 35,000-instruction pilot program had been developed for the Cape Cod system, but the difference between that and the full-scale SAGE program was "the difference between the model shop and production."[28] Whereas the Cape Cod system had been programmed as a one-of-a-kind system by a select group of engineers, for SAGE it would be necessary to use inexperienced programmers to develop a generalized program that could be reconfigured and rolled out to 20 or more sites. In the absence of a private-sector contractor willing to take on the programming challenge, the RAND Corporation, a nonprofit government-owned research organization, was given the task. RAND (a contraction of "research and development") had been incorporated in Santa Monica in 1948 as a "think tank" to provide the Air Force with research studies in the "techniques of air warfare," a spectrum of activities that ranged from secure communications to psychological studies of man-machine systems. RAND had already been involved with SAGE, training Air Force personnel for the Cape Cod system, and that activity continued alongside programming throughout RAND's 8 years of involvement with the SAGE project.

In December 1955, the RAND Corporation created an autonomous Systems Development Division to undertake the programming work. At that time, the corporation reckoned that it employed about 10 percent of the top programmers in the United States—but this amounted to only 25 people. It was estimated that there were no more than 200 "journeyman" programmers—those capable of the highest class of development work—although there were probably 6 times that number of professional programmers working on relatively simple commercial applications.[29]

It was clear that far more programmers would be needed, though RAND had no inkling of how many would be working on SAGE by the late 1950s. In February 1956, a recruitment manager was hired, and he began a national advertising campaign in trade journals, newspapers, and on radio. A recruiting office was opened in New York, and hiring teams trawled the country for talent. Hiring progressed at the rate of about 50 per month, but few of the individuals hired were actually programmers. It was necessary to recruit raw talent suitable for training in programming. The majority of applicants were men between the ages of 22 and 29, mostly college graduates in a wide variety of disciplines. ("Music teachers were particularly good subjects."[30]) Though there was no shortage of applicants, less than one-fourth passed through the initial screening and secured a job offer. The screening consisted of a three-day battery of psychological and mental aptitude tests that proved to be a fair predictor of programming ability.[31] Once hired, the recruits were given an 8-week training course organized by IBM for the Q-7 computer, then another 8-week course organized by RAND.

By October 1956, the staff of the System Development Division outnumbered that of the parent RAND Corporation, so it was decided to incorporate the division as a separate nonprofit organization: the System Development Corporation (SDC). The following year, SDC moved to a purpose-built 250,000-square-foot facility in Santa Monica that housed a burgeoning programming staff and a newly acquired Q-7 computer, whose "power supply equaled one-twelfth of the power consumption of the entire city."[32] In the fall of 1957, the programming staff of Lincoln Laboratory relocated to the new facility. In a passage evocative of John Steinbeck, SDC's historian wrote: "The bulk of the eastern contingent, consisting of five hundred families and their household goods, was brought to Santa Monica in special trains chartered by SDC. Some were returning home; most were seeing the land of sunshine and oranges for the first time. They settled in California, forming the core personnel in SAGE programming and augmenting the staff in systems training."[33] By 1959, there were more than 700 programmers working on SAGE, and more than 1,400 people supporting them. This was reckoned to be half of the entire programming manpower of the United States.

The SAGE program, when completed, contained more than a million instructions and was by far the largest program of its time. The operating programs consisted of 230,000 instructions, the utility and support programs 870,000. The programming regime was very sophisticated for its time, and much effort went into developing management disciplines and programming tools for writing and testing the hundreds of modules that made up the system. The project ran about a year late, and the cost was an astounding $50 per line of code.³⁴ As later became evident, the slippage and the high cost were not due to the boondoggle culture of a nonprofit corporation; they were quite typical of large-scale software development.

In June 1958, the first SAGE direction center went live. *Newsweek* reported the event enthusiastically:

Inside a windowless concrete blockhouse at McGuire Air Force Base, N. J., this week, huge electronic computer elements clattered, clinked, and blinked. Nearby a row of airman hunched over radarscopes as yellow-white images flickered across the screens.

Lightning fast and unerringly, the electronic unit of the SAGE air-defense system supplemented the fallible, comparatively slow-reacting mind and hand of man. For the first time, SAGE's electronic brains were tied into the network of early-warning radars, all-weather jet interceptors, missile sites, offshore picket ships, and ground observers protecting some 44 million Americans in the New York Air Defense Sector.³⁵

In all, 23 direction centers had to be equipped and integrated into the nationwide defense system.

From this point on, the nature of the activities at SDC changed from a pure development environment to one that included maintenance and roll-out. Direction centers came on stream at the rate of approximately one every 2 months, SDC supplying a program installation team of up to 100 programmers. Once the system was up and running, about eight technical personnel were permanently assigned, along with a few instructors to train military personnel in the use of the equipment. By 1959, about 400 programmers and 200 trainers were in the field. Back at SDC's Santa Monica facility, a production-line style of organization was adopted so that the different versions of the operating program could be produced for different computer configurations and operational requirements.³⁶ By 1962, when the system was fully deployed, the total cost of the software was $150 million—an impressive sum, but only 2 percent of the $8 billion total cost of SAGE.

The roll-out work was much less challenging than the original development activity, and there was an exodus of programming talent from SDC. Highly trained programmers found a ready market for their skills. In 1958 the attrition was 20 percent. "By the end of 1960, along with 3,500 SDC employees, there were already 4,000 ex-SDCers; by 1963, SDC had 4,300 employees while another 6,000 past employees were 'feeding the industry.'"³⁷ Only 50 percent of SDC programmers remained after 4 years in the job, only 30 percent after 7 years.

Hiring to replace lost talent was a serious concern for SDC, but its president later recalled: "Part of SDC's nonprofit role was to be a university for programmers. Hence, our policy in those days was not to oppose the recruiting of our personnel and not to match higher salary offers with an SDC raise."³⁸ It was later said that "the chances [were] reasonably high that on a large data-processing job in the 1970s you would find at least one person who had worked with the SAGE system."³⁹

By 1960, although the SAGE project was winding down, SDC continued to grow by taking on other military and public-sector projects. By 1963, SDC had a staff of 4,300 and offices in seven locations in addition to the headquarters in Santa Monica, and its computer center—with two SAGE Q-7s, four other hefty mainframes, and several mid-size IBM 1401s—was the largest in the world. It had an annual income of $57 million, generated by 45 projects, some for the Air Force, some for NASA, some for the Office of Civil Defense, and some for the Defense Advanced Research Projects Agency. SDC was by far the largest software enterprise in the United States.

SABRE, "The Kid's SAGE"

By 1960, both hardware and software technologies had improved dramatically. Unreliable first-generation vacuum-tube computers were giving way to smaller, cheaper, more reliable second-generation machines that used transistors, reliable core memory was now standard, and magnetic-tape storage could be augmented with random-access disk stores. Software technology had matured, allowing some de-skilling and some cost reduction through the use of programming languages and manufacturer-provided utilities.

Hence, most large and medium-size firms could now achieve routine computerization with in-house staff members, occasionally augmented by programming contractors. However, real-time projects pushed the technology to its limit. They required a computer to respond instantaneously to external inputs and to process many transactions simultaneously, a requirement for which IBM used the term "teleprocessing." Examples included airline reservations, bank automation, and retail systems. These leading-edge applications required one or more mainframe computers, novel terminal devices, and telecommunications equipment, all of which had to be integrated by means of software. Such systems cost many millions of dollars, were generally beyond the capabilities of even the most adventurous and deep-pocketed users, and were usually contracted out to mainframe manufacturers, to systems integrators, or to major software houses.

The first and classic civilian real-time project was the IBM–American Airlines SABRE airline reservation system. Though SABRE did not become fully operational until 1964, it was the outcome of more than 10 years of planning, technical assessment, and system building. IBM's initial involvement began in 1953, before it even had a computer on the market. At that time, the biggest problem facing American Airlines and its competitors was passenger reservations. Reservation operations were being run much as they had been in the 1930s, despite a big increase in the number of passengers. The reservation operation involved two main activities: maintaining an inventory of seats for flights and maintaining "passenger name records." Passenger name records included the personal details of passengers—contact details, itinerary, dietary requirements, and so on. Maintaining the seat inventory in real time was the most critical problem. An airline seat was the most perishable of products; once the plane had left the runway, it had no value. To maximize the number of occupied seats, it was necessary to maintain an accurate and up-to-date inventory, so that bookings and cancellations could be accepted up to the moment of takeoff. The heart of the manual system was an "availability board" on which reservations clerks kept track of the seat inventory. A vignette of an American Airlines reservation office in 1954 captures the scene:

A large cross-hatched board dominates one wall, its spaces filled with cryptic notes. At rows of desks sit busy men and women who continually glance from thick reference books to the wall display while continuously talking on the telephone and filling out cards. One man sitting in the back of the room is using field glasses to examine a change that has just been made on the display board. Clerks and messengers carrying cards and sheets of paper hurry from files to automatic machines. The chatter of teletype and sound of card sorting equipment fills the air. As the departure date for a flight nears, inventory control reconciles the seat inventory with the card file of passenger name records. Unconfirmed passengers are contacted before a final passenger list is sent to the departure gate at the airport. Immediately prior to take off, no-shows are removed from the inventory file and a message sent to downline stations canceling their space.[40]

Of all the carriers, American Airlines had the most innovative reservation operations. By 1952 it had already gone through two phases of mechanization. In 1946 the airline had commissioned the Teleregister Corporation (a manufacturer of brokerage display equipment) to develop the Reservisor, a machine that enabled travel agents to determine the availability of seats before making a booking by telephone or telex. In 1952, an improved system, the Magnetronic Reservisor, enabled agents to buy and cancel seats remotely, still confirming the transaction and giving passenger details by telephone or telex.[41]

The Reservisor-based systems had two major faults. The first had to do with inconsistent data. Once a seat had been sold and the inventory had been adjusted, the passenger details had to be recorded in a separate transaction. Reconciling the two separate transactions created discrepancies, and as many as one-twelfth of the transactions were erroneous. The result was either overbooking or underselling of seats. Because overbooking was unacceptable from a customer-relations standpoint, a precautionary "cushion" of unsold seats had to be maintained, with a consequent loss of revenues. The second major fault was that the system was too slow. For example, "completing a round-trip reservation between New York and Buffalo still required the efforts of 12 people, involved at least 15 procedural steps, and consumed as much as 3 hours."[42] With jetliners promising 5-hour transcontinental flights on the horizon, the information system would be slower than the flight time.

IBM's involvement with American Airlines came about through a happy accident. In the spring of 1953, on an American Airlines flight from Los Angeles to New York, R. Blair Smith of IBM's Santa Monica sales office chanced to take a seat next to C. R. Smith, the airline's president. The reservations problem was high on C. R. Smith's business agenda, and their conversation inevitably turned to the possibility of computerization. R. Blair Smith was invited to American Airlines' reservations office at La Guardia Airport to examine its operation. Sensing a major sales opportunity, Smith alerted Thomas Watson Jr., who had primary responsibility for computer developments at IBM. Watson was receptive to the idea. IBM had a number of real-time technologies under development, and a project code-named SABER had been established to look out for suitable business applications. SABER was an acronym for Semi-Automatic Business Environment Research—an oblique reference to SAGE (Semi-Automatic Ground Environment).

A small study group at IBM analyzed American Airlines' existing reservation system and, in June 1954, recommended development of "an integrated data system."[43] The initial response from IBM management was somewhat lukewarm, because the project threatened to be large and resource consuming for what was likely to be a limited and specialized market. This was not so much a conservative view as a rational one: the airline reservation system would require an amount of equipment comparable to that of a SAGE direction center, and, though the final costs of SAGE were not yet known, it was known that the Q-7 computer alone would cost $20 million and a single direction center $300 million.

Without a firm commitment to proceed, the study group continued to work with American Airlines on the feasibility of a computerized reservation system. While this was happening, computer technologies were maturing rapidly—IBM's 7000 series was under development, as were random-access disk stores and communications terminals. By late 1957, IBM felt confident enough to propose integrating all of American Airlines' operations—reservations, ticketing, check-in, and management reporting—in a single computer-based system. After some tough negotiations, the project, with an estimated total cost of $30 million, got the go-ahead in 1959. The airline's president, C. R. Smith, remarked: "You'd better make those black boxes do the job, because I could buy five or six Boeing 707s for the same capital expenditure."[44] Rather than "SABER," the American Airlines system was called "SABRE"—not an acronym but "only a name suggesting speed and ability to hit the mark."[45]

SABRE was to be organized along the lines of a scaled-down SAGE direction center, with an IBM 7090 computer (duplexed for automatic switch-over in the event of a system failure), 16 model 1301 disk stores with a combined capacity of 800 million characters, and 1,100 specially developed agent sets.[46] The agent set had a standard IBM Selectric typewriter with additional electronics and communication capabilities. Even so, each set cost $16,000, and the 1,100 sets accounted for about half of the system's cost.

Though SABRE was called "the kid's SAGE," it was by far the largest civilian computerization project to date, involving 200 technical personnel for 5 years.[47] Whereas IBM had been unwilling to undertake the programming for the SAGE system, by the time SABRE came along it had built up all the in-house skills that it needed to build the operating software. American Airlines was also fully involved: "We tapped almost all types of sources of programming manpower. The control (executive) program was written by IBM in accordance with our contract with them. We used some contract programmers from service organizations; we used our own experienced data processing people; we tested, trained and developed programmers from within American Airlines, and hired experienced programmers on the open market."[48]

The operational programs for SABRE contained half a million lines of code, but as much again was written for testing modules and simulating test environments for program debugging. Real-time software was much more difficult to construct than conventional programs. For example, it was not always possible to repeat the exact set of interactions that caused a program error, and this made bugs difficult to detect. Software bugs delayed deployment 6 months, and American Airlines had to issue a press statement: "Contrary to what you may have read . . . American Airlines SABRE system would not be ready for cut-over in New England in May. It is true that Hartford is a test city and will begin to do some testing in April and May, but at the present it is impossible for us to pinpoint any target date. We are hopeful that the system test period which began this month (March) will not take too long; however, we do not intend to install a system until it has been completely checked out."[49]

Parallel running of SABRE with the existing manual system began in late 1962, with the individual reservation offices coming on line one by one. The system was fully operational in December 1965, handling 85,000 phone calls, 40,000 passenger reservations, and 20,000 ticket sales per day. It was capable of handling 2,100 messages per minute with a response time of 3 seconds.[50] Analysts estimated that SABRE earned a 25 percent return on investment; the intangible benefits in terms of customer service and brand recognition may have been even greater.

Systems Integrators

SAGE and SABRE were the leading examples of computerized "systems integration" projects in the defense and civilian sectors, respectively. These projects, and those that followed, gave rise to the systems integrator, a new kind of company with the ability to engineer complex systems incorporating a range of electronic, computer, and software technologies.

Most of the systems integrators were pre-existing firms that caught the wave of Cold War defense expenditure on command-and-control systems. They included electronics and electrical engineering firms (General Electric, ITT, RCA, Western Electric, Westinghouse), computer and business machine manufacturers (IBM, Burroughs, NCR, Remington Rand), aerospace companies (Hughes, Martin Marietta, McDonnell-Douglas, Rockwell International, Sperry), and defense contractors (RAND, SDC, Aerospace, the Planning Research Corporation). There were also a few startup firms. For example, the Ramo-Wooldridge Corporation was founded as a defense contractor in 1953 by two scientists, Simon Ramo and Dean Wooldridge, with $300,000 in startup capital. Its first project was to design a digital computer for Westinghouse, then a prime contractor for a US Army ground-to-air missile project. Ramo-Wooldridge was renamed TRW in 1958 after a merger with Thompson Products (which greatly increased its capital base). TRW went on to play a major role in the development of "one of the most significant technologies of the twentieth century, the intercontinental ballistic missile."[51] Another startup was the nonprofit MITRE Corporation, established in 1958. MITRE grew directly out of the SAGE project and the difficulty Lincoln Laboratory encountered integrating systems from many subcontractors. As MITRE's historian explains it, in the late 1950s "the air-defense system was fragmented at this point in its development. The radars were off-the-shelf items procured through normal Air Force channels. Western Electric was responsible for the communications on which SAGE would rely. And the weapons that the system was to control were being developed under a variety of independent contracts. Quite simply, development authority was divided."[52] MITRE was created to take over the role of Lincoln Laboratory "as a new, independent, nonprofit corporation, to undertake a major advisory role in the systems engineering of the country's air defense."[53]

The scale of SAGE-type command-and-control systems in the late 1950s and the early 1960s was truly breathtaking:

SAGE—Air Force project 416L—became the pattern for at least *twenty-five* other major military command-control systems of the late 1950s and early 1960s (and, subsequently, many more). These were the so-called "Big L" systems, many built in response to the emerging threat of intercontinental ballistic missiles (ICBMs). They included 425L, the NORAD system; 438L, the Air Force Intelligence Data Handling System; and 474L, the Ballistic Missile Early Warning System (BMEWS). SAGE-like systems were also built for NATO (NADGE, the NATO Air Defense Ground Environment) and for Japan (BADGE, Base Air Defense Ground Environment).[54]

Prime contractors for the L-Systems included the major electronics firms (for example, RCA undertook BMEWS) and nonprofit corporations (including RAND and MITRE). One of the biggest projects, 465L, was the Strategic Air Command Control System (SACCS), for which about twenty proposals to act as "prime integration contractor" were solicited.[55] Proposals were submitted by IBM, Hughes, RCA, and ITT. The contract went to ITT, with SDC subcontracting for the software.

The software effort for the SACCS project was almost as heroic as that for the original SAGE.[56] Though it turned out to be considerably larger and more complex than the earlier system, SACCS was completed with far fewer human resources, owing mainly to the development of JOVIAL, a command-and-control programming language. Because the civilian sector was still firmly rooted in non-real-time computing, neither of the leading civilian languages, FORTRAN and COBOL, was suitable. JOVIAL, specially developed for SDC's real-time military applications, was in use by January 1960, and about 95 percent of the code for SACCS was written in the language. The software was completed in about 2 years, with 1,400 programmer-years of effort—less than half that for SAGE. JOVIAL went on to be used by thousands of programmers in the "closed world" of military computing well into the 1980s, although it was scarcely used in the civilian sector.

Other software technologies developed for military applications did, however, diffuse into civil computing. TRW, for example, developed impressive software capabilities, particularly in the management of large programming projects. The culmination of this activity was a 1972 joint project with McDonnell-Douglas for the $100 million Site Defense project, a part of the United States' anti-ballistic missile (ABM) defense system. According to TRW's historian:

These requirements resulted in the biggest and most complex software system in history, a feat so daunting that many experts doubted whether it could be achieved at all. Bell Labs, the original contractor for the program, abandoned the challenge as not feasible under existing requirements.

To develop the software, the TRW McDonnell-Douglas team pioneered many new techniques to ensure that the initial requirements were fully understood at all levels and that every piece of software met those requirements as they inevitably evolved during the project. The "waterfall" technique, for example, provided for comprehensive reviews at key points in the development of the software as it proceeded from requirements analysis and definition through coding to test and evaluation. . . .

Site Defense represented "a genuine landmark" in software history. Although the United States abandoned a full-blown ABM system, the software, which was developed on time and on budget, tested successfully against an actual ICBM in the Pacific missile range. The achievement sent shock waves through the software development community and helped position TRW for subsequent work in large-scale software projects.[57]

Here TRW's historian may overstate the influence of the Site Defense project. There were other large and influential developments at the same time. However, the catchy epithet "waterfall" certainly caught on for this style of managing a software project.

The United States' requirements for command-and-control systems established real-time technology through the expenditure of billions of dollars, an amount the private sector could not have contemplated spending. In the 1950s, the vast majority of civilian computers were used for mundane "batch" applications, such as payroll, inventory control, and customer billing. In these applications, the computer was used essentially as an evolutionary enhancement of mechanized accounting procedures that had existed long before computers.

The most direct spinoff from military command-and-control systems was the airline reservation system. When one compares the $8 billion cost of SAGE with the $30 million it took to build SABRE, one gets a sense of the magnitude of the cost of establishing real-time technology and the benefit to the civilian sector. This indirect form of "technology push" by government was unique to the United States, and it helped established US dominance of the computer industry.[58] Airline reservation systems were the most important civilian systems integration projects of the period 1955–1965. IBM, which pioneered in airline reservations with the SABRE project, created similar systems for Pan American (PANAMAC) and Delta Airlines (9072 SABRE). Univac, Burroughs, and Teleregister also developed reservation systems.[59] In each of these systems, the developer played a role similar to that of the prime contractor in a defense project, providing a bundle of services: an initial feasibility study and system design, the design and production of special-purpose agent sets and telecommunications equipment, the development of software, the rollout of the system into multiple locations, and the training of the client's staff. In the case of the SABRE, the entire activity was handled in house because IBM had all the necessary organizational and technical skills. In other cases, the prime contractor lacked some of the necessary skills (particularly in software), so a subcontractor was brought in.

After the airlines, the next industry to attempt real-time computing was banking, one of the most information-intensive businesses and a leader in information technology throughout the twentieth century. In the 1950s there were two major projects to develop a bank automation system, one unsuccessful (Chase Manhattan's Diana project) and one successful (the Bank of America's ERMA project).

In 1956 the Chase Manhattan Bank contracted with the Laboratory for Electronics (LFE), an MIT spinoff specializing in electronics systems, to develop a large-scale computer system.[60] At that time, although LFE had a staff of 700 (including 150 engineers) and annual revenues of $3 million, it lacked sufficient resources to develop a real-time computer. It subcontracted computer development to the British office machine company BTM, originally a manufacturer of IBM equipment under license in England.[61] BTM invested enormous resources into the project, largely as an act of faith in order to gain access to American technology. In 1957, because of cost overruns and time slippage, Chase Manhattan decided to withdraw from the project, and BTM derived little benefit from the experience.

The Bank of America's bank automation efforts went back to 1950, when it commissioned the Stanford Research Institute (SRI) to develop a prototype check-reading machine. That machine eventually evolved into a full-scale computer called ERMA (Electronic Recording Machine Accounting).[62] ERMA was a great success, and the Bank of America decided to contract for a total of 36 systems, to be installed in its major offices and integrated into a complete information system. Approximately 30 companies were invited to submit proposals, ranging from "such predictable contenders as International Business Machines (IBM) and Remington Rand to companies as remote from electronics as General Mills and United Shoe Machinery."[63] By late 1956 the bank had shortened the list to four companies: IBM, RCA, Texas Instruments, and General Electric. General Electric was awarded the contract, largely on the basis of its low bid ($31 million). General Electric subcontracted the check-reading machinery to NCR and the magnetic-tape equipment to AMPEX. The software was jointly developed with the Bank of America's data processing staff. The system was rolled out, with extensive publicity, over 2 years: "Opening ceremonies for ERMA, held in 1960 at three different locations connected by closed circuit television, were hosted with great fanfare by Ronald Reagan of *General Electric Theater*. With the installation of the last ERMA in June 1961, 13 ERMA centers, employing 32 computers, were servicing 2.3 million checking accounts at 238 branches. Conversion of the bank's 2,382,230 savings accounts was begun on January 11, 1962, and completed on February 23, 1962."[64]

After the early 1960s, with many command-and-control systems in place, the US government dramatically reduced spending on such systems. This left the systems integrators with the problem of turning swords into plowshares. Fortunately, in the second half of the 1960s the US government invested heavily in non-defense projects, particularly for healthcare programs during President Lyndon B. Johnson's "Great Society" period. The private sector also began to invest in large real-time systems. The hardware technology was becoming increasingly standardized and available off the shelf, so the main areas of activity for systems integrators were project management and software development. As a result, many of the systems integrators of the 1950s found themselves major players in the software industry of the 1960s, and many small software houses were spun off by former employees. Writing in 1971, an industry analyst observed:

> The software industry today still bears the unmistakable traces of its early beginnings in the military, aerospace, and university environments. The chief executives of most software companies have scientific and engineering backgrounds developed during the late 1950s and early 1960s, when difficult technological problems were dealt with by the massive application of "human resources." . . . The first pioneer independent software houses in the late 1950s and early 1960s spring, for the most part, from the big "systems analysis" organizations such as Systems Development Corporation, Planning Research Corporation, Rand, Mitre, etc. From a handful of companies in 1960, there were, at one point in 1968, estimated to be no less than 3,000 independent software and services companies.[65]

Software Contracting Startups

Some of the firms with software-writing competency were very large, others very small; few were of intermediate size. Large computer manufacturers and systems integrators were developing software skills so as to be able to undertake major computer projects. At the other end of the size range, about 40 startups—typically with six or fewer programmers—entered the new business of software contracting in the second half of the 1950s. A few of these startups—notably the Computer Usage Company (CUC), the Computer Sciences Corporation (CSC), the Council for Economic and Industrial Research (C-E-I-R), and Computer Applications Inc. (CAI)—grew rapidly and began to occupy the middle of the size range.[66] By the 1980s, it was difficult to distinguish systems integrators such as TRW from computer services firms such as CSC by size, by revenues, or by range of activities.

The trajectories of most of the successful early software startups were similar. They were established by entrepreneurially minded individuals from the technical computing community who, individually or severally, combined the skills of the technical expert and the business promoter. The skills of the technical expert were the most critical, since a high level of programming competence was the sine qua non of being in the business. In the 1950s, the only way to acquire these skills was to learn them as an employee of a computer manufacturer or user. However, if one had these skills, there were very few other barriers to entry into software business—"all you need is a coding pad and a sharp pencil."[67] Another source put it this way: "All you need is two programmers and a coffee pot. Many don't even have their own computer, but rent time to debug programs at a service bureau."[68] Even better, it was usually possible to use the client's computer for program development.

The Computer Usage Company

The Computer Usage Company (CUC), generally agreed to have been the first software contractor, was founded in New York in 1955 by John Sheldon and Elmer Kubie, then in their early thirties.[69] Sheldon and Kubie had both got their entrée into computing in IBM's Technical Computing Bureau. Sheldon had been the director of IBM's Technical Computing Bureau from 1949 to 1953, though his talents were more technical than managerial. In 1953, with a number of technical papers in the area of mathematical physics to his credit, he left IBM for graduate studies at Columbia University, supporting himself by technical consulting. Kubie, also a mathematically oriented programmer in the Technical Computing Bureau, was the more entrepreneurial of the two. Having played an important role in the development of the IBM 650, he was keen to enter IBM's management development program.[70] However, faced with a mandatory 2 years in sales, he decided to leave IBM in 1954 to become an operations research specialist with General Electric in New York. By this time, Sheldon and Kubie had excellent technical and managerial skills and a wide network of contacts in the user community.

Early in 1955, Sheldon approached Kubie with the idea of establishing a programming services company. Sheldon had raised $40,000 in startup capital from his personal and family resources—a significant sum, though not nearly enough to buy a computer. It was agreed that Kubie would be the public face of CUC, while Sheldon would be the technical manager. In April 1955 they set themselves up in business, working out of Sheldon's New York apartment. Their staff consisted of a secretary and four female programmers, each a recent science or mathematics graduate. Since the median salary for a programmer was $3,000–$4,000 a year, the startup funds gave Sheldon and Kubie about a year to make the business a success. In fact, CUC was successful from the very beginning. It was able to move into proper offices within 2 months, and in its first year of operation its income was $200,000 and its profits $10,000.

CUC's early contracts were for technical applications. The first, which was typical of those that followed, was obtained though personal connections with the California Research Association. The assignment was to simulate the radial flow of fluid in an oil well. The programming was done on an IBM 701, time on which Sheldon and Kubie rented from their former employer, IBM's Technical Computing Bureau. Within 2 years, CUC began to obtain lucrative contracts with NASA. Nonetheless, Kubie recognized that the market for technical applications was somewhat limited, so he began to broaden the firm's base by tendering for business applications and systems software. The portfolio of its early projects was impressively diverse: "Among these were oil reservoir simulation, nuclear reactor design, structural design, space technology, chemical engineering, supersonic air flow, circuit and logical design of complex electronic equipment, linear programming, inventory control, sales analysis, cost analysis, payroll, trust accounting, statistical analysis, information storage and retrieval, cataloguing, traffic analysis, production planning, insurance accounting, computer simulation, computer operating systems, and development of interpretive, compiler, and assembly programming languages."[71] Kubie estimated that "About 50–60 percent of our development work is in the business application area, 20–30 percent in [systems] software, and the remainder in scientific applications."[72]

By 1959, CUC had a staff of 59 and had opened a branch office in Washington. The company's initial public offering, in 1960, brought in $186,000 in new capital. Sheldon and Kubie's personal equity stakes were now worth about $300,000 each. Even with access to capital, Kubie remained reluctant to invest in computer facilities and continued to rely on access to clients' own machines and bureau facilities. (Interestingly, this led to a small activity in computer time brokering—reselling client firms' spare computer capacity.) By 1962, CUC had opened an office in Los Angeles. Now with a staff of 240, it had revenues of $2 million and profits of $60,000.

The Computer Sciences Corporation

The Computer Sciences Corporation (CSC) has been by far the most successful of the 1950s startups.[73] In 2000 it had annual revenues of $9.4 billion, nearly 60,000 employees, and about 600 facilities around the world. However, it began very modestly as a niche player in systems software.

CSC was founded in Los Angeles in 1959 by Fletcher Jones, Robert Patrick, and Roy Nutt.[74] The promoter was the charismatic Fletcher Jones, then not quite 30, "a Gary Cooper type—Texas-tall, lean and soft spoken." As a senior data processing manager with North American Aviation, Jones was a leading light in the Southern California aerospace computing community and a founding member of SHARE.[75] Patrick, another socially adept individual, had been a senior executive with

C-E-I-R and hence had firsthand knowledge of life in a software house. Nutt, the indispensable technical genius, was then head of automatic programming at the United Aircraft Corporation. An "introverted mathematician," he had already achieved mythical status in the small world of computing, having been a leading participant in IBM's FORTRAN development and the principal author of the SHARE assembly program (SAP).[76]

The origin of CSC was a classic example of a startup's seizing the day. In late 1958, Jones learned through his social network that Honeywell was intending to develop a business language compiler for its new model 800 computer. Knowing that Honeywell lacked the in-house skills for such an undertaking, he sensed an opening. He persuaded Patrick and Nutt to join him in a partnership to bid for the Honeywell project. Honeywell "accepted their proposal to specify, develop, and deliver this compiler . . . in 18 months with an effort of 17 man-years."[77] The work was to be done on a time-and-materials basis, with regular stage payments; this freed Fletcher and his associates from the need to raise startup capital. CSC was formally incorporated in May 1959.

Nutt's guru-like reputation was a "magnet for talent," and CSC quickly recruited a team of outstanding individuals to develop the Honeywell compiler, now known as FACT. The project turned out to be far too ambitious to complete within 18 months, and it quickly began to fall behind schedule. Whereas the original FORTRAN compiler had consisted of 18,000 lines of code, FACT would eventually contain 223,000 instructions. In 1962 it was said to be "the most extensive data-processing compiler in existence" and to be "more extensive that any other compiler on the drawing boards."[78] After CSC's initial contract terminated, Honeywell took over the project. The system was eventually delivered to customers in 1963, after 4 years of development and 100 programmer-years, "two-thirds of which was Honeywell effort."[79] The cost of FACT was never disclosed; however, CSC earned $300,000 from its first year on the project, so one can speculate the total cost was well in excess of $1 million. (Eclipsed by COBOL, FACT had little influence.)

Despite CSC's wild underestimation of the FACT project, there was no acrimony between it and Honeywell. The technology was still immature, and Honeywell was self-confessedly "naive."[80] Ultimately, Honeywell gained significant prestige in the development of data processing compilers from this project—prestige that enhanced the sales of its computers. Nonetheless, the cost overruns and delivery slippage in developing FACT were much talked about in the industry, and clients became more and more reluctant to offer open-ended time-and-materials contracts, preferring fixed price terms, ideally with penalties for late delivery. This gave CSC an enormous competitive advantage in its niche. Having passed through the baptism by fire that the FACT project represented, the company had developed project-management and cost-estimating skills in compiler development that no other firm possessed; as a result, it virtually cornered the market for compiler systems. Of course, that was a limited market, and by the mid 1960s CSC (like CUC) found it necessary to diversify into a broad range of computer services.

Summary

Throughout the 1950s, programs were perceived as objects without intrinsic value, or at best with value that were no market mechanisms to realize. Users got "free" software from a computer manufacturer, or they found it through a user group; more often, however, they wrote it in house.

The software provided by manufacturers grew from almost nothing in the early 1950s to an established and standardized set of tools and utilities by the end of that decade: assemblers, programming languages, subroutine libraries, input-output control systems, and simple operating systems. The SHARE user group, and groups like it, played a pivotal role in this process by acting as a conduit through which users' needs were channeled into the manufacturers' programming departments. It was through the efforts of user groups that a consensus was reached on the type of systems software users needed.

In the 1950s, and ever after, organizations tried to build systems that pushed at the boundaries of software-writing capabilities. SDC, the most influential software organization in the early history of the industry, was the first to develop the necessary technical and project-management capabilities for such ground-breaking projects. Because of its high visibility in the industry and the number of programmers it trained, SDC was, consciously or unconsciously, the model for the programming services firms that followed.

Although by the late 1950s computer manufacturers had begun to supply the systems software needed to run a computer installation, the creation of application programs remained the job of the computer user, and programming staffs of 20 or more remained the norm well into the 1960s. The advent of the programming languages FORTRAN and COBOL simplified the creation of programs greatly, and without them the world of computers would have evolved very differently. Even so, the prospect of transferring at least some program-development activities to third parties was very attractive, especially for a firm that lacked a particular software competence and did not want to develop it. Small software contractors sprang up spontaneously in many places to satisfy this need, and from their ranks came some of the major software houses of the 1960s.

The next time somebody asks you why the talent is moving to the Independent Software Houses, quote him Bauer's Second Law.

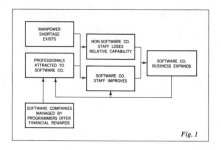

Fig. 1

Fig. 2: Dr. Bauer

Bauer's Second Law: Talent migrates from areas of well defined and stratified responsibility to areas of expanding activity at a rate proportional to the rate of expansion. Or, stated more simply: *talent goes where the action is.*

COROLLARY

Independent software companies, those not associated with a manufacturer or user group, are attracting an increasing percentage of available programming talent.

MANPOWER SHORTAGE EXISTS

Our basic premise is that talent, especially top talent, is in limited supply in any field. In the software industry the demand for top-rated specialists exceeds the supply. Consequently, software experts have a choice as to where they work. At present, and with increasing frequency, they choose to work for independent software companies. This is not to say that you can't find very talented people employed by computer manufacturers or user organizations. You certainly can. But more and more of them concentrate in independent companies.

SOFTWARE COMPANIES, MANAGED BY PROGRAMMERS, OFFER FINANCIAL REWARDS

It is true that part of the attraction is financial. Independent software companies depend on talent for their livelihood and consequently are willing to pay for it in several ways. Empiricists please call.

PROFESSIONALS ATTRACTED TO SOFTWARE COMPANY

But specialists are attracted to the independent company by more than money. A professional, given his choice, would rather work among his fellows. It is always best to work where your contribution is essential to the success of the enterprise, a place where you feel yourself in the mainstream of the business. Furthermore, when a man and his management are of the same discipline his needs are understood, his accomplishments rewarded, and his individual worth appreciated. Finally, working among top talent, a man can improve his own skills. This is especially true where people who have relatively narrow specialties within the basic discipline have a chance to exchange ideas and to learn from one another.

SOFTWARE COMPANY STAFF IMPROVES

For these reasons the staff of the independent software company improves, both in quality and in quantity. Since the talent pool is limited, it follows that the increased capability of the independents results in a decrease in the relative capability of non-independent software groups.

SOFTWARE COMPANY BUSINESS EXPANDS

This increase in capability brings more business to the independents. This in turn, makes it possible for the software company to offer more challenging work, more responsibility and more rewards. All this attracts still more talent. Thus the whole process repeats itself and becomes self-propagating.

IS THE INDEPENDENTS' GROWTH GOOD FOR YOU?

In five short years the independent software industry has grown from a meager $5,000,000 annual business, to $70,000,000 last year. And this year the figure is expected to double. Such growth must have sound economic reasons. There must be something the independents have to offer. There is. Stated in the simplest terms, the independent software firm can offer a pool of specialized talent which few users could afford to maintain for themselves. You can buy all this expert know-how, and use it for just as long as you need it to solve a given problem. And you will pay less than if you tried to solve the problem yourself. Furthermore, you will get the results on time.

HOW DOES INFORMATICS FIT IN?

Within our own organization (you knew the commercial was coming, didn't you?) we can call upon systems specialists, language specialists, experts in artificial intelligence, in data retrieval, in PERT, and many more. Without even leaving the building. Now, this kind of talent doesn't come cheap. (Look at our payroll and at our salary incentive plan, unique in the software industry.) We're always busy working on the latest problems. Right now, about 80% of our work is in the new field of on-line computing systems. We sponsored the first national symposium on the subject together with U.C.L.A. (We'll be happy to send you some of the papers presented in return for the coupon, below.)

THE MORAL:

If you have read this far, you might be interested in talking to us further about our services, capabilities, and opportunities. Simply call (213) 872-1220 and ask for me, for Frank Wagner, for Bob Rector or for anyone else on our staff. If more convenient call Werner Frank at our Washington office (301) 654-9190.

---

informatics inc®
Department F
5430 Van Nuys Boulevard
Sherman Oaks, California

Please send me the articles checked below:

☐ On-line Systems—Their Characteristics and Motivations
☐ On-line CRT Displays: User Technology and Software

Name_____

Company_____

Address_____

City_____

State_____ Zip_____

An Equal Opportunity Employer.
P.S. IN CASE YOU'RE WONDERING, BAUER'S FIRST LAW IS: IF THE PROGRAM HAS A BUG, THE COMPUTER WILL FIND IT.

In the 1960s, advertisements for software contractors emphasized their leadership in the evolving technology of software construction.

3

Programming Services, the 1960s

Until the late 1960s, the term "software industry" was not well defined. It tended to be used synonymously with "computer software and services" to signify the sector of the computer industry that supplied intangibles, to distinguish it from the rest of the industry that supplied hardware (computers, mass storage devices, peripherals, telecommunications equipment). The computer services and software industry constituted four main sectors: programming services, processing services, facilities management, and teleprocessing services. (These terms will be defined below.) By the end of the 1960s, however, the term software industry had largely taken on its present-day meaning, signifying commercial organizations engaged in the production of programming artifacts.

Computer Software and Services

In the 1960s there were thousands of entrants into the software industry. What began as a trickle in the early 1960s became a flood during the go-go years 1966–1969, when "few investments looked as exciting as the computer software houses[,] established companies were growing by 50 percent to 100 percent annually, and newer firms were going public weekly—all vying for a piece of the multi billion dollar market for computer programs and related services."[1] Most of the entrants came into programming services much as firms such as CUC and CSC had in the 1950s. They were typically founded by entrepreneurially minded individuals who had acquired technical skills with a computer manufacturer or a major user and who had a network of contacts that led to the crucial first contract. Three typical firms that went on to gain some prominence were Advanced Computer Techniques (ACT), Applied Data Research (ADR), and Informatics.

ACT was established in New York, in April 1962, by Charles P. Lecht, then in his late twenties.² Lecht had entered the computer field as a programmer with IBM in 1955. After working on the SAGE project, he joined the MITRE Corporation. Lecht started ACT as a one-man operation with "pocket money" of $800 and "one good contract." By the mid 1970s it had become "a 325-person worldwide mini conglomerate of consulting, processing, software development, and education services" with annual revenues exceeding $10 million.³ ACT was in fact an unremarkable firm that owed its prominence largely to Charles Lecht's flair for self-publicity, which was given substance by the fact that he had written several programming textbooks and a well-regarded survey of the industry.⁴

ADR, founded in Princeton, New Jersey, in 1959, had begun operations somewhat earlier than ACT. It too had an effective communicator, Martin Goetz, among its founders.⁵ ADR's first contract was for the development of a commercial programming language compiler and various other systems programs for RCA's 601 computer. However, ADR's place in the software hall of fame is due primarily to Autoflow. Introduced in 1965, Autoflow is generally credited with being the first software product. It was undoubtedly the first software product to lodge itself in most people's minds.

Informatics, established in Woodland Hills, California, in 1962, was another firm whose growth and prominence owed much to the personality of its principal founder. Walter F. Bauer, a mathematics PhD, succeeded in combining his entrepreneurial activities with his role as a leader in the technical computing community. Though Informatics was a major player in the early software contracting business, it made its mark on the history of software in 1967 by developing Mark IV, which would be the world's best-selling independent software product for 15 years.⁶

ACT, ADR, and Informatics were just three of the 40 or 50 software firms founded in the early 1960s; they happen to be the ones we know most about, largely because they were rather good at getting their stories in the media or because they took the trouble to preserve their history. In contrast, the great majority of software firms vanished, barely leaving a historical trace. For example, of the software firms listed in table 3.1, Data Systems Analysts (DSA) was an anonymous, low-profile firm that even in 1975 was described by an industry analyst as "a relic of the custom-software era of the mid 1960s" that had "survived primarily because of its unique technical competence in the design and programming of data-communications systems, particularly in the area of a message-switching systems for telegraph utilities, airlines, and the military."⁷ DSA did survive, however, and it remains a significant though somewhat anonymous company. Another firm listed in table 3.1, Programming Methods Inc., is more typical. It "vanished" because it merged with ADR in 1976, and ADR in turn was acquired by Computer Associates in 1987. Another prominent firm of the 1960s, CAI, went bankrupt after the 1970–71 computer recession; its history has been largely lost. This is not to say that the histories of these particular companies could not be recovered by dint of tracing the surviving founders and personnel; it is to say that recovering them would be an immense labor that would not greatly illuminate the larger story.

Indeed, if anything, the problem is that there are *too many* firms to analyze individually. Even if one could trace each of the thousands of software firms to its origin, the resulting mass of corporate biographies would not shed much light on the industry as a whole. Hence, probably the best one can do is select a group of firms and let them stand as proxy for the rest. This was, in fact, the approach taken by contemporary analysts. Table 3.1, taken from what is perhaps the earliest surviving analysts' report on the computer software and services industry, presents data on 17 firms. The analyst never explained how the table was derived, and some prominent players (including CUC and CAI) are missing. In all probability, this was the best selection of firms the analyst could locate that were willing to cooperate with his or her survey, which was no doubt done in a hurry and without posterity in mind.

In the 1960s the software industry was too small an entity to warrant analysis in its own right. For this reason, *programming services* was conflated with three other activities to form what was known as the Computer Software and Services Industry (CSSI). The other sectors were *processing services, facilities management,* and *teleprocessing,* the precise terminology varying over time and between analysts. (Table 3.1 also distinguishes a fifth sector, the leasing of computer equipment, which was a useful financial opportunity exploited by a number of firms. However, the only firm to make leasing a principal activity was Saul Steinberg's LEASCO, one of the notorious financial enterprises of the go-go years.[8])

Processing services was the first sector of the CSSI to emerge. The function of the processing services firm was to perform routine data processing for organizations that did not own a computer or other accounting machinery. Originally termed "service bureaus," processing services organizations had their origins in the 1930s, when IBM—then a manufacturer of accounting machines—was the dominant operator. Service bureaus were located in major cities, and clients were charged on the

Table 3.1
Revenues of leading US software and services firms, 1968–1970. ("Software" includes custom software, software packages, and consulting.)

	Revenues			Estimated distribution of revenues				
	1970	1969	1968	Software	Processing services	Facilities management	Teleprocessing	Leasing
Advanced Computer Techniques	$2,700,000	$3,200,000	$2,500,000	93%	5%	2%	—	—
Applied Data Research	$7,200,000	$6,200,000	$4,200,000	100%	—	—	—	—
Automatic Data Processing	$39,000,000	$29,000,000	$22,000,000	—	80%	—	10%	10%
Computer Sciences Corp.	$114,000,000	$109,000,000	$80,000,000	90%	—	—	5%	—
Computing and Software	$89,000,000	$79,000,000	$65,000,000	5%	75%	5%	5%	10%
Data Systems Analysts	$1,900,000	$2,100,000	$1,600,000	100%	—	—	—	—
Electronic Data Systems	$46,600,000	$16,000,000	$7,700,000	5%	—	90%	—	5%
Informatics	$18,400,000	$16,500,000	$11,500,000	74%	16%	10%	—	—
Leasco	$110,000,000	$84,000,000	$44,000,000	29%	—	—	2%	69%

National CSS	$7,800,000	$2,600,000	$140,000	1%	—	—	98%	1%
Planning Research Corp.	$65,000,000	$57,000,000	$23,000,000	88%	2%	—	10%	—
Programming Methods Inc.	$6,700,000	$4,300,000	$2,300,000	80%	—	20%	—	—
Rapidata	$4,400,000	$2,100,000	$470,000	1%	—	—	99%	—
Scantlin	$7,900,000	$8,700,000	$9,000,000	—	—	—	100%	—
Systems Development Corp.	$55,600,000	$60,800,000	$53,400,000	94%	1%	—	5%	—
Tymshare	$10,200,000	$6,400,000	$2,600,000	—	—	—	100%	—
University Computing Corporation	$114,900,000	$90,400,000	$47,600,000	20%	10%	10%	20%	40%

Source: Frost & Sullivan, *The Computer Software and Services Market*, p. 112.

basis of machine time and man-hours consumed. The customer's punched cards were delivered to the bureau by a courier service; the results, in the form of printouts and more punched cards, were returned by the same means. Because of the physical transportation of media, this was a highly localized business; it was often called "local processing." The concept of the service bureau came into its own with the arrival of the mainframe computer, whose capital cost ruled out individual purchases for all but the largest businesses. During the 1960s, IBM, NCR, Honeywell, General Electric, and CDC all developed extensive service-bureau operations.[9]

By far the biggest independent player in processing services was Automatic Data Processing (ADP),[10] established in 1949 by the entrepreneur Henry Taub as Automatic Payrolls Inc. in Clifton, New Jersey. Automatic Payrolls specialized in payroll processing, using non-electronic accounting machinery. In 1961 the company went public, changed its name to Automatic Data Processing, and obtained its first computer, an IBM 1401. Under its long-term president, Frank Lautenberg, the firm grew phenomenally. One of the best-known figures in the industry, Lautenberg (later a US senator) combined "an air of homeyness mixed with authority" and was an energetic president of ADAPSO in 1967–68.[11] ADP grew both organically and by acquisition. By 1970 it was "processing the payrolls of 7,000 firms, totaling $5 billion in wages."[12] After 50 years of growth and diversification, ADP has become one of the world's largest computer services organizations. In the 1960s and the 1970s, its principal competitor was Computing and Software Inc., which offered a portfolio of data processing services (e.g., credit reporting and the processing of insurance claims).

Facilities management was, in a sense, the opposite of processing services. A facilities management firm did not provide a computing service; it managed a data processing installation on behalf of the firm that owned it. The concept of facilities management predated the computer industry, having long been used in US military markets. The facilities manager provided personnel and expertise to operate a technical installation within the client organization, which usually retained ownership of the facility. The concept was brought to the computer industry by H. Ross Perot, who established Electronic Data Services (EDS) in 1962.[13] Perot's basic strategy was to seek out large and inefficiently run computer operations, which EDS would take over for a fixed fee. The more inefficient the original installation, the higher the potential profit. EDS was phenomenally successful, and through diversification, acquisition, and organic growth it became a global player in computer services. Although other computer services firms offered facilities management as one of their portfolio of services, no firm made such a success of the concept as a core strategy as Perot's.

Teleprocessing, the fourth sector of the CSSI, encompassed activities in which computing was supplied to the user by means of public or private telecommunications networks. This sector was also known as "remote processing," to distinguish it from the local processing of the older service bureaus and processing services firms. Activities ranged from routine data processing to database access. The sector came into its own in the mid 1960s with the advent of commercial time-sharing services. A time-sharing service consisted of a mainframe computer and typically about 30 terminals were connected to the mainframe by telephone lines. Users of the service could run library programs and develop their own software using the full power of a mainframe, typically at $10–$20 per hour. The late 1960s was a boom time for the time-sharing industry, with Tymshare, National CSS, Scantlin, and Rapidata among the major players. However, the arrival of inexpensive minicomputers in the 1970s devastated the industry. The survivors were firms, such as a Tymshare and Comshare, that refocused on network operations and on-line databases.

During the 1960s, the CSSI became a significant sector of the computer industry, growing somewhat faster than the industry as a whole (figure 3.1).[14] In 1960, software and services revenues of US firms amounted to $125 million, or 13 percent of the computer industry total of $960 million. By 1970 this had risen to $1.9 billion, or 24 percent, of an industry total of $7.9 billion. There is little reliable information on the number of firms in the industry, however. One indicator, perhaps, is the fact that ADAPSO went from 72 members in 1963 to 235 members by 1970, at which time it claimed to represent "34 percent of the industry's companies . . . and 48 percent of the estimated sales volume."[15] Though these numbers probably capture the number of processing services, facilities management, and teleprocessing firms accurately, they undoubtedly underestimate the number of programming services firms. Of the four sectors, programming services had by far the lowest barriers to entry; consequently, it attracted the most new entrants. All the evidence on the size of the industry, however, appears to be anecdotal. Perhaps the most reliable source was Larry Welke, founder of International Computer Programs Inc., who testified in the IBM antitrust suit that there were between 40 and 50 independent suppliers of

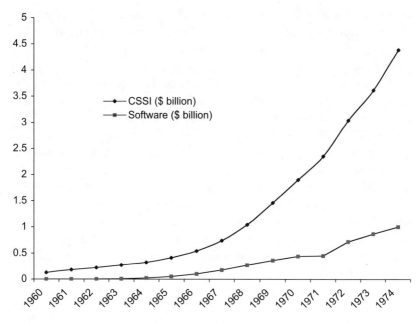

Figure 3.1
The US computer services and software industries, 1960–1974. Source: Phister, *Data Processing* (table II.1.26, p. 277).

software in 1965. By 1968, at the end of the period Welke described as "the flowering of the independent software industry," he estimated there were 2,800 software vendors.[16] Other estimates in the late 1960s varied from 1,500 to 3,000 firms, only about 30 of them exceeding $10 million in annual sales.

As was noted in chapter 1, statistics for the software industry are scarce and those that exist are difficult to interpret. Table 3.1 hints at one reason for this: There were relatively few "pure plays" in any one sector. Thus, although Informatics (say) was always classed as a software firm, it actually derived only 75 percent of its income from software activities, the remainder coming from processing services and facilities management. Most analysts classified firms according to their dominant activity, but this was not always possible. For example, the University Computing Corporation (UCC), with activities in all sectors, defied simple classification.[17] In 1970 its total software revenues (primarily from custom programming) exceeded $20 million, considerably more than that of any of the second-tier programming services firms, but it was rarely described as a software contractor.

Inside the Software Contractor

In the early 1960s, the primary market for programming services remained the US government and the computer manufacturers. Although the total market was small (only about $50 million in 1960), so was the number of firms competing in this market (about 40 or 50). Most contracts were given on a time-and-materials basis, and in the words of Informatics' historian it was a "highly profitable and comparatively risk-free line of business."[18]

The most interesting projects were the big real-time systems that were then at the leading edge of technology. The federal government, particularly the Department of Defense and NASA, remained the biggest source of such projects. With the important exception of airline reservation systems, the civilian sector was slow to embrace teleprocessing. Of the software contractors, only SDC had the capability of acting as a prime contractor on major real-time projects. The smaller software firms tend to be subcontractors, though as they gained experience they moved up the supply chain, securing bigger projects. In addition to these leading-edge projects, there was a more mundane but financially significant market: In the early 1960s, IBM and its seven competitors were all developing new computer models, for which the independents supplied both systems software and programming manpower. Supplying manpower (sometimes referred to in the industry as "body shopping') was lucrative. Even a lowly coder could command $10 an hour, about 3 times the direct labor cost.

Marketing

When it came to selling programming services to the federal government, big contractors such as SDC and MITRE and manufacturers such as IBM, RCA and Burroughs were in a league of their own. They were all inside the charmed circle of systems integrators who were invited as a matter of course to tender for government projects. For the entrepreneurial startups, breaking into this inner circle was difficult. The case of Informatics is typical.

Informatics had been established by Walter Bauer and his partners in 1962. Bauer, then in his late thirties, was head of TRW's computer division, which had a staff of 250. Bauer persuaded Werner Frank and Richard Hills, colleagues whose skills complemented his entrepreneurial and marketing talents, to join him, Frank in the technical area and Hills in management. A fourth principal, who joined a little later, was

Frank Wagner of North American Aviation, a founder and past president of SHARE. Bauer and his partners had a wide network of contacts, so that Informatics was about as well integrated into the West Coast computer community as was possible. Bauer commissioned a modest advertising campaign in the trade press, prepared a "brochure of capabilities," and awaited results. However, the early months produced only a trickle of small contracts, and the firm was losing several thousand dollars a month. Informatics was not in the network that heard about government projects. Bauer invested in a sales trip to the East Coast to make the company known to the National Security Agency, the Air Force Office of Scientific Research, the Office of Naval Research, and several other agencies. As a result of that trip, Informatics bid (unsuccessfully) on several large projects. In the next few months, Bauer and his colleagues improved their proposal writing skills in the direction of "comprehensiveness and perfection."[19] Almost a year after its founding, Informatics got a project, valued at $150,000, for the Rome Air Defense Center (RADC) in upstate New York. This led to further RADC contracts totaling several million dollars over the next 20 years. Other projects were secured during 1963 and 1964 with the Department of Defense's National Military Command Systems Support Center, the Navy's Pacific Missile Range, and the Advanced Naval Tactical Command and Control System. There were follow-on contracts, and by 1967 Informatics had revenues of $5 million a year and an estimated 3–4 percent of the market for custom software.

Informatics' marketing, "circumscribed by the personal circle of acquaintances of the principals," was typical of the software contracting industry, and it was extremely cost effective.[20] In the software contracting industry it was estimated that sales costs, when they were separately accounted for at all, amounted to as little as 2 percent of sales. Most advertising was aimed at enhancing a firm's reputation. Firms often touted their "star" creative employees (in this case, programmers), much as advertising agencies did. CSC, for example, ran an advertisement stating that its staff included "such well-known professionals as Roy Nutt, Owen Mock, Charles Swift, Lou Gatt, Joel Erdwinn, and others."[21] A client often insisted on having one of these "super programmers" on its project. Another technique for raising a firm's profile was to sponsor meetings and conferences that got the firm known by potential clients—for example, ADR sponsored symposia for the Association of Computing Machinery, Informatics sponsored the first national symposium on real-time systems with the University of California, and ACT organized the American Management

Association seminars on project management. Another method was for the founder to become a prominent personality. Martin Goetz of ADR was the most quoted software pundit of the 1960s, and his name was rarely out of the trade press. Charles Lecht of ACT was a prominent national lecturer to the computer-using community. This style of marketing was effective in taking "a company up to $2 million (sales per year) mark without difficulty."[22] However, getting beyond that level required a more organized and professional style of marketing, "straining the profitability of the company," and increasing sales costs to as much as 15 percent of revenues.

Project Management

For a software contractor, the ability to complete a programming project on time and within budget was the most critical competence. In the case of fixed-cost contracts, the profitability of a project depended on making a price-competitive bid that still left a margin of profit—margins were typically of the order of 8–15 percent, so even moderate slippage could easily turn a profit into a loss. In the case of a time-and-materials contracts, although there was no direct risk of loss to the contractor, follow-up contracts depended heavily on successful fulfillment of the initial contract. The common pattern was for the first contract to an individual supplier to be let on a time-and-materials basis, but for follow-on contracts to be fixed-cost, fixed-time.

Early software projects got an exaggerated reputation for cost and time overruns and poor reliability.[23] However, it was mainly leading-edge projects and unseasoned development teams that experienced these problems. The great majority of projects were completed within budget and on time. Indeed, it could hardly have been otherwise; the survival of the programming services industry depended on it.

During the 1960s, managing software projects was something of a black art. There were no formal methodologies until the late 1960s, and then they tended to be little more than prescriptive checklists. As is true of many aspects of large-scale software development, the origins of project management can be traced back to the SAGE project. At the Lincoln Laboratory, John F. Jacobs (later with MITRE) was head of programming for the early SAGE systems in 1955–56, when the programming staff numbered between 300 and 400. Jacobs borrowed the established techniques of engineering management and made them into a "systematic but ad hoc" sequence of stages, from initial concept to final program artifact.[24] He described these techniques in a seminal

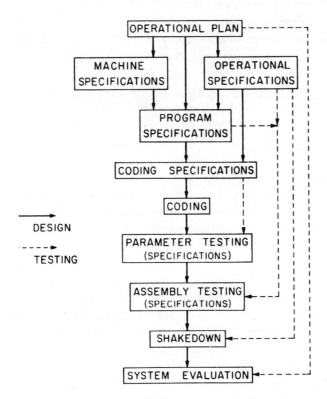

Figure 3.2
Project management in SAGE. Source: Jacobs, *The SAGE Air Defense System,* p. 184.

lecture at the Lincoln Labs in November 1956. In his memoirs, Jacobs recalls: "I called it 'The Romance of Programming,' and it seemed to have been the best thing I've ever done. More people remember me for that speech than for more significant contributions I made to the program."[25] In the seminar, Jacobs divided the programming project into a sequence of consecutive tasks (figure 3.2). First, the *machine specifications* and *operational specifications* phases determined the top-level requirements of the system. Once specified, these were frozen before embarking on the next stage, *program specifications.* In this stage, senior programmers divided the program into a set of separate, communicating subprograms. In the third stage, *coding specifications,* programmers determined the program logic for each of the subprograms. In the next stage, *coding*, coders (the lowest form of programming life) converted the coding specifications into annotated machine code, documented with flowcharts and ready for card punching. In the next stage, *parameter testing*, the subprograms were systematically tested to debug them. In *assembly testing*, the subprograms were integrated into the finished program. In the final stage, *shakedown*, the program was tested in the operational environment. This development technique set the pattern for all the programming activities of the SAGE project. The exodus of programmers from the SAGE project in the late 1950s caused this project-management style to diffuse throughout the software industry. For example, in the 1960s Charles Lecht, the founder of ACT and an alumnus of Lincoln Labs, SAGE, MITRE, and IBM, organized a set of American Management Association seminars on the "Management of Computer Programming Projects" and published what was probably the first book on the topic.[26]

Although the SAGE method was the first and best-documented project-management technique, it was not the only source of ideas. Computer manufacturers and computer users in aerospace and in other industries adapted their engineering-management practices to control the development of software. A common theme of these techniques was dividing the programming project into a temporal sequence of tasks, freezing specifications between them, and using a hierarchy of labor that cascaded down from systems analyst to senior programmer to programmer to coder. By 1970 at the latest, there was an almost universally accepted software-production style that included such tasks as problem analysis, system analysis, system design, coding, testing and debugging, acceptance, and maintenance. As was noted in chapter 2, at TRW this was called the "waterfall model" because of the irreversible cascading of tasks.

In the 1960s, systematic project management was a touchstone of certainty in an uncertain world, but it was a social as well as a rational construction. For example, the need for program flowcharts was an article of faith. In the 1950s and the 1960s, computer flowcharting created its own literature and culture. The American Standards Institute produced standard charting conventions, stationers produced stencils and drawing aids, and college programming instructors required their students to use flowcharts. By the second half of the 1960s, software packages such as Autoflow had been developed to produce flowcharts automatically. In the early 1970s, a fad for "structured programming" swept aside 20 years of flowcharting history in about 3 years. In a similar way, the hierarchy of labor was challenged, and the arbitrary distinctions between senior and junior programmers and between programmers and coders were lost. Indeed, the term "coder" fell out of use.

In practice, knowledge of the management of programming projects resided not in textbooks but in people. It took 10 years or longer for a programming manager to become fully seasoned in a particular application domain. The ability to estimate a programming task reliably and see it through to completion was a highly valued asset within a firm. Indeed, when the press talked of the vulnerability of software firms, whose only assets "could walk through the door," it was this project-management capability that was at issue. Rank-and-file programmers were relatively easy to find or train, but project managers had to be grown.

Economies of Scope

All the successful and fast-growing software contractors specialized in market niches in which, over a few years, they developed capabilities against which it was difficult for newcomers to compete.

Invariably, the niche selected by a firm was determined by the history of its founders. For example, as was noted in chapter 2, CSC came to specialize in developing compilers largely because one of its principles, Roy Nutt, had been involved in IBM's FORTRAN project. After CSC developed the Honeywell FACT compiler, it was awarded a fixed-cost contract to develop a FORTRAN system for the Univac LARC, Sperry Rand's high-performance computer primarily intended for defense contractors. CSC's compilers were recognized as the "fastest in the industry and . . . noted for their object code efficiency."[27] To broaden its capabilities, CSC subsequently brought Owen Mock, who had developed early operating systems with General Motors, into the firm. This enabled CSC to secure a contract to develop the entire software system for the Univac 1107—operating system, utilities, assembler, and compilers. That project "represented a significant state-of-the-art advance . . . which immediately brought industry-wide recognition of CSC as a major force in the software industry."[28] From that point on, CSC's growing reputation enabled it to dominate the market for systems software and to gain development contracts with almost every US computer manufacturer. After Honeywell and Univac, there were contracts with IBM, Burroughs, RCA, and Philco. There was also one with Phillips in Europe.

In a similar way, but in a completely different niche, Informatics built directly on the capabilities of its founders—capabilities that were largely gained while they were working on real-time defense contracts at TRW.

Recognizing that "the capabilities of Informatics are the capabilities of its employees," Bauer's initial marketing plan for 1962 stated:

> Clearly the area of interest of Informatics is in the most modern and most advanced information processing systems. Its area of specialty is frequently called the real-time applications. Programming systems for these real-time or on-line applications are considerably behind in development as compared with those of standard scientific and business applications. Automatic programming techniques in these areas need to be developed and applied. Informatics will offer the complete line of services for these systems, including programming.[29]

Having secured a single contract for one government client, Informatics found it increasingly less risky and more profitable to take on similar but usually larger projects for the same client. Though Informatics would pursue other technical areas, 70 percent of its income would come from its real-time market niche.

Thus, software contracting firms exploited economies of scope by becoming ever more expert and efficient (and therefore more profitable) in their market niche. As firms picked up more and more contracts in the same application domain, the knowledge was effectively captured by software tools and code assets that could be endlessly redeployed for different clients. In most cases, the development of these software assets was simply a rationalization of the software production process. Rather than build a piece of software from the ground up for each new contract, the firm began with a program that had been developed for a similar application and tailored it to the new one. In time this led to "pre-packaged" software, a path that is obvious only in hindsight.

"A Little Bit of Everything": Diversification of the Software Contractors

In the mid 1960s it became evident to the "big five" software contractors—CSC, SDC, CUC, PRC, and CAI—that custom programming was a low-profit business from which they would ultimately have to diversify.

Because of its low capital requirements and its low risk, custom programming had been an attractive proposition in the period 1955–1965. Although profit margins on time-and-materials contracts had always been low, rarely exceeding 10 percent, this had been compensated for by the negligible risk and the low investment required. Now, with the entry of hundreds of competitors, margins were being driven down, and maintaining a full order book was difficult. A team of idle programmers was extremely expensive to maintain, but letting programmers go had a disastrous effect on morale. Matters were made worse by a downturn in the software contractors' traditional markets. The number of new military systems declined in 1964, when the United States' deepening involvement in Vietnam diverted spending from advanced command-and-control systems to basic armaments. Computer manufacturers, another traditional market, had developed their own capabilities and needed far fewer software contractors. For example, between 1962 and 1967 IBM spent about $150 million on its System/360 software, of which only $4 million went to independent software companies. Another disappointment came from the commercial sector, where the majority of big users, such as banks and insurance companies, had chosen to develop in-house programming capabilities rather than rely on software contractors. The few crumbs that fell to the independent contractors tended to be for small-scale, specialized software, or for contract programmers to bridge a temporary manpower gap—the inglorious business of body shopping. Hence, to stabilize their revenues and achieve a measure of corporate growth, all the major software contractors diversified into other areas of computer services: large-scale systems integration, time-sharing services, facilities management, software products, and peripheral activities such as leasing and education.

CSC was now by far the most successful of the software contractors, and it diversified into every one of these activities. The company had gone public in 1963. In the next 3 years, the number of employees grew tenfold to 2,000. In 1968, CSC was the first of its peers to be listed on the New York Stock Exchange. In 1969, with 3,200 employees and revenues exceeding $100 million, it overtook SDC as the largest software firm. In the second half of the 1960s, CSC graduated from software contracting to full-scale systems integration, working as a prime contractor rather than as a programming subcontractor. Indeed, CSC was dismissive of the business that had given it its start: "Most software houses are doing programming, laying the bricks-and-mortar. . . . We are not in the programming business[; we are] systems *architects*."[30] About 70 percent of CSC's systems integration business was with the military and the federal government. In 1966, CSC was the prime contractor for NASA's $2 million NASTRAN project, a structural engineering program containing 150,000 lines of FORTRAN code; the MacNeal-Schwendler Corporation was a programming subcontractor. Other major projects were undertaken for the Goddard and Marshall Space Flight Centers. Another highly publicized project was for New York City's off-track betting system, for which CSC supplied a system with 462 on-line terminals. CSC also managed to pick up a good share of the meager number of private-sector projects that were available at the time, including one for General Electric and one for Travelers Insurance.[31]

In 1966, in a bold and arguably ill-considered move into processing services, CSC launched Computax and Computicket. In these ventures, CSC adopted the processing services model pioneered by ADP, creating software for which it would be able to charge on a per-transaction basis (rather than selling the software outright) and thereby ensuring a constant revenue stream. Computax (a tax preparation service) was moderately profitable, but Computicket (a theatre ticket reservation service) was a fiasco that led to a $12 million writeoff. CSC's boldest initiative was its Infonet computer network, initial planning for which was begun in 1968. The system was to cost more than $100 million—$50 million for a network of Univac's large 1108 mainframes located in 20 cities, $35 million for operating software, and $18 million for application software. Targeted at the thousands of firms with about 50 employees, and charging on a per-transaction basis, Infonet was intended to eventually contribute 40–50 percent of CSC's revenues. However, software delays and the computer recession of 1970–71 caused the project to die "a thousand deaths" before 1974, when it became profitable.[32]

CSC's major competitors went through similar experiences, though with differences of detail and emphasis. For example, PRC established the International Reservation Corporation, a hotel reservation network, and CAI created Speedata, a grocery warehouse service. Both were market failures; indeed, Speedata eventually precipitated CAI's bankruptcy. CUC also diversified in the second half of the 1960s. CUC's founder, Elmer Kubie, recalled: "Even though we had installed some equipment, we were still primarily a 'job shop' for analysis and programming. As with any job-shop operation, we suffered from slack times and times when clients asked for more help than we were staffed to provide. For this reason, it seemed sensible that we enter additional areas of service, which could lead to longer term contracts with greater stability and higher profit margins."[33] In 1965–66, CUC reorganized itself into four divisions: Development, Facilities Management, Business Services, and Education. The Development Division constituted the established software contracting heart of the business, while the Facilities Management Division aimed to compete with EDS and the Business Services Division with ADP.

SDC's position as one of the big five software contractors was anomalous, in that it was a government-owned, non-profit organization whose only customer was the US Air Force. Its problems, however, were similar to those of the other firms. By mid 1966 it was faced with a declining number of major command-and-control projects; "the sun had set on the heyday of the early 'L' systems."[34] SDC was permitted to seek contracts in the private sector, much to the chagrin of the hard-pressed independents. In contrast with the lavish Air Force projects, those from the private sector were puny and relatively uninteresting. By 1968 the size of the average contract was down to $65,000, and the 387 contracts SDC had underway generated only $53 million in revenues. SDC also ventured into time sharing and software products. Growth remained minimal, however, while SDC's competitors were diversifying rapidly and growing at up to 50 percent a year. The decline of challenging leading-edge defense projects, the lack of equity sharing, and low government salaries all contributed to a hemorrhage of programming talent from SDC. One manager recalled: "This was the heyday of go-go stocks, with the market paying fantastic multiples of earnings for software companies. It was demoralizing to our managers to see their counterparts in industry becoming wealthy overnight."[35] In 1969 the government agreed to allow SDC to privatize in order to compete on equal terms with the rest of the industry. Again the hard-pressed private-sector companies reacted strongly, organizing themselves into the Association of Independent Software Companies in order to lobby against the conversion.[36]

While the larger independents were moving into computer services, the second-wave software contractors (those established in the early 1960s) were also reaching the limits of custom programming and, albeit from a smaller base, were seeking to diversify for the same reasons. Many of these firms went public between 1965 and 1967, using the proceeds to diversify. ADR and Informatics diversified into software products. Informatics also invested in information services, and by 1970 it was deriving only half of its revenues from software contracting. Another startup of the early 1960s, Charles Lecht's ACT, also saw its once-lucrative software contracting work dry up: "During the mid sixties ACT, while still a small company, acquired a prestigious customer list and a reputation for delivering good work on time. Still, Lecht was becoming aware of the problems endemic to the custom software business. Droves of small companies—started, like ACT, with the founder's pocket money and one good contract—had moved in to fill the demand for programmers. Many of them hadn't the mettle to survive, but the intense competition was driving prices down."[37] ACT went public in 1968, using the injection of new capital to diversify into "a little bit of everything."[38]

The International Scene

In the 1960s, the European computer scene lagged the American scene by 3–5 years in number of computers, size of computers, and development of computer and software industries.[39] The primary reason for the US computer industry's lead was the government's lavish spending in the 1950s and the 1960s on defense projects such as the L-Systems, on the space program, and on federal information systems. By 1962, US manufacturers had captured 75 percent of the European computer market.[40] In the absence of a thriving domestic computer industry and large-scale government spending, it was difficult for software contractors to get started in Europe. Writing in 1967, a *Datamation* journalist noted:

> The software market in Europe today has more potential than profit. In England and on the Continent, there's neither a huge military machine nor an affluent government that's willing to spend money on the development of advanced computerized systems. There's no space program that can generate millions of dollars of contracts for programming services and computing time. And the few mainframe manufacturers who might contract out for compilers are, through mergers, becoming fewer in number each year.... In England and Europe, it is not a glamour industry—not a twinkle in the eye of investors.[41]

Several of the leading US software houses set up European subsidiaries in the early 1960s. C-E-I-R was probably the first to do so. Established in London in 1960, C-E-I-R's subsidiary overestimated the strength of the European software market and ended up as a "a bit of a shambles."[42] After being acquired by British Petroleum, it became a part of BP's Scicon computer services subsidiary. Some of the smaller US software houses also established European beachheads early on. For example, Informatics entered the Dutch market in 1964, securing custom software contracts from Univac, Phillips, and Electrologica. In 1965, Auerbach established a European base. In each of these cases, however, the company's founder was atypically international in outlook. Walter Bauer and Issac Auerbach were active in organizations such as the International Federation for Information Processing.

By the mid 1960s, there was great political concern, particularly in Britain and France, over the "American invasion" of high-tech industries such as aerospace, nuclear power, and computers.[43] In France, concern over computing led to the Plan Calcul for the establishment of indigenous computer and software industries. In Britain, the parlous state of the computer industry was a major concern of the incoming Labour government in 1964. In addition to restructuring the computer industry into a single national player, International Computers Limited (ICL), the government established a National Computer Centre to foster the diffusion of computer technology through training and software development.[44]

In 1967 the American invasion stepped up a gear when the Computer Sciences Corporation began European operations. CSC, by then America's second-largest software company, established Computer Sciences International, with headquarters in Brussels and subsidiaries in the Netherlands and London. The Dutch subsidiary was a joint venture with Phillips, for which CSI developed systems software for its computers. In London, CSI became the center of a minor political storm in 1968 when it was awarded the systems development contract for the £5 million, 500-man-year LACES (London Airport Customs Entry System) project at Heathrow Airport.[45] In the wake of this and similar American incursions, London's *Times* warned the European computer industry to brace itself for a "second invasion from America"—this time from the software firms.[46]

At the risk of overgeneralization, it is fair to assert that the various European countries developed distinctive software industries that reflected their particular business cultures and market opportunities. France, with its tendency toward centralization and nationalization, developed large, bureaucratically inclined software houses. Britain developed much more in the American pattern of the entrepreneurial startup. Germany's custom software industry was handicapped by its federal structure, although it developed a strong packaged software industry in the 1970s. Italy and the Netherlands were the only other European countries to develop significant software industries; these too were shaped by the countries' particular economic and cultural circumstances.

The most prominent European software firm was Société d'Economie et de Mathématiques Appliquées, founded in 1958.[47] SEMA started operations (about 4 years after PRC and C-E-I-R, with which it was most frequently compared) as a "think tank" specializing in economics and mathematics. Like PRC and C-E-I-R, SEMA acquired a computer for technical calculations. In 1962 it set up a subsidiary computer bureau operation. SEMA quickly established computer competencies, and by 1967 it was a fully capable systems integrator with a staff of 1,500 and annual revenues of $20 million. It undertook projects for major nationalized organizations, including Electricité de France and Orly Airport, and its subsidiaries in Spain, Italy, and Germany contributed 40 percent of its total revenues. SEMA, however, was in a class by itself, being 3–4 times the size of its nearest European rival. Its existence and its structure owed a great deal to France's distinctive bureaucratic structures, high defense spending, and nationalized utilities.

Britain's software industry was the most dynamic in any European country and the one most like the American industry. It included both entrepreneurial startups and spinoffs from major computer users and manufacturers.[48] The first significant British software firm was Computer Analysts and Programmers Limited, formed by three programmers in 1962. By 1967, CAP had 250 programmers on its staff. Eventually it became a cornerstone of Cap Gemini.[49] Many of the British startups had trajectories remarkably similar those of American firms, such as CUC and CSC, though lagging them by a few years. For example, the Computer Machinery Group (CMG), established by Bryan Mills in 1964, might almost have been the Computer Usage Company transplanted to Britain a decade later: "Bryan Mills took two colleagues from Burroughs, and set up in a basement at home, with little more than a thousand pounds and his determination to get into a fast-growing market. . . . [T]he three founders of the business bought time on a computer leased by someone else—initially all they needed was half an hour a week. It's astonishing how much work a computer can do in 30 minutes."[50] CMG specialized in accounting software, and was one of the first European firms to move from custom to packaged software. Of the world of custom software in 1964, Mills recalled: "There was a lot of talk around then of packages—the word used to describe a set of programs to do a given job. What happened was that a software house wrote a payroll program for company A. When company B came along asking for its payroll to be computerized, the original program was taken down, dusted off and adapted. It got to look more and more like a hand-me-down suit—good quality, but a terrible fit."[51] CMG developed a packaged solution that was very successful, though it fell short of a software product as the term was later understood. By the late 1970s, CMG had diversified into computer services and had 17 subsidiaries in Europe. It remains one of the major European computer services firms. Other British startups of the 1960s included the Hoskyns Group (established by John Hoskyns, a former IBM salesman) and Systems and Programs Ltd. (established by Alan Benjamin, later chief executive of the Computer Services Association, the British equivalent of ADAPSO). Both companies eventually fell under American ownership in the early 1990s; the Hoskyns Group was acquired by Martin Marietta and SPL by EDS. Logica, the last major startup before the 1970–71 computer recession, obtained 25 percent of its initial funding from PRC.[52]

In Britain, as in the United States, several leading computer users established software and service subsidiaries in the late 1960s and the early 1970s. Among these subsidiaries were British Petroleum's Scicon, the British Oxygen Company's Datasolv, and the Barclays Bank–ICL joint venture BARIC. In 1970, the British mainframe manufacturer ICL (which employed 2,000 programmers, the most of any company in Europe, to develop its mainframe software) established a programming services operation called Dataskil.

Although Britain had more software firms than any other European country, it never experienced the epidemic of firm formation that occurred in America. *Computer Weekly* estimated that as late as 1974 there were only 80 significant software firms in the United Kingdom (table 3.2); another contemporary source suggested there were only "several dozen."[53] As in the United States, most of the software firms were quite small, with a few tens of employees at most. Some of the firms survived for many years in a market niche before fading into obscurity as the founder moved into retirement. A typical example was Vaughan Systems and Programming Ltd., established in 1959 by the husband-and-wife team of Diana Vaughan and Andrew St. Johnston. In 1980 the firm was still thriving, but with only 33 employees.[54]

By far the most interesting of the UK startups was Freelance Programmers Limited, established by Stephanie ("Steve") Shirley.[55] Born in 1933, Shirley was a graduate mathematician with one of the smaller British computer firms. She left that firm in 1962 so that she could start a family and work at home as a freelancer. At that time, she was the single employee of Freelance Programmers Limited. However, she attracted

Table 3.2
Data on European software industry in 1974.

	Computers	Software companies
UK	4,500	80
Germany	7,100	47
France	4,400	40
Netherlands	1,300	30
Belgium, Luxembourg	1,000	25
Italy	2,600	20

Adapted from p. 264 of Gannon, *Trojan Horses and National Champions*. Original data attributed to *Computer Weekly*.

many other women in a similar situation, and by 1966 FPL had evolved into a cooperative of 70. In 1971 the company was renamed F International. Its business model was tried out in Denmark, in the Netherlands, and in the United States, although in none of those countries did it meet with as much success as in Britain. After some management changes in the 1980s, F International matured into a conventional computer services operation. It was renamed the FI Group in 1988.[56]

If the story of the FI Group illustrates anything, it is that no two software firms were quite alike. At one level, there were great similarities—for example, the great majority of startups were established by entrepreneurs and technicians trained by a computer user or in the computer industry. At the strategic level, however, firms were highly individual and idiosyncratic, and they exhibited strong path dependencies related to the founders' experience, financial constraints, and business networks. For example, FPL's success owed everything to Steve Shirley's personality and her individual circumstances in 1962.

The Go-Go Years: From Boom to Bust

The classic history of the 1960s' stock market boom is *The Go-Go Years*, written by John Brooks, a distinguished financial journalist. As is typical, the boom of the 1960s took 2 or 3 years to be recognized and to acquire an identity. Brooke explains the origin of the term "go-go" as follows: "Sometime in the middle nineteen sixties, probably in late 1965 or early 1966, the expression as used in the United States came to have a connotation that the dictionaries would not catch up with until after the phenomenon that it described was already over. The term 'go-go' came to designate a method of operating in the stock market. . . . The method was characterized by rapid in-and-out trading of huge blocks of stock, with an eye to large profits taken quickly, and the term was used specifically to apply to the operation of certain mutual funds, none of which had ever previously operated in anything like such a free, fast, or lively manner."[57]

Late 1965 and early 1966 was the very moment at which many of the second-wave software companies, including ADR, Informatics, and UCC, made their initial public offerings. On the whole, these offerings were undramatic, even conservative, and intended to fund expansion and reward their founding entrepreneurs. By the summer of 1968, however, both investors and entrepreneurs had lost their prudence and had been caught up in a frenzy for computer stocks that seemed to know no bounds. In 1968, *Fortune* ran an article titled "The Computer Industry's

Great Expectations." Its opening paragraph caught the mood of the day exactly:

Never before has the stock market shown quite so much enthusiasm about an industry as it has lately about the computer industry. Recent prices of computer stocks represent some of the highest price-earnings ratios ever recorded. Even shares of giant I.B.M., which increased sixfold between 1957 and late 1966, have doubled since. In July the market valued I.B.M., whose physical assets amount to less than $6 billion, at more than $40 billion—more than any other company in the world, actually as much as the gross national product of Italy. And the market value of smaller and newer companies in the industry has gone up even more steeply than I.B.M.'s. In less than three years the price of University Computing Co. of Dallas rose from $1.50 a share (adjusted for splits) to $155. The stock market valued this newcomer, who sales last year were less than $17 million, at more than $600 million.[58]

Though the case of the University Computing Co. (UCC) was a little more dramatic than most, it is worth describing in some detail, as it illustrates the general phenomenon.

UCC was founded in Dallas in 1963 as a service bureau and software company by 28-year-old Sam Wyly, a former Honeywell computer salesman. Wyly started the business with a CDC 1604 computer, which he persuaded the Southern Methodist University to house in exchange for computer time. The relationship with the university gave the company its name. UCC went public in 1965, initially trading at $4.50 a share on the over-the-counter market. The investing public was "so eager for new and speculative issues" that the stock quickly shot up to $7.50.[59] For the next 2½ years, UCC proved a solid investment, with its shares trading at between $10 and $20. Wyly was able to go to the market time and again to finance further expansion. In 1967, overseas operations were begun in the United Kingdom, in Ireland, and in the Netherlands, and two subsidiary companies were established: the Computer Leasing Co., a highly profitable equipment leasing operation, and Computer Industries Inc., a manufacturing operation. By 1968, UCC stock was trading at a price-earnings ratio of more than 100. According to *Fortune*: "Sam [Wyly], whose own UCC shares are worth about $60 million, was asked if the stock wasn't overpriced. 'Several analysts I know meet every 60 days to discuss that question. Each time they decide to postpone their decision until the stock gets more overpriced.'"[60] Ben Voth, UCC's chairman, recalled in his memoirs: "There was no deliberate market manipulation. The inflation of the stock price was due to the nature of the computer business and the mood of investors during 1964 to 1968. . . . Of the many new companies in the business, somehow University Computing Co. caught the public fancy, and slap-happy plungers pushed the quoted value of its stock up to $1 billion by 1968."[61] In 1969, UCC obtained a listing on the American Stock Exchange. On August 28, 1969, after a 3:1 split, its shares opened at $73. UCC shares eventually reached a peak of $186.

UCC was a classic go-go stock, and perhaps its fall from grace was only a matter of time. The fate of Informatics was less deserved. Informatics was regarded as positively staid within the industry; it was even criticized for its excessive bureaucracy. Informatics had been formed in 1962 as a wholly owned subsidiary of the Data Products Corporation, a Californian manufacturer of computer equipment. In 1965, Walter Bauer and his co-founders were granted 7.5 percent of the stock, and in May 1966 an IPO was made at $7.50 per share. At that time, Informatics had 300 employees, annual revenues of $4.5 million, and a net income of $171,000. The $3.5 million realized by the offering represented a price-earnings ratio of about 25. During 1968, however, "the data processing market was booming with the stock market and investors expressing keen interest in the stocks of software products."[62] In the first half of 1968, Informatics' share price rose to $72, representing a price-earnings ratio in excess of 200. It was an opportunity too good to miss, and Informatics issued 30,000 shares at $65, producing $1.8 million in new capital that was used to repay short-term borrowings and to fund the development of a software product called Mark IV. Early in 1969, after a 2:1 stock split, another 73,000 shares were sold at $25.50. At this point, one could argue, investors had become detached from reality, for Informatics' 1968 earnings had been only $40,000, representing a price-earnings ratio in excess of 600. In effect, investors were buying Informatics for its potential rather than for its present earnings. Informatics was now valued at $45 million, and the shareholdings of its three founders were worth $3 million. Informatics invested the proceeds of this final offering before the end of the boom to develop a Data Services Division.

In the US economy as a whole, the go-go market began to slow in 1967–68. Software stocks, however, remained extremely buoyant. As late as February 1970, *Forbes* observed: "In this environment, small software companies are going public at the rate of one or two a week. Despite a decline in the stock market, some computer-software companies still sell for 50 or 100 times earnings. An index of software stocks kept by *Computer World*, a trade publication, is up 35 percent from March 1, 1968, while the S&P 500 is about even for the same period."[63]

Computer services and software stocks, already attractive owing to the allure of high-tech, had been boosted by IBM's decision to unbundle, announced in June 1969. IBM had stated that, beginning January 1, 1970, it would charge users directly for computer services and software instead of supplying them "free." This decision was to have its biggest impact on the packaged software industry, but it portended a bright future for the software industry in general.

The US economy fell into recession in late 1969, and by the spring of 1970 gloom had overtaken the computer industry. Computer deliveries in the United States fell by 20 percent. Even IBM, the bellwether of the industry, saw its revenues stall and recorded its first-ever decline in staff numbers (about 10,000 out of 160,000 employees). IBM did not resume growing until 1973. In a phrase of the day, when IBM sneezed the rest of the industry caught a cold. In May 1970, General Electric decided to sell its loss-making computer division to Honeywell. The next year, RCA also threw in the towel. It was now no longer IBM and the seven dwarfs; now it was IBM and the bunch (Burroughs, Univac, NCR, CDC, and Honeywell).

Users not only deferred computer purchases in the 1970–71 computer recession; it also deferred spending on software and services. As a result, customers for the investments initiated in the heady days of 1968 never arrived. The downturn in the computer software and services industry can be timed with considerable precision.

John Brooks's book *The Go-Go Years* began with this observation:

On April 22, 1970, Henry Ross Perot of Dallas, Texas, one of the half-dozen richest men in United States, was so new to wealth, at 40, that he was not listed in *Poor's Register* and had just appeared for the first time in *Who's Who in America*. . . . Yet that day Perot made a landmark in the financial history of the United States and perhaps of the Western world. It was hardly a landmark to be envied, but it was certainly one to be remembered. That day, he suffered a paper stock market loss of about $450 million.[64]

Brooks noted that Perot still had, on paper, almost a billion dollars left. Relative to UCC, whose stock was 80 percent below its peak, Perot's EDS was doing well.

In the weeks that followed, the financial press began to carry reports of software houses beset with problems and falling stock prices. On May 16, the *New York Times* reported: "Tight money claimed its first major victim in the computer software field, as John A. DeVries, president and chairman of Computer Applications, Inc., the nation's second largest computer software concern resigned yesterday from the company he founded."[65] The story went on to explain that CAI had lost $9.8 million in 6 months' trading and had had to abandon the "several years and more than $10 million" it had invested in its Speedata grocery system. DeVries was quoted as follows: ". . . we reached the stage where we needed nine months more and $5 million in capital to make the system completely operational and revenue producing. There was just no money to be had and so we had to scrap the whole deal." Six months later, on October 1, 1970, CAI filed a petition for reorganization under Chapter X of the Federal Bankruptcy Act. Trading in CAI shares were suspended the same day at 3⅝; in 1967 they had stood at 47⅜.[66]

On July 30, the *New York Times* carried a story about the dumping of half a million shares of CSC stock.[67] No one knew who had sold the shares, although some speculated that it had been Bernie Cornfeld, the infamous promoter of the troubled I.O.S. Investment Fund, who was known to be a heavy investor in computer-related companies.

EDS, UCC, CAI, CSC—the list of troubled software firms went on. These were not mere startups; they were the largest firms in the industry. The software firms hit worst were those that had diversified into teleprocessing services in 1968 and 1969, including CAI, CSC, and UCC. Each of these firms had effectively bet its survival on the emergence of the "computer utility." This concept was always somewhat nebulous, its exact meaning depending on who was doing the promoting. In essence, however, the idea was that eventually users would own neither computers nor software; they would run applications software on remote machines through private or public data communications networks. This concept reached its peak of popularity between 1966 and 1968 with the publication of books such as *The Challenge of the Computer Utility* and *The Future of the Computer Utility*.[68] *Fortune* enthused: "Some computer industry prophets regard the service bureaus as the key to the industry's future. The computer business, they say, will evolve into a problem-solving service. What they envision is a kind of computer utility that will serve anybody and everybody, big and small, with thousands of remote terminals connected to the appropriate central processors."[69]

Two things went wrong with the computer utility concept. The first was the time and cost it took to develop the operating software. The second was that the market never materialized. The reason the computer utility market never materialized to the extent anticipated was the arrival of small-business systems and minicomputers in 1969 and 1970. These small computers, and the software packages that went with them, finally gave a medium-size firm the opportunity to own its own computer, which most preferred to a remote computing service. Sadly, this development became apparent only after the computer services firms had invested huge sums in computer and software development.

Because CSC was the largest software company, its problems with its Infonet utility attracted the most publicity: "Its $80 to $100 million Infonet project overhangs the whole software industry like a threatening but magnificent cornice of snow on an Alpine peak. Every company in the business is looking anxiously up, hoping that the cornice is firmly anchored to the mountain. If Infonet crashes down, the ensuing avalanche is likely to be destructive far beyond the limits of its own company."[70]

By 1970, Infonet's 1968 budget of $50 million had swollen to $100 million, half of that for software. In March 1972, CSC took an "almost fatal" $62.8 million writeoff in development costs, which resulted in a $35.7 million net loss for the year.[71] From a high of $60 in 1968, CSC's stock fell to $2; it hovered around the $6 mark for another 5 years. CSC survived largely because software contracting, the core of its business, still generated 90 percent of its revenues. Infonet did not turn a profit until 1974, when it generated sales of $50 million. However, that was far short of the $300 million a year that CSC had originally expected from Infonet. CSC was widely regarded as the best-managed company in the industry, and it was one of the few to eventually rescue something from the computer utility fad.

UCC was less fortunate. In 1969, at the height of the boom, UCC's Sam Wyly decided to enter the computer utility market. He planned a completely vertically integrated operation—hardware, communications, and software. In terms of hardware, UCC introduced its own terminal device, called COPE (Communications Oriented Processing Equipment), which was to be manufactured by its Computer Industries Inc. subsidiary. To establish a communications infrastructure, UCC attempted to merge with Western Union. This fell through when CSC made overtures to the company for the same purpose.[72] In November 1969, UCC therefore created a subsidiary, the Data Transmission Co., to develop a data network called Datran that would "compete head on with AT&T."[73] It was anticipated that Datran would eventually require an investment of $200 million. Somehow, while all this was going on, UCC also acquired the Gulf Insurance Company as a wholly owned subsidiary, giving it access to capital and a degree of financial stability. This was a remarkable performance for a 6-year-old firm whose 1968 revenues had been $30 million. The Datran project was the start of a decade-long disaster story. To keep the project afloat during the downturn, UCC sold off its manufacturing and other subsidiaries one by one. In 1976 it was forced to obtain a $55 million refinancing package from the "reclusive Swiss billionaire" Walter Haefner in exchange for 58 percent of the company.[74] Wyly was ousted in 1978, and UCC was subsequently renamed Uccel.[75]

Even the conservatively managed Informatics did not escape the computer utility debacle. It had used to the proceeds of its 1968 stock issue to establish a Data Service Division, described as "a unique computer utility."[76] To expand rapidly and obtain a customer base it acquired a number of service bureaus in the spring of 1969 at a cost of $3.6 million. Primarily because of software problems, however, many customers left, others suspended payments, and one sued. Informatics came to regret its hasty purchases and conceded that it had failed to exercise "due diligence" when acquiring the bureaus. By 1970, the operating losses of the Data Services Division exceeded $100,000 a month. In May 1970, "after operating in the black since 1963," Informatics announced an annual loss of $4.2 million.[77] The computer utility boom was truly over. It had been an "illusion" and one of the "computer myths of the sixties."[78] Informatics' fortunes did not recover for several years. In 1974, unable to obtain investment capital, it became a wholly owned subsidiary of Equitable Life Insurance, one of its major customers.

Another casualty of the recession was SDC's privatization plan. SDC's July 1969 announcement of its intention to become make a public offering[79] immediately provoked intense lobbying by the Association of Independent Software Companies, led by Informatics, PRC, and ADR. Already under intense competitive pressure, AISC's members "contended that the unleashing of a non-profit built up from sole-source government contracts was unfair competition."[80] As a result of the lobbying, the privatization was postponed. Then the window of opportunity shut for a decade. Within a few months, SDC was overtaken by the recession. Its time-sharing service, launched in 1969, had to be closed down in a matter of months, with a loss of $3 million. In 1970, SDC lost $550,000 on sales of $56 million. This precipitated a change of management and a major restructuring. SDC made further attempts at a public offering in 1973 and 1976, but each of these had to be called off because investors' confidence had failed to recover. SDC was acquired by the Burroughs Corporation in 1981.

By 1971, half of the estimated 3,000 computer software and services firms that had been in business at the height of the 1966–1969 stock market boom had gone out of business or "simply faded away."[81] The programming shortage was over. A Boston software executive was quoted as saying "This is certainly the best time in history for hiring" and "I estimate that 15 percent of US programmers are out of work. I am getting about 50 résumés a day in my own specialty, and some are very good people. They frequently offer to take 25 percent or 30 percent salary cuts."[82]

The computer software and services industry would experience and survive many economic downturns in the next 30 years. They were not crucial in shaping the industry; they were simply features of the ever-shifting economic landscape. The 1970–71 computer recession was of a different order, however. It profoundly shaped the industry and haunted it for a decade. Within software firms, it was a baptism by fire for largely unseasoned managements that had known only good times. *Business Week* reported: "Many software houses simply expanded beyond their capability to manage. . . . The software companies were not sufficiently market-oriented. They either overestimated the size of the market or misjudged what the user needed."[83] Within the investment community, confidence was undermined for nearly a decade. Not until 1978 would another software firm, the Cullinane Corporation, successfully make an initial public offering. And for firms that were already publicly owned, it would be a similar period of time before they could go back to the market for fresh capital.

Summary

The 1960s saw several thousand programming services firms started. Many of these entered the industry on the wave of the 1965–1969 US boom in computer stocks. There was, however, a high attrition rate, due to the combined effects of intense competition and an exceptionally difficult economic environment. Fewer than half of the firms established in the 1960s survived the 1970s. A programming services industry also became established in Europe and Japan, though because of the less developed entrepreneurial culture there were far fewer players. The firms that survived best against American competition did so by focusing on domestic markets, such as defense and finance, in which they had an indigenous advantage.

The most successful firms evolved capabilities in large-scale software development that were difficult for new entrants to emulate. These capabilities included estimation and project-management skills, the development of specialized programming competencies, and the creation and exploitation of software assets. Software contracting was a cyclical industry, however. To stabilize their incomes, large firms were forced to diversify into other computer-related activities, such as processing services and facilities management. Buoyed by funds easily obtained from the stock market during the go-go years, some of the biggest firms were caught up in the computer utility fad. Unfortunately, the computer utility concept was rendered uneconomical by the advent of small, low-cost computers and the escalating costs of developing software development. All the computer utility ventures generated huge losses and writeoffs. Only the best-managed firms survived the triple whammy of the failure of the computer utility, the crash in computer stocks, and the computer recession. The irrational euphoria for computer stocks and the failure of the computer utility were nearly unique events in the computer industry, and it was a generation before a comparable phenomenon occurred with the Internet boom at the end of the century.

Software contacting has remained an activity with low barriers to entry, and with a correspondingly a high mortality rate because of intense price competition and cyclical demand. Each year brings, like the flowers in the spring, a new crop of thousands software contractors worldwide, and the demise of fewer firms, but still several thousand. Software contracting remains the most popular way of participating in the software industry, programming services enterprises outnumbering software products firms by 2 or 3 to 1.

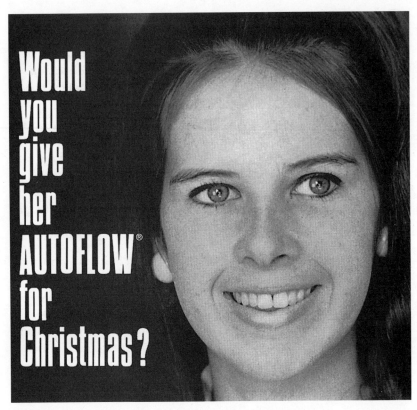

You Should!
Most programmers and managers agree that AUTOFLOW is "great", Christmas time, any time, or all year 'round.
Your favorite programmer would appreciate AUTOFLOW because AUTOFLOW users at over 300 installations agree that they're enjoying ultimate performance in automatic flowcharting — with AUTOFLOW. ADR's AUTOFLOW, a Computer Documentation System, is a proprietary, automated flowchart system that accepts FORTRAN, COBOL, PL-1 or Assembly language and produces two dimensional flowcharts accurately, completely, and effortlessly.

AUTOFLOW is available on lease or service basis for IBM System/360, 1400 and 7090 series, RCA Spectra 70, and Honeywell 200 Series.
A new, speedy 360/DOS AUTOFLOW is now available to new and existing users.
As always, AUTOFLOW can be yours for a free 30 day trial. You should review all of the merits of AUTOFLOW beyond speed, economy and dependability. Then—ask yourself, "Would you give her AUTOFLOW for Christmas?"
. . . Or why not at least a "*ring*"—at *609-921-8550*?

 APPLIED DATA RESEARCH, INC.
ROUTE 206 CENTER, PRINCETON, N.J. 08540 • PHONE: 609-921-8550
Offices in principal cities throughout the world.

Early software advertisements often portrayed a software product as a tangible artifact, to distinguish it from earlier programming services.

4
Origins of the Software Products Industry, 1965–1970

In the second half of the 1960s, the way in which most computer users acquired most of their software began to change. Up to that time, almost all of a user's software was custom written, either by in-house data processing personnel or by a programming services company. The only software a user did not write or commission was the few thousand lines of basic operating software and programming aids supplied by the computer manufacturer. The practice of custom writing software for each computer installation inevitably had to change, if only because of the proliferation of computers. As the number of computers soared, the world would simply run out of programmers. By the mid 1970s, only the largest and most traditional computer users continued to rely heavily on custom-written software. For the rest, especially medium-size companies, the software problem had become an information-searching task: finding the best software product for a particular application.

The Rise of the Software Product

Four factors can be identified as having led to the emergence of software products in the second half of the 1960s: the proliferation and growing capabilities of computers, the changing balance of hardware and software costs, the so-called software crisis, and the introduction of a standard platform (IBM's System/360 computer range).

In 1960 there were 4,400 computers in the United States; by 1965 there were 21,600; by 1970 there were 48,500 (table 4.1). This growth led to widely reported shortages of programming manpower. For example, in 1966 *Business Week* reported that, although there were 120,000 programmers in the United States, there was a demand for half again that number.[1] The 120,000 programmers amounted to fewer than six per computer, and the stock of computers was growing much faster than the population of

Table 4.1
Stock of general-purpose computers, 1955–1974.

	US	World
1955	240	246
1956	700	755
1957	1,260	1,460
1958	2,100	2,500
1959	3,110	3,800
1960	4,400	5,500
1961	6,150	7,750
1962	8,100	10,500
1963	11,700	15,200
1964	16,700	21,900
1965	21,600	29,600
1966	28,300	40,300
1967	35,600	53,100
1968	41,000	63,000
1969	46,000	72,000
1970	48,500	79,000
1971	54,400	90,600
1972	57,730	98,520
1973	62,250	106,800
1974	65,040	111,840

Source: Phister, *Data Processing*, p. 251. General-purpose computers were defined (ibid., p. 8) as "the larger machines, mostly used for business data processing and scientific calculations"; minicomputers were excluded.

programmers. These differential growth rates alone would have made the emergence of the software product something of a historical necessity, but improvements in computer performance exacerbated the problem.

The performance of computer hardware improved spectacularly between 1955 and 1965. Table 4.2 illustrates this improvement with respect to IBM's three most popular EDP computers: the model 650 (announced in 1953), the model 1401 (announced in 1959), and the smallest member of the System/360 range, the model 30 (announced in 1964). The 360/30 had 66 times the memory of the model 650 and 43 times the speed, while the relative cost per instruction executed had fallen by a factor of 40. Unfortunately, because of manufacturing and marketing economics, there was no possibility of delivering a version of a 360/30 with the power of a 650 for one-fortieth the cost. Indeed, com-

Table 4.2
Comparative performance of IBM computers (memory in kilobytes or kilocharacters; relative purchase price per instruction executed per second). Here and in other tables, NA means not available.

Model	Memory	Relative processing speed	Relative purchase price
650 (announced 1953)	1	1	1
1401 (announced 1959)	4	7	NA
360/30 (announced 1964)	66	43	0.025

Based on pp. 645–646 of Pugh et al., *IBM's 360 and Early 370 Systems*. Similar statistics appear on p. 191 of Moreau, *The Computer Comes of Age* (MIT Press, 1984).

puters had become somewhat more expensive—for example, in 1955 a month's rental of a model 650 cost about $3,500; in 1965 a month's rental of a 360/30 cost twice as much.

Big computers required big software. All users of third-generation computers were faced with a software vacuum: high-speed processors with large memories that had to be filled somehow so that users could get the best from the machine. The best supply-side evidence for the growth in the quantity of software per computer is the system software supplied for IBM's more popular computers (figure 4.1). The basic measure of software size was the total number of lines of code. The model 650 came with about 10^4 lines of code, the 1401 with 10^5. The new System/360 came with 10^6 lines of code initially, and with 10^7 by 1970. Thus, the amount of code supplied by IBM was increasing by a factor of 10 every 5 years, and it was estimated that the cost of developing systems programs had increased from about 10 percent of the total R&D budget in 1960 to 40 percent by 1965.[2]

For more than 15 years, one of the most persistent beliefs of the software community was that there would be a dramatic shift in the balance of hardware and software costs of running a computer installation. The graphs in figure 4.2 all tell much the same story: In the mid 1950s, 80 percent of the cost of running a computer had been for hardware, 20 percent for programming; at some point in the future, it would be 20 percent for hardware and 80 percent for software. Few of these graphs were based on empirical data, and they were plagiarized, with random embellishments from one author to the next, well into the 1980s.

The graphs shown in figure 4.2 were derivative of a graph by Werner Frank, one of the principals of Informatics, published in *Datamation* in 1968.[3] The graph was largely impressionistic, as Frank later admitted: "In

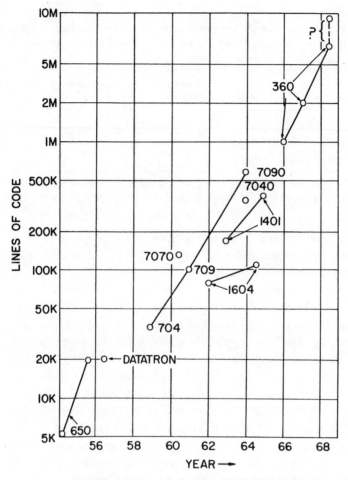

Figure 4.1
Growth in software requirements. Source: Naur and Randell, *Software Engineering*, p. 66.

those days, being unaware of Pareto's law, I was not sophisticated enough to apply the now famous 80-20 rule in drawing the curve. Careful inspection of the figure reveals that it was hand drawn and not precise. Curiously subsequent figures produced by other authors, in different contexts, somehow retained the crudeness of this first rendition."[4]

The rendition best grounded in data appeared in "Software and Its Impact: A Quantitative Assessment" by Barry Boehm of the RAND Corporation, also published in *Datamation*.[5] Writing in 1973, Boehm predicted that the

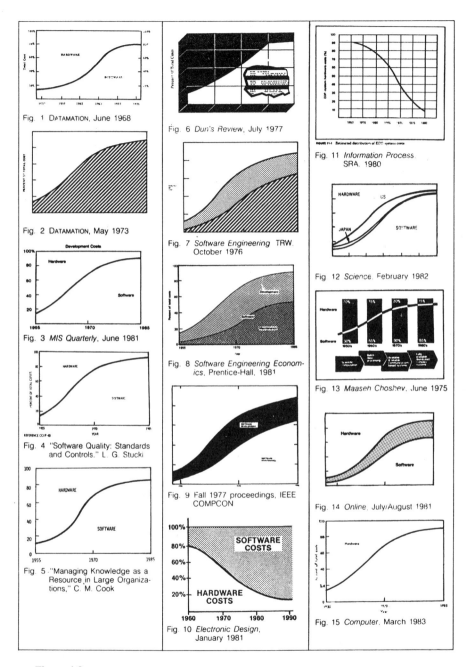

Figure 4.2
The rhetoric of the shifting balance of hardware and software costs, as illustrated in a number of similar graphs. Source: W. L. Frank, "The History of Myth No. 1," *Datamation,* May 1983, p. 252.

80-20 point would be reached in the mid 1980s. When Frank revisited the issue in 1983, however, he had concluded that his original projection had not been borne out by the facts. It was a "software myth," and he gave a variety of contemporary estimates showing that the software proportion of the costs of running a computer rarely amounted to 50 percent, much less 80 percent. Nonetheless, while the myth was believed it portended a bright and confident future for the software industry. It was one of the factors, along with soaring stock prices, that tempted many new entrants into the software products business.

There were several software crises in the 1960s, apart from the programmer shortage: the low productivity of programmers, the poor reliability of programs, and cost overruns. Like the reports of manpower shortages and of the shifting balance between hardware and software costs, the evidence was more anecdotal than empirical, and it was exaggerated. The best historical source on the extent of the software crisis is the proceedings of the first-ever conference on software engineering, sponsored by NATO and held in Garmisch, Germany, in October 1968. (The fact that NATO sponsored the conference is evidence of the military significance and the international dimension of the software crisis.) The proceedings of the conference give a revealing glimpse of the nature of the debate—almost all the evidence was anecdotal, often in the form of compelling "war stories" from leading software practitioners.[6]

While computer power had increased a hundredfold in a decade, it was estimated that programmer productivity had increased by a factor of 2 or 3 at most. The best contemporary evidence on programmer productivity came from a RAND Corporation report in 1965, which was based on data from 169 completed software projects.[7] The data showed a huge variance, ranging from 100 to 1,000 finished instructions per programmer-month, indicating costs ranging from $1 to $10 per line of code. Other estimates at about the same time indicated a range from $2.50 to $25 per line. The lower costs tended to be for routine, straightforward applications, such as payroll programs, the higher costs for the more challenging systems programs, such as real-time reservation systems. The most extreme costs were reported for manufacturers' operating systems. For example, IBM's OS/360, the major operating system for the new System/360 range, was said to have cost at least $50 million to develop. The first release in 1967 had 220,000 instructions, which works out to a cost of $225 per line of code. Other estimates put the cost even higher. Most of the software costs for large projects were not in fact for writing the code per se but for testing and debugging.[8]

Thus, the second aspect of the software crisis was the number of errors in software. For obvious commercial reasons, suppliers were not eager to disclose figures; however, the anecdotal evidence was massive. For example, Boehm stated: "Each new release of OS/360 contains roughly 1,000 new software errors. On one large real-time system containing about 2,700,000 instructions and undergoing continuous modifications, an average of one software error per day is discovered."[9] The lesson that was being learned in the 1960s, simultaneously in many projects, was that software writing did not scale up linearly. This was particularly true of the transition from custom-written to packaged software. At the Garmisch conference, one individual commented:

Regardless of how brave or cowardly the system planners happen to be, they do face difficulties in undertaking large software projects. These have been called "problems of scale," and the uninitiated sometimes assume that the word "scale" refers entirely to the size of code; for example, any project of more than 50,000 source statements. This dimension is indeed a contributory factor to the magnitude of the problems, but there are others. One of increasing importance is the number of different, non-identical situations which the software must fit. Such demands complicate the tasks of software design and implementation, since an individually programmed system for each case is impractical.[10]

Cost overruns were closely related to these problems. Estimates of programming manpower required—and therefore project duration and cost—often proved optimistic, and often by "a factor of 2.5 to 4."[11] Again, the most celebrated example was IBM's OS/360. That project started in 1965 with a team of 150 programmers, quickly slipped 6 months behind schedule, so more programmers were added; this happened time and again, and eventually a thousand people were working on the project. Writing after the event, Fred Brooks, the manager of the project, concluded: "Like dousing a fire with gasoline, this makes matters worse, much worse. More fire requires more gasoline, and thus begins a regenerative cycle which ends in disaster."[12] In all, 5,000 man-years were expended on OS/360. It was eventually released in mid 1967, a full year late, unacceptably slow and with many residual errors. (Later releases were 10 times as efficient.) Together, OS/360 and the rest of the System/360 software cost $500 million, over the original budget by a factor of 4 and "the single largest expenditure in company history."[13]

At a time when there were many computer designs on the market, there was relatively little scope for software packages. For example, in 1960 IBM had no fewer than seven software-incompatible computer architectures. Unique operating systems and utilities had to be developed for these platforms, and each language processor had to be recoded. As for application programs, it was too costly to produce packaged software except for a few major industries, such as banking and insurance.

In 1960, IBM established a "SPREAD Task Group" to determine its future computer strategy. SPREAD was a contrived acronym that stood for Systems, Programming, Review, Engineering, And Development but was really meant to connote the magnitude of the challenge of specifying a computer that would likely reshape the industry. In December 1961 the task group issued its famous SPREAD Report,[14] which noted that the "explosive growth" of software had been central to the task group's discussions.[15] The report recommended replacing IBM's existing computer models with a single series of machines, software compatible throughout the range from the smallest to the largest. The recommendation was adopted, and the announcement of System/360 in April 1964 was a watershed in the software industry.

From Software Package to Software Product

One of the abiding legacies of user groups was the perception of software as a free good. Walter Bauer, president of Informatics, recalled of the early SHARE meetings that "everybody who developed a piece of software was only too happy and flattered to have somebody else use it."[16] This perception was a brake on the development of software products as a traded good until 1970, when IBM itself began to sell packages.

In the 1950s and the 1960s, IBM went to considerable lengths to facilitate the distribution and exchange of software packages, but in those days the activity was a non-remunerative marketing service to users. With the announcement of the model 1401 in 1959, IBM established a Program Applications Library for the cataloging and distribution of software written by the company and its customers. Though the systems programs in the Program Applications Library were essential for computer operation and were universally used, this was not true of the applications programs, which largely served as exemplars and were "seldom if ever used as programs; rather, the overall system design was usually modified and then recoded to the particular user's requirements."[17]

With the announcement of System/360 in 1964, IBM refined the Program Applications Library by way of a four-way classification. The first two classes were for IBM supported programs: Type I for systems software and Type II for application programs. The distinction between systems software and application programs was a natural division that was used in all subsequent software classification schemes. Systems software included the operating systems and programming aids that all users needed to run their computer installations efficiently and to create their own programs. Application programs were provided for particular industries (such as financial services) and for common business tasks (such as payroll processing). Most application programs required customization by an IBM sales engineer or the user. The two remaining program classes were for "contributed" software. Type III comprised programs developed by IBM's systems engineers (typically for a customer in order to secure a sale). Type IV comprised programs submitted by IBM users. Type III and Type IV programs were supplied "as is," without warranty or support, and, like the old Program Applications Library, they were useful primarily as exemplars.

Around 1960, IBM and the other computer companies began to produce program packages for some of the major industries. One of the best-known and widest-reaching of IBM's industry packages was '62 CFO (short for Consolidated Functions Ordinary, 1962), developed for the insurance industry, a market second only to banking in importance to the company.[18] This was the first widely used application "package." It was actually a set of programs. The software was developed for the model 1401 in order to capture the hundreds of medium-size insurance companies then contemplating computerization for whom the cost of custom-written software would be prohibitive. Based on the US Society of Actuaries' recommendations for handling insurance billing, accounting and related functions, the software was developed by an IBM team that worked with a consulting actuary and with several potential users. In 1962 the package was incorporated in the 1401 Program Application Library so that it could "be used broadly throughout the Life Insurance Industry as operational computer programs or as a guide in the development of personalized total systems on an individual company basis."[19]

At first, few if any users adopted '62 CFO in its entirety, although a claim by one user that it saved his company "about 35 man-years of research, planning, and programming" was typical. By 1964, when the package had been expanded and refined in the field, '62 CFO was "well on its way to widespread acceptance." Three regional '62 CFO user groups were established along the lines of SHARE, enabling users to compare experiences and trade gossip. A typical success story, reported at one meeting of a user group, was of an insurer's being able to install and modify the package in just a month and a half, with a team of three. Another user stated that his firm "decided to order an IBM 1401, specifically to utilize the IBM 'package' of programs for Life Insurance known as '62 CFO." This was of, course, exactly the outcome that IBM had intended. The story was repeated many times over. Contemporary estimates indicate that between 200 and 300 US insurance companies eventually used the package, many with multiple 1401s. Increasingly, users were becoming aware of the advantage of making their operations conform to an available package, rather than making a package conform to their historical bureaucratic idiosyncrasies. Only the biggest firms, such as Prudential, could still justify the assertion that they liked to do things their own way.

As the 1960s unfolded, software packages for all the major industries were developed by IBM and the other mainframe makers. These software packages, like other customer services, were "bundled" with the cost of the computer. Software development was always perceived simply as a marketing expense; there was no thought of recovering costs by selling or leasing packages.

While IBM covered the entire application spectrum, its competitors focused on a few applications, usually in the areas of their historical competence. Burroughs, which had supplied accounting machinery to banks since the turn of the century, specialized in financial-sector software; NCR, the former cash register king, targeted the banking and retail sectors. In 1965, NCR developed a banking package for its model 315 computer and ran an advertising campaign offering "a complete EDP system for banks" and promising "You'll have complete software packages of programs ready and waiting for your command, plus trained NCR EDP personnel to assist in site preparation and installation."[20]

When Honeywell had designed its 200 series family of computers (announced in December 1963), its vice-president of Planning and Engineering, J. Chuan Chu, had made applications software central.[21] Some 35 applications packages were supplied for twelve major industries. In a path-breaking tactic, the 200 series was also made software compatible with the IBM 1401, so that users could readily migrate to the more price-competitive Honeywell series, taking their existing applications (whether written by the users or developed using IBM's software packages) with them. Honeywell even supplied a provocatively named "Liberator" program, which was the basis of an entire advertising campaign: "Liberator lets you convert your old 1401 programs . . . and run them at high new speeds on a new Honeywell Series 200 computer. Automatically, permanently, painlessly. No reprogramming. No retraining."[22] This was possible only because IBM's program packages were supplied free of charge, with no intellectual property protection.

Hence, software packages were a major aspect of marketing strategy for all the mainframe companies by the mid 1960s. Packaged software was the principal way IBM expanded the sales of the 1401 to medium-size companies that could not afford (or even find) a programming work force. For the other mainframe producers, software packages were the principal means of differentiating their products from IBM's and targeting niche markets.

The First Software Products

The term "software product" firmly entered the computing lexicon in 1966. In that year, Larry Welke resigned his post as vice-president for data services at the Merchants National Bank and Trust Company of Indianapolis in order to found International Computer Programs Inc., a software products information service. The new firm's principal activity was to publish *ICP Quarterly*, a catalog of software products in which vendors could list their software packages free of charge. As was noted above, the emergence of software products transformed the software problem from a programming activity into an information-searching task. International Computer Programs played a crucial and largely unsung role in this process. Without ICP, the software products industry might have taken much longer to become established.[23] With a modest annual subscription of $25 for four issues, *ICP Quarterly* sold widely, but primarily in the computer using community, which was just beginning to see software products as a viable alternative to in-house programming or the hiring of a programming services company.[24] The first issue of *ICP Quarterly* appeared in January 1967. Around the same time, some of the key industry associations also began to produce program catalogs. For example, in the banking industry alone, the American Banking Association produced the *Abacus* listing, the US Savings and Loan League offered a *Software Exchange* service to members, and IBM published a *Finance Program Exchange Directory*.[25]

The distinction between a software package and a software product was that the latter was a traded capital good. A software product was a discrete software artifact that required little or no customization, either by the vendor or by the buyer; it was actively marketed, it was sold or leased to a computer user, and the vendor was contractually obligated to provide training, documentation, and after-sale service. In contrast, the manufacturer's "free" software package was supplied with few contractual obligations, and no matter how well the customer might be supported in practice, there was no legal requirement that the manufacturer supply a fully operational application.

The most prominent if not actually the first program to meet all the criteria of a software product was Applied Data Research's Autoflow. ADR had been approached by RCA in 1964 to develop a flowcharting program for its model 501 computer.[26] Until the mid 1970s, as has been mentioned, flowcharting was the most common method of documenting a program. A flowchart was a diagram that showed the logical flow of a program, and a good flowchart considerably eased the burden of software maintenance. Unfortunately, producing a final flowchart at the end of a system development project was famously the most hated part of the programmer's task, and for this reason it often did not get done—particularly if there was pressure to get on with the next development project. RCA wanted a software package that would enable programmers to produce flowcharts mechanically as a by-product of punching their decks of program cards, thus minimizing the documentation chore for the programmer and maximizing the chance that it would get done.

ADR had a long-standing relationship with RCA, going back to the development of its first COBOL compiler in 1960. As with its previous programming contracts, RCA offered to pay ADR a fixed fee for developing a flowcharting program, which it would then supply to users free of charge. Somewhat misjudging the strength of RCA's interest, ADR developed the program without a formal contract and offered it to RCA for $25,000. Unfortunately, RCA declined, as did other computer manufacturers ADR tried subsequently. Having invested $10,000 in the project, Martin Goetz, president of ADR, decided to market it direct to RCA 501 users. He named the program Autoflow, priced it at $2,400, prepared some direct marketing literature, and did a mail shot. This resulted in just two sales, the equivalent of about 2 percent of the RCA 501 customer base. This was disappointing; however, Goetz reasoned that if he had achieved 2 percent take-up by users of the IBM 1401, which had an installed base of about 10,000 machines, the story would have been quite different. ADR therefore rewrote the system for the 1401. This produced a mild flurry of interest but few sales. Follow-up inquiries suggested that, though the package was useful, what users were "desperate" for was a flowcharting package that would produce flowcharts for their *existing* software. This would be a major aid for installations trying to maintain software whose original authors had long since departed, leaving little or no documentation. Again, Goetz decide to take the risk. The new version of Autoflow required a significant further investment because the program was much more sophisticated than the previous version; indeed, it was comparable in complexity to a compiler.

The new Autoflow package sold quite well, in the original 1401 version and also in subsequent versions for the IBM System/360, the 360-compatible RCA Spectra 70, and the 1401-compatible Honeywell 200. By the end of 1968, 300 copies had been delivered. Though this was a success in itself, sales remained far below a 1 percent market take-up. Goetz was convinced that the major reason for the limited number of sales was the fact that IBM had begun to offer a package called Flowcharter for free. Although Flowcharter was not fully automatic, as Autoflow was, there was an expectation among users that IBM would eventually produce a program of comparable capability to the ADR package, and this kept them from paying for what they would eventually get for free. Moreover, Goetz had evidence that IBM salesmen had exaggerated Flowcharter's capabilities, suggesting that it was automatic when it was not. In retaliation, Goetz applied for patent protection and initiated a lawsuit against IBM, alleging that IBM had misrepresented the capabilities of its Flowcharter to the detriment of Autoflow.[27] This happened simultaneously with the IBM antitrust lawsuit.

As was noted, Autoflow has often been called the first software product. However, to focus on its priority is misleading. Autoflow is better viewed as simply one example of an idea that was occurring more or less simultaneously to many people in many places. The only common traits among these individuals were an entrepreneurial streak and experience in the computer industry. In September 2000, the Software History Center, an organization that seeks to record the origins of the software industry, organized a reunion of about a dozen software products pioneers, all of whom subsequently became major players in the industry.[28] Present at the reunion were the founders of Whitlow Computer Systems, Boole and Babbage, and Pansophic. Although there were differences in timing and in product offerings, there was a surprising uniformity to their business stories.

Whitlow Computer Systems was formed by Duane Whitlow in 1968. Whitlow had left a senior technical position in TWA's computer department to form a software products company in Englewood Cliffs, New Jersey. Wanting to go into software products rather than to step onto the treadmill of programming services or consulting, he had initially thought of developing a database system. However, while exploring this possibility he had become aware that IBM's sorting programs for the System/360 were notoriously slow. He therefore decided to see if he could write an improved version. While doing so, he stumbled across an undocumented machine instruction that provided a great improvement in sorting speed. (That IBM was unaware of this instruction may seem surprising today, but many early computers had instructions that the designers had not planned and which simply "fell out of the wiring." Rarely did they do anything useful, however.) Whitlow was able to get a patent on the undocumented instruction, which prevented even IBM from using it in its own programs. He then proceeded to write SyncSort, a program that operated much faster than IBM's "free" sorting programs. By 1975, Whitlow Computer Systems was able to advertise that SyncSort was used by more than half of the *Fortune*'s top 50 companies. Since IBM was seventh in the top 50, the advertisement noted wryly that it would never be 100 percent, because "no matter how good we make the sort—and improving it is a fetish with us—we're never going to be able to sell it to our Large Competitor, Old No. 7."[29]

Boole and Babbage was formed in San Jose in October 1967 by Ken Kolence and David Katch, both of whom worked with CDC's Palo Alto facility. Kolence had expertise in performance-monitoring software, which enabled the resources consumed by a user's programs to be measured so that their performance could be optimized. After starting their company, Kolence and Katch set about building software products for performance monitoring on IBM's new System/360 computers. To finance the development, they obtained $50,000 in venture capital from Franklin "Pitch" Johnson's Asset Management Company in exchange for a 95 percent stake of the company. Boole and Babbage's first two products, PPE (Problem Program Evaluator) and CUE (Configuration Utilization Evaluator), were launched in 1968. Applying the packages typically enabled a user to achieve a 10–20 percent saving in computer resources and either to get more work out of an existing installation or to defer the leasing of additional resources. For the next 30 years, Boole and Babbage dominated the niche in performance-monitoring software, and the firm was regularly among the top 20 independent software vendors (though not the top 10). In 1998m BMC Software acquired Boole and Babbage and its 950 employees for $900 million.

Pansophic was established by Joe Piscopo in 1969 in the Oak Brook suburb of Chicago. Piscopo, a trained computer scientist, was then studying for an MBA but was not sure of his career direction. As Piscopo has told it, he was attending a family gathering when, in a scene reminiscent of the "plastics" incident in *The Graduate*, a wealthy relative encouraged him to come up with a business plan to enter the computer business.

After a process of elimination, Piscopo decided that software products would be the most promising business. With $150,000 obtained from relatives, he first had to decide what product to make. "I finally came up with the idea of a library system. I had carried trays of punched cards in the back seat of my convertible from Montgomery Ward's to the Midwest Stock Exchange on many occasions with the constant worry that I would get the cards out of sequence. So we decided there would be a market for a library system that would make it possible to store programs on tape. It turned out that the $150,000 was more than enough. We were down to $10,000 by the time we finished the product. But that was plenty. With $10,000, we could have gone 3 more months."[30]

As a small company (ten people), Pansophic lacked marketing resources. It therefore tried to sell the program, called PANVALET, outright to one of the leading computer services firms. They all declined, even before a price had been mentioned, and Pansophic had to develop its own marketing capability. Piscopo, certain that PANVALET would be a winner, had hoped to sell it to a major computer services operation for between $5 million and $10 million. Although this says something about the exuberance of youth, PANVALET in fact went on to achieve sales of $100 million, and Pansophic was regularly among the top 10 independent software firms. It was acquired by Computer Associates in 1991 for $290 million.

Pioneering in the Software Products Industry: Informatics Mark IV

Although packages such as Autoflow, SyncSort and PANVALET heralded the software products industry, they were not the products that ultimately shaped it. That distinction must go to Informatics' Mark IV file-management system. Computer users could take or leave Autoflow, SyncSort, or PANVALET, essentially convenience products that were useful but not mission critical. In contrast, at its peak, Mark IV enabled more than 4,000 organizations to computerize their businesses. It was an expensive $30,000 product that was central to a computerized business operation. In importance it was second only to the choice of the computer—and when plug-compatible mainframes became available, in the mid 1970s, it was even more important than that.

From its launch in November 1967 until its peak in the early 1980s, Mark IV was the most important and best-selling product from an independent software vendor. It was the first software product to reach $1 million in cumulative sales, then $10 million, and eventually $100 million. At the annual "million dollar" awards ceremony organized by International Computer Programs Inc., it was often named the best-selling product from an independent software vendor.

More than any other product, Mark IV shaped the fiscal, marketing, and after-sale practices of the independent software products industry. The software products that followed were constrained by the price model established by Informatics, sold in a competitive environment based on the marketing policies established by Informatics, and provided with after-sale support in the form of maintenance and upgrades on the lines that evolved first with Mark IV.[31]

Product and Platform

As early as 1965, Walter Bauer, Informatics' president, had been keen to develop "proprietary programming items."[32] However, his concept then was of a permanent body of code that could substitute for the conventional writing of custom software for clients. Bauer had no concept of the high-volume, stand-alone software product that was to come—at that time, he has said, "we were only thinking 10s or 20s."[33]

If one person in Informatics can be said to have had a concrete vision of the software product, it was John Postley, Mark IV's inventor and its product champion.[34] Postley, a 20-year veteran of computing, had graduated in mathematics in 1948 and had subsequently been employed by the Institute for Numerical Analysis at the University of California in Los Angeles. He had then held computing posts with a number of aerospace companies, eventually landing at the RAND Corporation. While at RAND, he had become interested in the information-handling capabilities of computers, as opposed to their scientific uses. In 1960, he co-founded a small consulting firm, Advanced Information Systems, to pursue opportunities in non-numerical computing. That firm soon became the AIS Division of Hughes Dynamics, which Howard Hughes had established in order to enter the computer services market. It was one of about 30 small firms acquired by Hughes Dynamics. After a year or two of disappointing results, the capricious Hughes decided to withdraw from computer services and sold off the various Hughes Dynamics divisions. As a result, Informatics acquired, at essentially no cost, the AIS Division, including John Postley and a dozen programmers.

Along with AIS, Informatics also acquired Mark III, a file-management system invented by Postley. A file-management system performed two major functions. First, it took over the routine housekeeping chores of data management, thus eliminating the need for special-purpose file-management programs. Second, it provided a "report generation" facility that simplified the production of management reports and statistics, thereby reducing programming requirements from days to hours, or even eliminating the programmer altogether through the provision of standardized forms that could be completed by executives.

The Mark III file-management system, had been evolving since 1960. When Informatics acquired it, it was in its fourth incarnation. (The first three versions had been called RR10, Mark I, and Mark II.) It had been developed not as a product but as a software asset that could be customized for different clients. By 1967, however, software products were in the air, particularly since the launch of ADR's Autoflow, and Postley suggested that Mark III be turned into a product. Among its advantages as a product were the fact that it was complex and therefore not easily imitated and the fact that it was already highly developed. Although Postley was aware of several similar systems being developed by computer manufacturers and other software houses, Mark III was farther along than any of them. But the decision to go to market with a file-management system was not an entirely rational one, and Postley's role as a product champion should not be underestimated.

Postley estimated that it would take a year and $500,000 to "productize" Mark III—that is, to make it a self-contained and robust product. Not surprisingly, the product was to be called Mark IV. Much of its success was due to Postley's deep understanding of the difference between Mark III (an in-house program used for fulfilling custom-software contracts) and Mark IV (which was to be sold to heterogeneous clients for myriad applications). Mark IV would have to be exceptionally robust and essentially bug free—fixing bugs in the field would not be economical, and any early unreliability would quickly destroy the product's reputation. And Mark IV would have to be supported by manuals and documentation so that it could be used in domestic and overseas markets without day-to-day support from Informatics. The $500,000 that Postley requested was probably more than the cumulative cost of developing Mark III, which in any case would mostly have to be rewritten. Thus, what Mark III represented as a software asset was 7 years of organizational learning and market awareness, rather than a chunk of code.

Mark IV was to be developed for the IBM System/360, then fairly new to the market. In the long run System/360 was sure to dominate the mainframe market, and Informatics intended sales of Mark IV to grow in tandem with sales of System/360. The System/360 platform was seen to offer the greatest potential reward for the fixed development cost.

Financing

Assuming modestly optimistic sales to half a dozen or so customers, Postley estimated that the $500,000 needed to develop Mark IV would be returned in about 2 years. Informatics' board, however, felt that such a large investment was an unacceptable risk for a company whose annual revenues were then only $4.2 million. Bauer therefore encouraged Postley to seek out some customers who would be willing to fund the development by buying the product in stage payments before delivery. This was not as strange as it now sounds; pre-payment was quite common in the programming services industry, and the software products industry had not yet arrived. After several months of effort, Postley secured five sponsors, each willing to put up about $100,000. The first was Sun Oil, which signed on in November 1966. By July 1967, four other sponsors had signed on: National Dairy Industries, Allen-Bradley Inc., Getty Oil, and Prudential. The sponsors expected some royalties on future sales, but they left the technical direction entirely to Informatics.

The market for Mark IV would be determined largely by its selling price. This was a wholly novel situation in the computer industry, because—as was often commented upon—the incremental cost of producing additional copies of a piece of software was extremely small relative to the development cost. In the case of Mark IV, the incremental cost would be no more than $200. Although the existence of low incremental costs was not unprecedented in the business world (movies, book publishing, and pharmaceuticals had long been accustomed to it), it was novel in the computer services industry. However, closer examination indicated that the *selling* costs were likely to be not dissimilar to those of a medium-size computer, and this suggested a minimum selling price of $15,000. On the other hand, Postley's negotiations while he was pre-selling the product suggested that the market would bear a price of $100,000. Another view making the rounds of the industry (a view that ignored selling costs, incidentally) was that software products should be priced at one-fifth or one-tenth of the development costs, in order to recoup the outlay with the first five or ten sales.[35] To get a second opinion, Informatics commissioned a consultancy to undertake a market analysis. The analysis suggested that a selling price between $25,000 and $40,000 would maximize revenues. Informatics settled on a price of $30,000. The company got just one detail wrong: It greatly underestimated the cost of maintenance and the potential revenue from upgrades. It took several years to build these factors into its price structure.

Bauer later recalled that potential buyers were "astounded" by the price of Mark IV.[36] In a world accustomed to free software, the price of $30,000 was indeed astounding. It set a precedent for software pricing that rapidly diffused across the software industry. With time, pricing would become more nuanced and formulaic; however, the software product was always priced high.

Intellectual Property Protection

It was a commonplace in the software industry that programs were trivially easy to replicate. This had never been a practical problem; custom programs were generally too specific to an organization to be attractive to another user, and manufacturers' software packages were free. However, with software generally perceived as a free good, legal and/or physical protection of programs was important for the software products industry.

The issues had been widely aired in the computer trade press and in the legal profession since the mid 1960s.[37] The law offered several forms of intellectual property protection: patents, copyright, trade secrets, and trademarks. Patents were initially the most appealing form of protection for software, and in 1967 patents were applied for by ADR for Autoflow and by Informatics for Mark IV. Though ADR succeeded in getting a patent, this was achieved by claiming that the program was a "process" akin to a manufacturing process. It was not clear that the patent would stand up if challenged. Informatics failed to get a patent in the United States, though it was successful in Great Britain and Canada. The software patent situation was exceedingly complex, with a backlog of about 75 applications in 1968 and with five appeals in progress. In addition, there were complex public policy issues regarding the validity of software patents, insofar as patents were designed to protect tangible artifacts rather than "ideas." Exactly what kind of an artifact software represented was an open question.

Copyright law also had significant limitations for protecting software. In 1964 the US Congress passed legislation allowing a program to be copyrighted provided that it could be said to be a "literary expression" and provided that a human-readable copy was deposited in the Copyright Office. The latter condition made copyright protection unsuitable. Informatics took the view that, although copyright protection would make actual copying of a product illegal, it would not protect the algorithms and the know-how that the literary expression encapsulated. Disclosing the code would make it relatively easy to re-engineer the software and produce a functional replica perfectly legally. For this reason, Informatics decided to rely on trade secret law for protection. It required its customers and its employees to sign non-disclosure agreements. The programs were kept secret, only machine-readable binary code being supplied to the customer. The product was supplied by a perpetual license, which could be revoked should the user make unauthorized copies for other installations or for non-purchasers.

Informatics did, however, fully utilize copyright law to protect the supporting documentation for Mark IV as conventional literary works. This afforded considerable protection in itself, as the software would be difficult to use without the supporting documentation. In practice, the copying of software was never a problem in the early software products industry. User firms tended to be more likely to stay within the law than individuals (a situation much different from the one that would later develop with videogame and personal computer software).

Informatics also made good use of trademarks. The trade name "Mark IV" was registered, along with its distinctive visual expression, which was liberally imprinted on documentation and packaging, on the labels for magnetic tapes, and on the card decks. Trademarks were widely used by all the software products firms to foster brand images.

Marketing

For the marketing of its software products, Informatics adopted the customs of capital goods selling in general and those of mainframe marketing in particular. Indeed, the biggest single source of early software products salespeople was IBM.

Before the launch of Mark IV, Informatics had never run a sales force. Sales of custom software had generally been achieved through personal contacts at the board level, or by responding to requests for quotes for government contracts. The early sales of Mark IV were all closed by John Postley, largely on the strength of his enthusiasm as a product champion. Now, however, a more traditional type of sales force would have to be created. Postley decided to copy IBM's selling techniques "chapter and verse."[38]

A national sales manager was recruited to build establish a small IBM-style sales force. To Informatics' surprise and pleasure, 100 sales were made in the first 3 months. This encouraged the company to finance a rapid buildup of the sales force and to recruit field engineers to install the software on customer sites and instructors to train users in Mark IV techniques. A European sales office was opened. By mid 1969, the original 12-person Mark IV development group had become the 100-strong Software Products Division of Informatics, and the majority of these people were involved in marketing, installation, and training.

A marketing idea with some panache was to create a user group along the lines of SHARE. This idea came from Frank Wagner, a founding vice-president of Informatics and a founder of SHARE in 1955. The user group, cutely named the IV League, held its first meeting in Los Angeles. It was attended by about 50 people, each from one or another of Mark IV's original five sponsors. Postley held a barbecue at his father's Bel Air home. The IV League's first full-scale meeting, in February 1969, was attended by about 130 delegates representing 80 companies. Following the practice of SHARE, the IV League was completely independent of Informatics.

By the spring of 1969 there were 121 installations of Mark IV in the United States and 16 in Europe. This was wildly in excess of anything anyone at Informatics had dreamed, except perhaps John Postley. But the best was yet to be. IBM was about to start selling software, thereby greatly expanding the market for software products.

IBM's Unbundling Decision

In 1967 the Antitrust Division of the US Department of Justice began an investigation of IBM, which was widely reported to have a 70 percent share of the domestic computer market. Since a high market share was not of itself a violation of the Antitrust Act, attention focused on IBM's commercial conduct. Among IBM's alleged offenses was its practice of "bundling," defined in court as "the offering of a number of elements that are considered to be interrelated and necessary from a customer's point of view, in the computer field, under a single pricing plan, without detailing the pricing of the component elements themselves."[39] It was argued that by this practice IBM was competing unfairly with other manufacturers by supplying software and other services, without regard to their true cost, in order to win an account. In effect, IBM was selling below cost—unquestionably an antitrust violation. Other alleged malpractices included premature product announcements and predatory pricing, subjects of a private lawsuit by CDC in December 1968.

Against this background, on December 6, 1968 (only days before the CDC lawsuit), IBM announced its intention to unbundle its prices and to charge separately for the five service elements then included in the overall price charged to the customer: systems engineering, education and training, field engineering, programming services, and software packages. IBM committed to announcing the full details of unbundling by the end of June 1969.

IBM has never officially admitted that the unbundling and antitrust issues were connected. At the trial, IBM's defense counsel argued that in the 1950s and most of the 1960s bundling had been necessary in order to make the unpredictable costs of running a computer—particularly the software costs—more certain. However, by the late 1960s many users were leasing their second or third computer, no longer had need of these bundled services, and were beginning to resent the fact that they were effectively cross-subsidizing less sophisticated users. Thomas Watson Jr. testified: "We had some very sophisticated customers by this time, Lockheed, Boeing and others, who felt that they were better at performing some of these services than we were. They felt it onerous to pay for them when they, themselves, could do it in their opinion better."[40] In their published account of the trial, IBM's defense team wrote: ". . . it was natural for IBM to consider unbundling in the late 1960s, but not before. IBM's announcement on December 6, 1968, of its intention to unbundle came before the filing of the government complaint in early 1969, but after the Antitrust Division's investigation had been underway for some time, as had consideration of unbundling. No evidence in the trial record suggests that the two events were related."[41] This may have been true, but at the operational level within IBM there is no doubt that the looming antitrust action was uppermost in the minds of those charged with determining the unbundling strategy. According to Burton Grad, a member of the unbundling task force: "IBM believed that it could prevent a US government suit by announcing that it was going to unbundle its services and by doing so promptly. This seemed far better than waiting for a suit and then negotiating a settlement or consent decree."[42] However, time was not on IBM's side. On January 17, 1969, the last day of the Johnson administration, the Attorney General of the United States filed an antitrust suit against IBM.

To explore how best to organize the separate pricing of software and services, IBM created an unbundling task force of almost 100. The members of the task force were on full-time assignment, with no other duties, and were not permitted to have business contact with other IBM employees except to collect information. The task force was divided into teams, each dealing with one of the five bundled services (systems engineering, education and training, field engineering, programming services, and software packages).

All the unbundling decisions were made in the light of the ongoing antitrust action. There were two conflicting objectives: the need for a gracious and positive acceptance of unbundling in order to present the best possible face to the antitrust trial, and the desire not to give up the marketing benefits of bundling in closing sales. The problem facing the unbundling task force was "how to minimize the programs which could be separately priced," and "the expressions used were to 'draw the line' and 'hold the line.'"[43]

The team considered Type I software, systems programs, and Type II applications programs separately. The Type I software consisted of the very large monolithic operating systems such as OS/360, on the one hand, and a large number of programming tools and utilities—language processors, sorting programs, file-management systems, data communications software, etc.—on the other hand. It was rather easy to justify keeping the operating systems bundled, on the technical grounds that "hardware and software implementation of certain technological functions were closely coupled" and "the decision on how to implement a function should be based on engineering and economic considerations, not on pricing polices."[44] Another reason was that the development expenses—especially those of OS/360—had been so enormous that no realistic level of pricing could begin to recover the sunk costs. Other than the operating systems, however, there was no technical case for keeping the other systems programs bundled, and they were priced separately.

The decision not to charge separately for the operating systems—the most controversial of the unbundling outcomes—maintained IBM's operating-system monopoly into the late 1970s, when at last the operating systems were priced separately. Critics argued that bundling discouraged innovation in operating systems and made it difficult for manufacturers of IBM-compatible equipment to compete. There was also suspicion of a darker motive: that IBM benefited from an inefficient operating system because it forced users to lease more powerful computers than they would otherwise need. Martin Goetz, the president of ADR, asked rhetorically: "In 1970, the operating systems were the *one* piece of software that IBM chose not to unbundle. Was it an act of kindness?" He then answered his own question: "IBM doesn't charge for them simply because it is significantly more profitable in the long run to lock customers into inefficient operating systems and lock out competition."[45]

The task force was also charged with recommending pricing strategies and policies regarding intellectual property protection. Prices were to be determined by the standard pricing methods that IBM used for its hardware. This was a time-honored process, dating from at least the 1930s, that involved predicting the revenue that would be derived at different price levels, with the need to recover development and marketing costs taken into account. The price selected was the one that maximized the sales opportunity and net income over time. To ensure intellectual property protection, the task force recommended that software be leased rather than sold, as it would be easy to revoke a leasing agreement should a user violate the terms of the contract. Thus, software packages would be subject to a monthly leasing charge added to that for the hardware. This had a further benefit for IBM: It would make it much harder for independent software firms to compete, because they generally did not have access to leasing funds.

The situation with regard to Type II application programs was much more complex, mainly because of the hazy distinction between a software package and a software product. Some application packages—especially the more scientific and mathematical ones, such as linear programming and project-management packages—presented little difficulty. These were already self-contained packages that required negligible support and were therefore easy to unbundle. However, most of the industry-specific programs had long been supplied as illustrative examples rather than finished products, and they usually required extensive customization by IBM's systems engineers. (For example, IBM's PARS airline reservation system, though described as a package, was more exactly a framework for application development that cost about $5 million to customize.[46])

Another complication was that some bundled application programs were legitimately being supplied as turnkey products—products with which a complete package of software and hardware was supplied for a dedicated task. (One of IBM's most successful products in this class was the 1130-based typesetting system.) Yet another difficulty was that some potential products were the results of ongoing developments with customers, and their development costs could not be reliably estimated.

As a result of these complications, each actual or potential software package had to be treated on its merits—whether it should be reengineered into a product, left as forever-free legacy software, or removed from the catalog. Once the broad unbundling strategy had been determined, by May 1969, the original task force was expanded by several hundred staffers to resolve the tactical issues for each of the bundled services.

On June 23, 1969, IBM announced its unbundling plan, which was to take effect January 1, 1970. Systems engineering, field engineering, and education and training were all to be priced separately. Custom programming would be charged on a cost-plus basis, with the option to bid for fixed-price contracts; thus, IBM would be free to compete head on with the programming services industry. Finally, all systems programs (excluding operating systems) and application programs would be subject to a monthly leasing charge, which would include support and product maintenance. Seventeen products were initially announced, including language processors (Assembler, FORTRAN, COBOL, and PL/1), the GIS file-management system, the IMS database, and the CICS data communications program. Where programs were already in the public domain (as was true of the language processors), leasing charges were kept modest and were to be reviewed as successive releases were issued. At the same time, IBM reduced its hardware prices by an average of 3 percent.

The measly 3 percent price reduction was roundly criticized by IBM users and watchers. Before IBM's announcement, there had been numerous speculative estimates of the potential size of the software products market that would be liberated by unbundling. Estimates had varied from as little as 2 percent of the total computer market to as much as 50 percent.

Probably the most optimistic view of the potential market for software products was that of ADR's president, Martin Goetz. In April 1969, ADR had filed its David-versus-Goliath suit against IBM. This tiny company, with annual revenues of $4.2 million, had claimed damages of $300 million in a catalog of complaints, some specific and some general. Most specifically, ADR had alleged that IBM had damaged the sales of Autoflow by making exaggerated claims for its rival Flowcharter package. More generally, ADR had argued that, by the practice of bundling, IBM had suppressed a huge potential software market for a decade. ADR argued that software accounted for 35–50 percent of IBM's sales, that in 1968 alone the software market that IBM had "thus foreclosed to ADR and all others" was valued between $1.95 billion and $2.79 billion, and that cumulatively over the period 1959–1968 ADR and others had been deprived of a market of between $9.6 billion and $13.75 billion.[47] On its own account, ADR claimed damages of $300 million, representing $97.5 million in lost sales to IBM users, $32.5 million in lost sales to non-IBM markets (since the universal practice of bundling had been legitimated by IBM), plus $150 million because IBM had "retarded [ADR's] economic and technological growth." Besides these damages, ADR argued that IBM should divest itself of all software activity and create a $3.57 billion fund to compensate users of bundled software, which they had paid for but had not taken advantage of. Shortly after initiating the lawsuit, ADR acquired Programmatics Inc. of Los Angeles, another software products firm that had filed a suit along similar lines. The two lawsuits were now combined into one, with a total value of $900 million. IBM subsequently settled out of court for $2 million, dropped its Flowcharter package, and agreed to market some of ADR's products. The $2 million was said to be just about enough to cover ADR's legal expenses, but ADR gained enormous publicity from the case, establishing a reputation as a scourge of IBM.

Although in retrospect ADR's action seems close to frivolous, it did reflect a commonly held belief that the potential market for software products was huge. Watts Humphrey, the IBM manager in charge of systems software, was only too aware of a "public perception that as a result of pricing the software, the hardware price would be reduced by 25 to 40 percent."[48] IBM went to some lengths to disabuse users of this notion, participating in panels on separate pricing both at the 1968 Spring Joint Computer Conference and at the fall meeting of IFIP. "While I could not tell them what the price impact would be," Humphrey explained, " I said that only about 3 percent of IBM's employees then worked on software. A much larger price change would thus seem unlikely."[49]

IBM's 3 percent reduction in the price of hardware was widely viewed as conservative in the extreme, and indeed as a covert way of raising prices. In fact, the price reductions were in line with informed economic analysis that pre-dated the announcement.[50] The main reason for the confusion was a failure to distinguish between R&D costs and manufacturing and sales costs. For 2 or 3 years, computer users had been assailed by press reports that software was the dominant cost of System/360, with percentages ranging from 40 to 60 commonly expressed. What these reports neglected was that R&D accounted for only about 5 percent of the cost of an IBM mainframe; manufacturing, marketing, and profits accounted for the remaining 95 percent. For bundled software, the manufacturing costs were negligible and no marketing was done. If that software development constituted between 40 and 60 percent of the R&D budget, the cost of software supplied to users actually amounted to between 2 and 3 percent of the total.[51]

After Unbundling

The historian JoAnne Yates has described unbundling as the "crucial inflection point" in the development of the software products industry.[52] This is an apt characterization. Although the software products industry was viable and growing before IBM's decision, unbundling resulted in a marked increase in the rate of growth. In 1969 the total revenues of software products companies had been variously estimated between $20 million and $50 million; after unbundling, analysts predicted the industry would grow much faster than the computer market as a whole, to $400 million by 1975. This prediction turned out to be substantially correct. In May 1969, in the run-up to the announcement, *Datamation* enthused: "As the great day nears when IBM will make the awaited announcement of some form or other of separate pricing of hardware and software, the independent software companies and packages proliferate and the usual chaos continues. No one is quite sure what the effect of separate pricing will be, but there is overwhelming confidence of a bright future for all and activity is increasing and bustling, prompting the conclusion that when the shakeout comes, it will, indeed, be a major one. Something is happening everywhere."[53]

The immediate beneficiaries of unbundling were the existing computer services and software firms—not only those with products, such as ADR and Informatics, but also those with unrealized software assets. For example, CSC (the dominant computer services firm, with a staff of 3,200) immediately moved into software products. In 1969 it announced twelve packages, including the Cogent II file-management system, the Compuflight II reservation system for regional airlines, and Exodus, a program for converting IBM 1401 software to the System/360 platform. These pre-existing programs, developed as in-house programming tools or for custom-written software contracts, represented an opportunistic exploitation of sunk development costs rather than a complete change of direction. CSC remained (and remains) a computer services business at its core. Many of the other computer services firms—Boeing Computer Services, McDonnell Douglas Automation, TRW, SDC, and Westinghouse, to mention only a few of the largest—also capitalized on their existing software assets by quickly bringing software products to market.

The two most prominent firms already selling software products in the 1960s—ADR, with its Autoflow flowcharting package, and Informatics, with its Mark IV file-management system—both experienced dramatic increases in sales after IBM announced unbundling.

By late 1968, ADR had sold only 300 units of Autoflow, the revenues from which had contributed no more than 10 percent of its $4.2 million annual turnover. Once IBM's decision to unbundle was announced, sales rose significantly. The first quarter of 1969 produced 100 sales, and ADR felt confident enough to install an IBM 360/50 in its Princeton headquarters for future product development. Fueled by the sales of Autoflow and of its other major package (Librarian, a utility for managing program development files), ADR grew at a rate of 20 percent for the next several years. In 1970, these two products alone accounted for $3 million of ADR's annual sales of $7.2 million.

In 1973, by which time Autoflow had about 2,000 users, a revised version, Autoflow II, was released. ADR had opened eight sales offices in the United States and was represented in 31 other countries. It now specialized in programming tools and aids, offering seven products (including MetaCOBOL, Look, and Roscoe) in addition to Autoflow and Librarian. In 1978, ADR entered the database market with a product called DATACOM/DB. By this time, ADR was deriving the great majority of its revenues from software products and had relegated its custom programming business to a subsidiary. Autoflow continued to flourish. It was the first software product to reach 10,000 unit sales. In 1976 it still brought in $700,000, about 5 percent of ADR's annual sales. However, by the early 1980s flowcharts (and flowchart packages) were old technology, and sales fell off quickly. In all, Autoflow had a life of about 15 years. That turned out to be typical of the software products industry, and it was much longer than early entrants had expected.

Informatics' Mark IV was well on the way to becoming a successful product well before IBM announced its decision to unbundle. Orders for 117 systems had already been received, and 1968 sales amounted to $1.8 million. Before unbundling, sales of Mark IV had been projected, in the corporate 5-year plan, to rise from $1.2 million in 1969, peak at $1.6 million in 1971, and decline to $0.8 million in 1973 as the market saturated (figure 4.3). In fact, sales widely exceeded this projection. In 1969 sales were $2.8 million, and in 1973 they soared to $7.5 million.

Mark IV sales rapidly became the single biggest source of income for Informatics, reaching 25 percent of its annual revenues. The high revenues were matched by unanticipated sales costs of an astonishing 55 percent, and it typically took between 6 and 12 months to close a sale. This was a new kind of business, and it transformed Informatics' outlook and culture. Once primarily a contractor of programming services to the federal government, Informatics was now a supplier of software products to private industry. By 1972, there were 500 installations, and Mark IV had taken on a life of its own. Meetings of the IV League user group drew as many as 750 participants, and there were chapters in Europe and Japan. In addition, a large number of imitators had come on to the

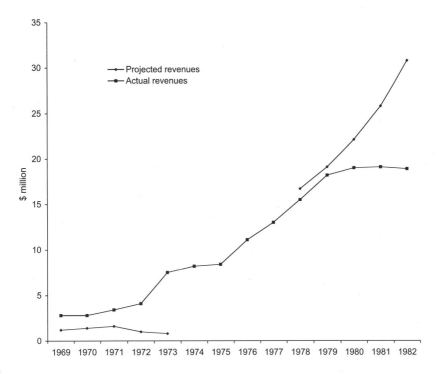

Figure 4.3
Mark IV revenues, 1968–1982. Source: Forman, *Fulfilling the Computer's Promise*, pp. 9/40–9/41.

market, including Application Software's ASI-ST, TRW's GIM, CCA's IFAM, CSC's Cogent II, Infodata's INQUIRE, Mathematica's RAMIS, and SDC's ORBIT—not mention IBM's GIS.[54] Each of these products sold or leased for a similar price to Mark IV and offered comparable facilities. In order to protect Mark IV's market share, Informatics decided to invest heavily in further product development, sales, training, and support facilities. Six sales offices were opened in the United States, and there were offices in Geneva, Copenhagen, London, Paris, Rome, and Germany.

In 1973, Informatics decided to make Mark IV its major activity. The sales force was doubled, and a heavy investment was made in a major new release: Mark IV/2. Complementary products were introduced, including Accounting IV, ICS IV, Bibpro IV and Recon IV (all developed internally) and CL*IV, Shrink, Score, and Monitor IV (acquired by purchase).

By 1975, annual sales of Mark IV amounted to $7.4 million (of which $2.2 million came from Europe), accounting for 32 percent of Informatics' revenues. The five-year plan for 1978–1982 projected continuing growth in Mark IV sales, from $16.7 million in 1978 to $30.8 million in 1982. In 1983, Mark IV's cumulative sales passed $100 million, a first for the independent software products industry. However, sales had hit a plateau in 1980 and had begun to decline. In 1982, sales were just $18.9 million, only 60 percent of the projection. Mark IV was reaching the end of its product life. After 15 years in the field, it was old technology.

Summary

The concept of a software package evolved gradually during the first half of the 1960s. At first, "pre-packaged" software was a technological response by computer manufacturers to a changing environment in which the number of computers was increasing logarithmically while the number of programmers was growing linearly. Packaged software was the only way to alleviate the programming bottleneck.

A market for software packages developed in the mid 1960s as it became clear that "free" software packages from manufacturers could never satisfy the burgeoning demand for software. Existing programming services firms were the first to exploit this opportunity, and they introduced the term "software product" in place of "software package." The term "product" was consciously adopted by vendors to imply a new kind of software artifact for which the vendor took contractual responsibility for performance and reliability in exchange for the license fees paid by users.

The manufacturing cost of software was negligible, and this initially led software pundits to assume that software would be cheap and would have a business model quite unlike that of hardware. In fact, as is perhaps obvious only in hindsight, corporate software products, with high marketing expenses, costly pre-and post-sales support, and ongoing maintenance and enhancement, turned out to differ little from other capital goods. Whatever their preconceptions, the early suppliers of software products quickly recognized the nature of the business they were in and established account managers in the IBM style, technical sales forces, user training, product documentation, and after-sale support. These business practices, fully shaped by the end of the 1960s, still persist among today's enterprise software vendors.

Although IBM's decision to unbundle is one of the landmarks of software history, the evidence suggests its effect was smaller than is commonly believed. One reason the real effect of unbundling is hard to discern is that its was swamped by larger economic forces: the crash of software stocks in the spring of 1970 and the computer recession of 1970–71. The software products industry did not walk into an optimistic, boundless future, but into the industry's most turbulent and difficult decade.

CORRECTION

"Is it true that almost half the 50 largest firms sort with SyncSort?"

"No. More than half."

(When you're running a bandwagon, it's hard to count the passengers.)

Call (201) 568-9700.

Find out what "Sort Support" really means.

OVERSEAS REPRESENTATIVES —
CAP-Gemini/CES, Brussels; CAP-Gemini GMBLL, Dusseldorf; CAP-SOGETI, Geneva; CAP-Gemini/Pandata, London; Syntax, Milano; CAP-Sogeti Prodvis, Paris; BRA, Stockholm; Advanced Technology, Tel Aviv; Shell Oil Co./of Australia; Melbourne; Deltacom Do Brasil, Sao Paulo.

WHITLOW COMPUTER SYSTEMS Inc. 560 Sylvan Ave., Englewood Cliffs, N.J. 07632

Our business is growing so fast we may have to hire a house statistician!

As an example of what we mean, take a look at the list below. It includes the names of the crème de la crème—the 50 largest companies, ranked by sales, on Fortune's famous 500 list of industrial corporations.

RANK	COMPANY '75		
1	Exxon	26	Phillips Petroleum
2	General Motors	27	Union Oil of California
3	Texaco	28	Bethlehem Steel
4	Ford Motor	29	Caterpillar Tractor
5	Mobil Oil	30	Eastman Kodak
6	Standard Oil of California	31	Rockwell International
7	International Business Machines	32	Dow Chemical
		33	Kraftco
8	Gulf Oil	34	RCA
9	General Electric	35	Esmark
10	Chrysler	36	Sun Oil
11	International Tel. & Tel.	37	LTV
12	Standard Oil (Ind.)	38	Beatrice Foods
13	U.S. Steel	39	Xerox
14	Shell Oil	40	United Technologies
15	Atlantic Richfield	41	Greyhound
16	Continental Oil	42	Firestone Tire & Rubber
17	E. I. du Pont de Nemours	43	Boeing
18	Western Electric	44	General Foods
19	Procter & Gamble	45	Ashland Oil
20	Westinghouse Electric	46	Monsanto
21	Union Carbide	47	W. R. Grace
22	Tenneco	48	R. J. Reynolds Industries
23	Goodyear Tire & Rubber		
24	International Harvester	49	Litton Industries
25	Occidental Petroleum	50	Lockheed Aircraft

Toss a dart at the list today, and you've got a better than 50-50 chance of finding a SyncSort user. Wait a few days or weeks—and who knows? Maybe 80-20?

It will never be 100-0, of course. Because no matter how good we make the sort—and improving it is a fetish with us—we're never going to be able to sell it to our Large Competitor. Old No. 7, above.

However, our disappointment is somewhat lessened by the knowledge that we've got a similarly high number of satisfied users among the other 450 companies on Fortune's list.

The reasons for SyncSort's success are fairly simple:

1. Performance—SyncSort outperforms any other sort on the market in all areas: Elapsed Time, Total CPU Time and I/O Activity.

2. Sort Support—Ours is better than our large competitor's. Proximity is one reason. Our sort support people sit right next to the people who develop the sort—not in another town, state or country.

If you'd like to confirm the last point, try this test. Make an intentional "zap" in any IBM sort and then call them for help. Then make a similar "zap" on SyncSort and give us a ring. Compare the answers you receive—and the time it took you to get them.

If we don't beat No. 7, we'll eat the whole town of Armonk.

SyncSort, a classic complementary product, enhanced IBM's System/360 operating system.

5
The Shaping of the Software Products Industry, the 1970s

For all the euphoria that IBM's unbundling announcement created, the 1970s was a disappointing decade for software products. Nonetheless, it was during the 1970s that the industry was shaped and acquired the characteristics that persist to the present day.

The software products industry is best understood though two vectors of analysis, with the organizations that produced software on one axis and their products on the other. This might suggest that the industry could be visualized in the form of a matrix, with vendors along the rows and products across the columns. However, there are thousands of software firms and tens of thousands of products, so such an image cannot be represented graphically. Instead, industry watchers have used these two vectors of analysis by considering each as a separate typology. This is the approach taken in this chapter. It leaves the reader to perform the necessary mental integration, but that has always been the case.

This chapter gives an overview of the business environment, the various types of software suppliers, and the their markets. It then explores the "product space" and describes representative examples of the various software genres.

The Business Environment

As figure 5.1 shows, sales of software products grew steadily but undramatically during the 1970s. The primary reason for the slow growth of the software products industry was a capital shortage that made it difficult for incumbents to expand and even more difficult for new firms to become established. Every one of the top 10 software products firms in 1979 had existed before the decade began. The lack of confidence among investors was due primarily to the hangover from the go-go years, compounded by the 1970–71 computer recession. Stocks hammered in

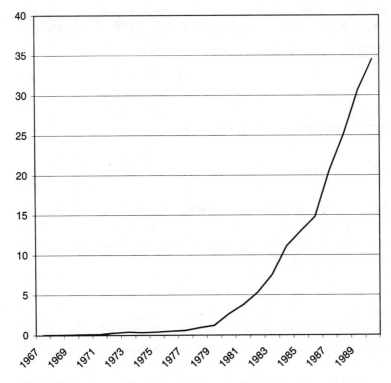

Figure 5.1
Worldwide sales of US software product firms (billions of dollars), 1967–1990. Data for 1967–1973 from Phister, *Data Processing,* p. 277; data for 1974–1990 from Association of Data Processing Service Organizations (attributed to IDC and INPUT).

the early 1970s took years to recover. For example, the software products pioneer ADR was one of the "software stock disasters of 1969–70," when it was "one of the $40-plus kites that plummeted like a squadron of bricks."[1] ADR's share price did not reach its low point of $1.50 until 1975. ADR at least managed to stay in the black. Informatics ended 1970 with a $4.2 million loss, the first since its founding in 1963. Investor confidence took a long time to recover. Not until 1978 did another software firm (Cullinane) go public.

Despite the prevailing gloom among investors, the 1970s offered bright new opportunities with modest capital requirements for software products startups. A vast new market for software was created by the rise of small computers. These systems sold in the tens of thousands to medium-size firms that did not employ computer programmers and were therefore entirely reliant on packaged software. The most popular small-business system was the IBM System/3, of which an estimated 13,000 had been installed by 1973. Although IBM dominated the System/3 software market by developing applications packages for the major industries, it also created "unique opportunities for small independent companies." Scores of startups rushed in, "most having less than a dozen employees, and several [were] one-man the firms."[2] Minicomputers, which had originated in technical environments, also created a new market in which startups could flourish out of the shadow of the major software firms. Some of these startups, including Oracle and ASK, became major players in the 1980s.

By the early 1970s it was well understood that the software products industry was a classic capital goods business. Its products were expensive to develop, sold in a relatively small numbers, and required a great deal of technical and after-sale support. Martin Goetz, founder of ADR, was quoted as saying: "The once-current publishing or record company analogy has been shunted aside in the industry today. This is not a get rich quick scheme. There is a minimum wait of three to five years for a return on investment on a successful product: 1.5 years for development; 1.5 years for product acceptance and refinement; one to two years during which recovery of investment begins, although technical costs are still incurred for continuing enhancement and support."[3] For example, it was estimated that the median cost of developing a database system was between $500,000 and $1 million, and that it might take up to 5 years to achieve 500 sales at a unit price between $25,000 and $50,000.[4] Development expenses, far from being a negligible part of the selling price, dominated costs in the early years, and it could be 2 or 3 years before a product became profitable. Marketing was an even bigger expense. Goetz noted: "On every dollar of revenue about 30 percent to 50 percent of costs are expended in sales; 30 percent is on the low side, generally it's 40 percent to 45 percent. Another 10 percent is in non-sales marketing costs, about 20 percent in product development, and about 10 percent in support. The margins of many software companies are between 10 percent and 30 percent, with overhead rising to support your more successful products."[5] Somewhat less anecdotally, Werner Frank of Informatics came up with similar data (table 5.1).

In fact, Goetz and Frank turned out to have been pessimistic. This was due largely to the unanticipated longevity of the IBM System/360 architecture—a longevity not expected even by IBM, which had planned to

Table 5.1
Five-year revenues and expenses of a typical software product.

Revenues	
Product sales	$19,500,000 (70%)
Maintenance	$8,500,000 (30%)
Total	$28,000,000 (100%)
Expenses	
Engineering	$4,000,000 (14%)
Marketing	$13,700,000 (49%)
Total	$17,700,000 (63%)
General and administrative (10%)	$1,800,000 (6.3%)
Gross profit	$8,500,000 (30.7%)

Source: Frank, *Critical Issues in Software*, p. 194.

replace its mainframes with the Future System in the second half of the 1970s.[6] However, in 1975—partly because of technical problems but mainly because of the extent to which people had become locked into System/360-based software—IBM abandoned its plans and stayed with the System/360 architecture. Until then, the conventional wisdom had been that the life of a software product was about 5 years, and software was priced to recover development costs in that period. When the life of System/360 was extended, the life of its software packages was extended too. With regular product upgrades, a life of 10 years, 15 years, or longer turned out to be not uncommon. This was perhaps the most significant reason why the major vendors of software products were able to survive the 1970s.

Suppliers of Software Products

Table 5.2 lists the sources of software products in the 1970s. The three major sources were computer manufacturers, independent software vendors, and turnkey suppliers. Whereas computer manufacturers and independent software vendors supplied programs plain and simple, turnkey suppliers delivered complete packages of software and hardware. Two further sources of supply in the 1970s were software brokers and time-sharing services. A software broker mediated between users and package suppliers who lacked marketing capability; a time-sharing service enabled a user to run a program from an on-line terminal. Brokers and time-sharing services had both faded away by the end of the 1970s.

Table 5.2
Sources of software products in the 1970s.

	Examples
Computer manufacturers	
IBM and the "bunch"	IBM, Univac, Honeywell
Minicomputers	DEC, HP, Prime
Small-business systems	IBM, Singer, Plus-Four
Independent software vendors	
Software products	ADR, MSA, Cullinane
Computer services	UCC, CSC
Turnkey vendors	Triad, Wang Laboratories, Computer Vision
Software brokers	COSMIC, INSAC
Time-sharing services	Tymshare, GE Time-Sharing System

Computer Manufacturers and Captive Revenues

Immediately after IBM unbundled, computer manufacturers were the main sources of packaged software. IBM was, not surprisingly, the biggest supplier, with three-fourths of the worldwide computer market. Initially only 17 software packages had been unbundled, but as IBM's programs were upgraded they were designated as paid-for System Products, usually with the suffix SP added to the program's name.[7] The early free-of-charge releases were stabilized, and educational and technical support was gradually withdrawn. By 1975 it was estimated that IBM had 40–45 percent of the market for packaged software.[8]

Historically, manufacturers had supplied both systems programs and applications programs. Systems programs were an essential operational component of a computer; application programs were primarily a marketing aid. After unbundling, manufacturers continued to supply most of the systems software for their computers (particularly operating systems, utilities, and programming tools), and it was difficult for the independent software vendors to compete in this market. As a result, users had little choice in their basic operating software, and for this reason manufacturers' software package income was usually referred to as "captive revenues."

Despite the lack of choice in systems software, users perceived unbundling as advantageous. It made pricing more transparent, and users no longer had to passively tolerate the take-it-or-leave-it attitude when "free" systems software was supplied solely on the manufacturer's terms, sometimes with unacceptable delays and bugs.

Independent Software Vendors

The structure of the software industry has been characterized as "boulders, pebbles, and sand"—that is, a few large firms at the top, a moderate number of medium-size firms, and a vast number of small firms.

In an authoritative 1980 report, industry analysts identified 6,104 firms that claimed to supply software packages—a number they regarded as "patently ridiculous": "In our opinion, any firm within an income under $500,000 per year is marginal. In a business dominated by regional suppliers, such firms are truly local. We cannot deny that small firms do play a role in the industry, but that role is limited."[9] For that reason, the aforementioned analysts considered the 940 software products firms with verifiable annual incomes exceeding $500,000 to constitute the core software industry. These 940 firms (table 5.3) accounted for 68 percent of the $2.5 billion package software market, the remainder being shared by perhaps 5,000 firms with incomes below $500,000. Then, as now, the term "software product" had cachet, and arguably a firm that had sold a product to a dozen firms in a small geographic area was really in the custom software business.

The most interesting firms, perhaps, were the big software products vendors established in the late 1960s that grew to prominence in the 1970s—e.g., Cincom, Cullinane, Information Sciences, and Pansophic. Another group of firms had longer histories as programming services and computer services firms but had also become significant players in software products—e.g., Informatics, University Computing Corp., and MSA. As figure 5.2 shows, packaged software remained well under half of the business of some firms (e.g., Informatics and UCC), whereas it became the whole of the business of others (e.g., MSA).

The 1970s brought many new entrants into the software products industry. Typically, for lack of capital, they started out as programming contractors so as to maintain their cash flow in the early years, but their medium-term aim was to move into products. A few of these firms, including Computer Associates, SAP, and Hogan Systems, became major players in the 1980s. Of course, the great majority of startups did not become global players. It was quite possible—indeed it was the norm—for a software firm to coast along for 10, 20, or 30 years with a single product and a small number of customers, eventually fading away as the owner approached retirement. Because of their size and their low profiles, the histories of these firms are elusive. However, there is one well-documented example: the MacNeal-Schwendler Corporation (MSC), whose founder Richard MacNeal privately published a detailed company

Table 5.3
Revenue distribution of suppliers of packaged software, 1980.

Revenues	Number of suppliers	Market share Size	Percentage
≥$20,000,000	3	$200,000,000	8%
$10,000,000–$19,999,000	8	$113,000,000	5%
$5,000,000–$9,999,000	25	$188,000,000	8%
$2,500,000–$4,999,000	90	$339,000,000	14%
$1,000,000–$2,499,000	238	$417,000,000	17%
$500,000–$999,000	576	$432,000,000	17%
Total of top 940	940	$1,689,000,000	68%
$250,000–$499,000	899	$337,000,000	14%
$100,000–$249,000	1,577	$276,000,000	11%
0–$99,000	2,688	$168,000,000	7%
Total	6,104	$2,470,000,000	100%

Source: Business Communications Corp., *Software Packages,* p. 87.

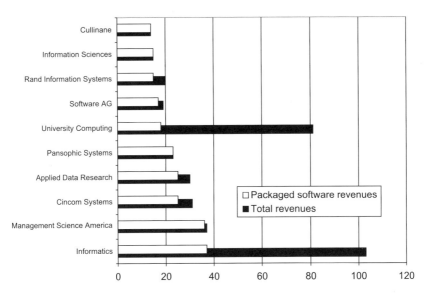

Figure 5.2
The revenue mix of the top ten software products firms, 1979. Source: "A Rush of New Companies to Mass-Produce Software," *Business Week,* September 1, 1980: 54–56.

history.[10] Founded in 1968, MSC was a subcontractor to CSC in the development of a stress-analysis program for NASA called NASTRAN. In 1972, MSC turned this program into a product called MSC/NASTRAN. Not until 1980 were 100 sales achieved; MSC's revenues in that year were still well under $5 million, and it had only 60 employees. Thus, MSC is perhaps more "average" than the larger, more visible companies.[11]

From a historical perspective, the boulders and the pebbles are easier to deal with. The boulders either still exist or were taken over by other firms in the consolidations of the 1980s and the 1990s; in any case, there are strong historical traces. The pebbles—especially those, like MSC, that became significant players—have some visibility. But the sand—the thousands of firms with a few employees each—are simply lost in the mists of time. Even in the aggregate they defy informed analysis.

Turnkey Suppliers

Users of small-business systems and minicomputers were confronted by two problems. First, they usually had neither the technical expertise nor the staff resources to operate a computer day to day. Second, they found it difficult to integrate software products with other software packages and with the hardware. A turnkey vendor addressed these two problems by supplying a complete package of hardware, software, and ongoing support. A prominent late-1970s example is Triad Systems. Founded in 1972 to supply an industry-specific turnkey package for auto parts suppliers, by 1984 it had become "the largest and most lucrative turnkey company in the business":

> The system Triad sells is complete, from soup to nuts: hardware, software, service, training, and future software enhancements. The entry-level price for a Triad system is $50,000. By reducing receivables and inventory levels, and accelerating inventory turnover, the typical retail auto parts store recoups its investment in 9 to 18 months. The Triad system is well suited to its market. It better be. When a small retail business puts all its financial, accounting, inventory, and order entry procedures on one system, that system better work right. . . . By the end of 1983, Triad had . . . 75 percent of the auto parts computer market.[12]

Turnkey suppliers flourished in vertical markets such as retail and professional practices. There were also two major cross-industry markets: office automation and computer-aided design. The market leaders were Wang Laboratories and Computer Vision, respectively. Both went beyond writing software and integrating it with hardware; they also manufactured their own hardware.

The turnkey supplier epitomizes the problem of measuring the size of the software industry. For example, Wang, the leading producer of word processors, was usually classified as a hardware supplier, whereas Computer Vision was regarded as a software vendor, yet these firms provided similar packages of hardware and software.

Software Brokers
In the 1970s, software brokers sprang up to mediate between software developers and software users. Some brokers concentrated on small software products firms that lacked marketing resources; others specialized in selling packages developed in house by computer users, who hoped to recover sunk development costs. In most cases, brokers did a little more than provide a catalog of programs and supply the product on a reel of magnetic tape or a deck of cards; users were then left with the difficult task of installing poorly productized software. This form of software brokerage proved unviable because of users' unhappy experiences. Most of the brokers that did not go out of business metamorphosed into conventional software products firms that provided full customer support.

A typical example of an early broker was Atlantic Software, established by Walter Brown and Richard Thatcher in Philadelphia in 1967.[13] Like many other individuals in the computer industry, Brown and Thatcher had realized that user-developed programs were potentially valuable assets. Their business plan was to identify such programs and sell them on a royalty basis, without significant support. To solicit candidate packages, advertisements were placed in the *Wall Street Journal* inviting computer owners to submit programs for resale. The response was overwhelming. Some enthusiastic computers users even sent magnetic tapes and operating manuals. Brown and Thatcher selected the ten packages they thought would do best and went to market without testing them—in some cases, without even having witnessed a demonstration. Sales were extremely slow—enthusiastic sellers weren't much interested in buying their neighbors' programs. Only one of the ten products sold at all well. The slow-selling nine were all dropped at the end of 1968. The one that succeeded—SCORE, an early report writing program—sold about 20 copies by the end of 1969. By that time, however, Atlantic had changed its businesses model, becoming a conventional supplier of software products, acquiring the full rights to SCORE and assuming responsibility for ongoing development and support.

Besides these small private brokers, there were two major public brokering initiatives: COSMIC and INSAC.

COSMIC was established in 1966—still in the era of free software—to make programs that had been developed for NASA at public expense available as a national resource for industry and education. Programs were distributed for nominal charges that bore no relationship to their original development costs. One of COSMIC's major packages was NASTRAN (the source of MSC/NASTRAN). Consisting of 150,000 lines of FORTRAN and having cost $3 million to develop, it was sold for $1,750. COSMIC was a spectacular failure. A contemporary report commented: "[COSMIC's] major problem—like any giveaway program—has been in giving it away."[14] Generally the programs were so poorly documented and productized that they could not be used without major investments in programming time. Even the Association of Independent Software Companies, which polled its members about COSMIC, agreed that it presented no threat to the industry.

INSAC was a much more significant venture. It was formed by the British government in 1977.[15] By that time, the major US software products firms had begun to set up sales operations in Europe, and Britain and other European countries had begun to establish their own software products industries. With the single exception of Germany's Software AG, however, there were no international players in Europe. The United States was the world's biggest market for software products, but none of the British firms had sufficient resources to market their products there. To address this problem, and (it was hoped) to improve Britain's balance of trade in software products, the Labour government's National Enterprise Board established INSAC to market the products of the leading British software vendors. By 1979, INSAC had seven US offices, a sales staff of 40, and a portfolio of 500 products. Despite negligible sales, INSAC—sarcastically referred to as the "software Titanic" by one business writer[16]—continued, in a triumph of hope over experience, to build up its sales force to a peak of 165. Sales remained negligible, and in 1982 INSAC was a finally disposed of by the newly elected Conservative government. In its 5 years of operation, INSAC incurred a loss of £6.8 million (about $16.5 million).

The reason all these attempts at software brokerage failed was that competitive pricing could not compensate for poor products and lack of customer support. Software was not a price-sensitive good; it was a small part of any data processing budget (typically less than 5 percent), and quality assurance and continuity of supply of mission-critical software were overwhelmingly more important than a factor of 2 or 3 in price.

Time-Sharing Services

A recurring issue in the software industry has been the charging model—for example, whether software should be supplied as a product that the user owns absolutely, whether the user should be granted a license, or whether the user should be charged on a per-use basis. Most often the middle course of a perpetual but revocable license was favored by software vendors because it represented the best compromise between recovery of development costs and control of the product's use. Time-sharing services offered a per-use charging mechanism.

Time-sharing services were mostly used by engineers and scientists as "super calculators" for stress analysis, linear programming, matrix analysis, and other mathematical applications. Financial workers also made use of time-sharing services—particularly financial analysis packages, which were the real forerunners of spreadsheet programs for personal computers.[17] The program catalogs of the major time-sharing companies contained hundreds of programs. Though some of these were developed by their salespeople for customers, most were developed by third parties. Time-sharing services typically charged $10–$20 an hour for their services, and the more programs they had the more computer time they sold. Authors were paid royalties, typically 20 percent of the cost of the on-line time consumed using the program. For the time-sharing firms, programs were simply a way of selling computer time; they had no real interest in software per se. Most of the programs were small (a few hundred lines at most) and were developed by freelance or moonlighting programmers. This offerred entrée to many small software companies, because it overcame the capital barriers of product development and marketing. The most important firm established in this way was ASK Computer Systems. For most program authors, however, writing programs for time-sharing services never amounted to more than a hobby or a part-time job. It was, however, the most accessible route into the software industry before the emergence of the personal computer. The concept of time sharing died with the arrival of the personal computer, and most of the software died with it.

Market Structure and Taxonomy

In July 1974, *ICP Quarterly* listed 2,928 software products from 740 vendors. Journalists and industry analysts used a variety of classification schemes to impose some order and coherence on this fragmented market.

At first, packages were usually classified by machine type or application. The trade press, particularly *Datamation*, routinely carried features with titles such as "File Management Systems: A Current Summary," "Programs for the IBM System/3," and "The DBMS Market Is Booming."[18]

For industry watchers with deeper pockets, about ten information providers published regular reports on the industry and its products by subscription. The most prominent of these providers were INPUT, Auerbach, IDC, Frost & Sullivan, International Resource Development, Business Communications, Data Pro, and Data Decisions. The reports were used by strategic decision makers within the software industry and by major computer users trying to navigate the software products market, and were sometimes distilled into articles by journalists in the computer and business press.

By the late 1970s, there was a consensus among analysts and journalists that the software products market was best represented by a taxonomy. The particular taxonomy shown here in table 5.4 is from the US

Table 5.4
Taxonomy of software products. Examples in right column give name of product, publisher, and date published.

Systems	
Operating systems	OS/360 family (IBM, 1965)
	EXEC-8 (Univac, 1965)
	Unix (AT&T, c. 1971)
Database management systems	IMS, DL/1 (IBM, 1969, 1973)
	TOTAL (Cincom, 1968)
	DATACOM/DB (CIM/Insyte/ADR/CAI, c. 1971)
Teleprocessing monitors	CICS (IBM, c. 1970)
	ENVIRON/1 (Cincom, c. 1969)
Programming aids	AutoFlow (ADR, 1965)
	Mark IV (Informatics, 1967)
	SCORE (Atlantic Software, 1968)
	MetaCOBOL (ADR, 1970)
Utilities	SyncSort (SyncSort Inc., 1968)
	PANVALET (Pansophic, 1969)
	CA-SORT (Computer Associates, 1976)
Applications	
Industry-specific	Manufacturing System (MSA, c. 1972)
	BankVision (Hogan Systems, c. 1980)
	CIS (Anacomp, c. 1980)
Cross-industry	Pay (CSC, 1967)
	General Ledger (MSA, 1972)
	MANMAN (ASK, 1974)

Department of Commerce's 1984 study *A Competitive Assessment of the United States Software Industry*, which (though not always acknowledged) became one of the primary inputs to several non-US governmental and quasi-governmental reports and so became the most widely accepted view of the market. Taxonomies from different sources varied in their details, but there were two invariants. The first was that the market was divided into two main sectors: systems packages and applications packages. (This was essentially the distinction that IBM had made in the 1960s between its Type I and Type II programs for the System/360 program library.) The division was both intuitive and historically grounded. By 1980 the systems software sector was typically subdivided into five sectors: operating systems, database management systems, teleprocessing monitors, programming aids, and utilities. The second invariant was that applications were divided into industry-specific packages and cross-industry packages. Industry-specific packages were those designed for a particular type of enterprise, such as banking, manufacturing, or health care. Packages designed for one industry sector were generally not applicable to another. For example, a package for processing insurance claims could be used only by an insurance company, a bill-of-materials program was only applicable to a manufacturer. Cross-industry applications, in contrast, performed administrative and financial functions that could be used in different industries—prominent examples were payroll programs, text processing systems, and project-management software. These were potentially salable to any firm that needed to pay its staff, to create documents, or to schedule projects.

In 1974, the 2,928 software products listed by *ICP Quarterly* were classified as follows: 922 systems packages, 1,229 industry-specific packages, and 777 cross-industry packages. Roughly speaking, the ratio of systems packages and application packages was 1:2. This was true whether measured by the number of packages, by the number of firms, or by the total value of the market (table 5.5), and it has remained approximately true over time.

At its deeper levels, the taxonomy grew more lush as more packages came onto the market and finer divisions were recognized in the product space. There were also occasional regroupings to reflect changing market relevance and the emergence of new technologies and applications. For example, in 1974 Informatics' Mark IV file-management system was a leaf on the branch of data management systems, but by 1980 it had been reclassified as a programming aid, because the database systems branch now included only modern disk-based databases with full

Table 5.5
Revenue split between systems and applications products, 1980. ("Firms" excludes approximately 10 percent of firms supplying both systems and applications.)

	Firms	Revenues
Systems	222 (23.5%)	$798,000,000 (28.8%)
Applications	721 (76.5%)	$1,972,000,000 (71.2%)
Total	943 (100%)	2,770,000,000 (100%)

Derived from Business Communications Corp., *Software Packages*, pp. 90–91.

random-access facilities. (This was, perhaps, a telling indicator that by 1980 Mark IV's salad days were long gone.)

Even in the 1970s, the taxonomy tree was luxuriant, with dozens of classifications at the secondary and tertiary levels. Every one of the thousands of software products occupied a position somewhere on the tree, and each had its own history.

Systems Software Packages

Table 5.6 shows the revenue split in 1980 for the five divisions of the systems software package market: operating systems, database systems, teleprocessing monitors, programming aids, and utilities.

The market for operating systems, the most visible of the five systems software sectors, was dominated by the computer manufacturers. Revenues in this sector were largely captive, reflecting the fact that users had little option in the choice of supplier. With the important exception of Unix, users had the choice of exactly one operating system for any particular computer configuration and application. The much more competitive market for database systems was split approximately equally between independent software vendors and computer manufacturers. The main reason there was a strong independent sector was that database technology was in its infancy at the time of unbundling, and the database systems produced by manufacturers were immature products against which it was relatively easy for the independents to compete. The third major market in systems software was for teleprocessing monitors. A teleprocessing monitor was a program that sat between the operating system provided by the manufacturer and the users' application program in a real-time on-line system. The market for teleprocessing monitors was utterly dominated by IBM's CICS, though there were 20 competing packages.

The two remaining sectors of the systems software market, programming aids and utilities, were much more fragmented.

Table 5.6
Expenditures on systems software products, 1980.

Operating systems	$103,000,000 (12.9%)
Database systems	$308,000,000 (38.6%)
Teleprocessing monitors	$103,000,000 (12.9%)
Programming aids	$217,000,000 (27.2%)
Utilities	$67,000,000 (8.4%)
Total	$798,000,000 (100%)

Derived from Business Communications Corp., *Software Packages*, p. 93.

Programming aids comprised software packages that enabled the programming process to be made more productive and efficient. They included assemblers and compilers, almost all of which were produced by manufacturers; as a captive market, this was second only to operating systems. There were few assemblers or compilers from independent software vendors; as with operating systems, the manufacturers had captured the market long before unbundling. The primary role for the independent software vendor was to produce complementary products that enhanced the standard programming tools; two prominent examples were Autoflow and MetaCOBOL, introduced in 1966 and 1970 respectively. Autoflow automated the production of flowcharts for a user's program; MetaCOBOL provided a set of high-level additions to the standard COBOL language. Programming aids were relatively inexpensive to develop. They were often produced to enhance a software house's own programming capability and marketed on an opportunistic basis when they proved to be particularly useful. There were numerous categories of such programming tools—program generators, application generators, query languages, fourth-generation languages, report program generators, decision table processors, test harnesses, and so on. The market was extremely fragmented, and products that sold as well as Autoflow were quite exceptional. Most products had fewer than 100 sales.

Utility software was another fragmented market, with hundreds of products, nearly all of them from independent software vendors. Most were complementary products that enhanced the manufacturers' operating systems in some particular way. Major operating systems, such as IBM's OS/360, were designed to run on thousands of machines with a vast range of possible processor speeds, memory sizes, and peripheral configurations. They thus had the inefficiencies inherent in a one-size-fits-all product, and the complementary products eliminated individual inefficiencies. There were products for systems management, resource accounting, diagnostics, simulation, file security, terminal display systems, and dozens of other functions. One of the classics of the genre was SyncSort, a sorting program introduced in 1968. By 1985, when it had 8,500 customers, SyncSort was one of the first half-dozen products to receive the *ICP Business Software Review* Hundred Million Dollar Award for aggregate sales.[19] In 1976, Computer Associates introduced CA-SORT, a competitor to SyncSort. There was room in the market for both products, and CA-SORT turned out to be the springboard for the growth that made Computer Associates the second most significant independent software vendor (after Microsoft) of the 1990s.

Industry-Specific Applications

Measured either by number of packages or by market value, industry-specific applications constituted the largest sector of the software products industry. In 1974, of the 2,928 packages listed in *ICP Quarterly*, 1,229 were industry-specific programs—42 percent of the total. Financial services (including banking, insurance, and brokerage) was overwhelmingly the biggest single industry category, with 575 packages—nearly half of the total. Within financial services, banking predominated, with 379 packages. Financial services, the most information-intensive sector of the economy, had the most to gain from computerization and was one of the earliest adopters. A 1975 analysis by International Resource Development reported: "It seems that the use of application packages in the banking industry has now reached a mature state, at which most bank EDP managers are generally aware of the availability of packages oriented to their needs. The same awareness is now . . . becoming evident in certain parts of the manufacturing segment, the health care industry and in some utilities. Most portions of the US economy, however, must be classed as being in an early state of development with regards to their awareness and utilization of packaged software."[20] By 1980, finance, utilities, and manufacturing had emerged as the most computerized areas of the economy. Distribution and public agencies were still relatively undercomputerized, but during the 1980s they began to catch up.

In the 1970s, the acceptance of software products was accelerated by the rise of small-business systems, which created a vibrant market for packages. For example, after IBM introduced its System/32 small-business system, it developed several application packages priced at about $3,000 for small operators in major markets (construction, hospitals,

Table 5.7
Industry-specific application software products, 1974.

Type of package	Number	Median quoted price
Banking industry		
Demand deposit accounting	40	$10,000
Savings systems	59	$7,500
Loans (commercial, instalment, mortgage, amortization schedules)	97	$10,000
Central information file	22	$17,000
Trust, accounting, and credit card systems	35	$12,000
Bond and portfolio accounting	20	$7,300
General ledger and financial information	24	$10,000
Other	82	$8,000
Communications, chemical, and petroleum industry	38	$10,500
Construction industry	38	$3,500
Education	40	$7,500
Engineering		
Civil	75	$500
Electrical and electronic	50	$6,000
Mechanical	97	$1,000
Chemical	24	$3,800
Other	32	$10,500
Financial		
Mutual Fund Management/Brokerage Operations	12	$20,000
Other	26	$15,000
Insurance	124	$11,000
Manufacturing industry		
Requirements planning	20	$6,640
Process control	23	$8,000
Numerical control, quality control	19	$5,000
Pharmaceutical, medical, and health care	57	$9,200
Retail/wholesale industry	42	$6,500
Transportation industry	31	$7,200
Utilities	44	$4,500
Property management/real estate	23	$6,000
Government/legal	18	$17,100
Printing, publishing and library management	17	$8,500
Total number of packages	1,229	

Source: *ICP Quarterly*, July 1974, as analyzed on p. 65 of International Resource Development Inc., *Computer Services and Software Markets*.

membership organizations, etc.). These packages sold in the thousands.[21] However, this still left space for independent software vendors to develop niche products for hundreds of specialized firms. Sales were typically in very small numbers, predominantly to local customers. The firms did not have the resources to develop truly productized software, so packages had to be installed, customized, and debugged on the customers' premises. When a software products firm had fewer than ten customers, as was often the case, it was hard to distinguish it from a programming services operation. The difficulty of developing well-productized software for very small markets was a primary reason for the fragmentation and wide geographical distribution of packaged-software firms. Not until the 1980s did things began to change:

> This industry is a classical example of a competitive economic market place. It contains firms of all sizes, yet none is dominant. To a great extent, we believe geography is responsible for this.
>
> With the techniques available in the past, a software supplier had to go out to his customer's site to install and service products. IBM is showing the industry the way to overcome this problem. They are now staffing centers that can be contacted at any time, not only by telephone, but also through the computer. Self-checking and self-correcting software may be on the horizon. We believe that such a capability will become a reality in the later part of this decade. Such capability would permit a vendor to sell software anyplace, let the user himself install it and, should problems develop, contact the vendor by computer "for a fix." Such a capability, however, costs money to build. This favors the firms that have already established a sizable position in the business. By 1989, it is possible that some true leaders will finally emerge from the mass of firms now fighting for their share of the software dollar.[22]

Industry-specific software was both the most competitive and the most fragmented sector of the industry. However, it was also the easiest sector in which to establish a niche, and there were many one- or two-person operations. The primary competence needed by a new entrant was an in-depth knowledge of the target industry; program-writing skills were relatively unimportant. (In the systems software sector, in-depth programming skills were critical.) The typical new firm was established by one or two former programmers or systems analysts from a user firm in the target industry. The industry-specific software sector, incidentally, was the one sector of the software products industry in which there were significant non-US players. As with custom software, indigenous firms had a competitive advantage in their knowledge industry and their familiarity with local fiscal conventions. However, except for one or two sectors (e.g., financial services in United Kingdom) there are no reliable statistics on the number or size of firms, other than that they numbered in their hundreds and that, as in United States, most of them had fewer than a dozen employees.

Cross-Industry Applications

In 1974, *ICP Quarterly* listed 777 cross-industry applications, the second largest category (table 5.8). (This table has a rather arbitrary quality: the taxonomy is plainly flawed, and some of the categories are strange—for example, it is hard to understand what was in the mind of the analyst who came up with the category "leisure time/ecology." The table is reproduced here as an artifact of its time.)

By the 1980s, cross-industry software taxonomies incorporated the categories shown in table 5.8, and many new ones, under five portfolio terms: Accounting and Finance, Human Resources, Planning and Analysis, Engineering and Science, and Office Automation. Thus, whereas in table 5.8 payroll occupies a category of its own, in later taxonomies it was subsumed under human-resources packages. In 1974, however, payroll packages, with a total of 93 products, constituted the biggest single category. Payroll was one of the first and easiest tasks to computerize, and hence it had a disproportionate presence in the early software taxonomy. Similarly, in 1974, accounting and finance packages were well developed and accounted for 325 packages (42 percent of the total). The next largest group was mathematical and operations-research software, which accounted for another 120 packages. In general, the cross-industry applications of the mid 1970s were dominated by routine, batch-oriented back-office tasks that were easy to computerize.

By the late 1970s, the market for cross-industry packages was divided between big players (such as MSA and University Computing) that supplied integrated suites of mission-critical applications to major firms and smaller vendors that supplied narrow, monolithic applications, mostly to medium-size firms. The big firms tended to specialize in integrated accounting software; the small firms tended to supply packages that helped to optimize particular tasks, such as organizing mailing lists and managing projects. The big firms, such as MSA in accounting software, represented the state of the art in cross-industry packages.

In the 1970s there were two major markets for turnkey equipment: the market for word processors and that for computer-aided design systems. In the case of word processing, integration was achieved between computers, software, terminals, and printers, and sometimes with telephone and facsimile equipment. In the case of CAD, integration was achieved

Table 5.8
Cross-industry application software products, 1974.

Type of package	Packages	Median quoted price
Accounting		
General ledger	46	$5,000
Accounts receivable/payable	96	$6,000
Integrated accounting	28	$8,400
Billing, budgeting, costing	24	$7,000
Responsibility accounting, asset accounting, and CPA packages	32	$5,000
Other	27	$5,000
Financial		
Financial analysis	44	$2,800
Stock transfers and shareholder accounting	10	$16,000
Credit union and profit-sharing systems	18	$6,000
Inventory management, distribution, and warehousing	50	$7,000
Leisure time/ecology	14	$300
Mailing activities	42	$2,000
Marketing	26	$10,000
Mathematics and statistics	36	$3,000
Operations research and management science		
Mathematical programming	34	$12,500
Simulation, models, and games	37	$9,200
Other	5	$10,000
Payroll	93	$5,600
Project management and control	50	$6,600
Personnel and pension	26	$10,000
Purchasing and order processing	22	$12,400
Other	17	$9,000
Total number of packages	777	

Source: *ICP Quarterly,* July 1974, as analyzed on p. 64 of International Resource Development Inc., *Computer Services and Software Markets.*

between computers and software, and between graphics display equipment and flatbed drawing and plotting machines. A turnkey supplier assumed the task of integrating software and machinery, which typically came from several different suppliers.

In the second part of this chapter, the software "product space" is explored by means of a historical account of the evolution of the most important genres. The narrative is organized in accordance with the product taxonomy of table 5.4. Though far from exhaustive, the examples have been chosen to give a representative view of the world of software products.

Operating Systems

Operating systems are the most sophisticated mass-produced software artifacts, both in size and in logical complexity. Because operating systems are so costly to develop and difficult to debug, the launch of a wholly new one is a rare and usually traumatic event. Most operating systems have evolved over 10, 20, or 30 years, constantly growing in "functionality" and reliability. This does not mean that a twenty-first-century operating system necessarily contains code from the 1960s; it means that over time the code has been replaced and re-engineered piece by piece.

Table 5.9
Evolution of principal IBM operating systems, 1965–1994.

Mainframe range	Small and medium models	Medium and large models	Time-sharing models
System/360 (1964)	DOS (1965)	OS/360 (1966) OS/MFT (1967) OS/MVT (1967)	CP/40 (1965) CMS (1967)
System/370 (1970)	DOS/VS (1972) DOS/VSE (1979)	OS/VS2 (1972) MVS (1974)	VM/370 (1972)
370/XA (1981)	VSE/SP (1985)	MVS/XA (1983)	VM/SP (1981) VM/XA (1983)
ESA/370 (1988) ESA 390 (1991)	VSE/ESA (1990)	MVS/ESA (1988) MVS/ESA SP4	VM/ESA (1990)

Sources: M. S. Auslander et al., "The Evolution of the MVS Operating System," *IBM Journal of Research and Development* 25 (1981): 471–482; R. J. Creasy, "The Origin of the VM/370 Time-Sharing System," ibid. 25 (1981): 483–490; Jim Hoskins, *IBM ES/9000: A Business Perspective* (Wiley, 1994).

In this sense, an operating system is analogous to a human organization: workers constantly enter and exit; eventually, none of the original workers are left, yet the organization itself lives on. (A cuter analogy, perhaps, is "Hepplewhite's Hammer." In that apocryphal story, a rural museum claims to exhibit the hammer of the famous furniture maker in a glass case. When challenged, the curator explains that it is entirely original, except for the fact that it has had its head replaced twice and its handle three times.)

The Captive Operating System

In the years that followed unbundling, IBM set the pattern for the supply of operating systems. Because of IBM's decision not to charge for software packages already in the field, the three major System/360 operating systems were supplied free of charge. The smallest of the operating systems was DOS, a disk operating system for small to medium-size machines. The strategic operating system was OS/360, which from 1967 on was supplied in two variants: OS/MFT and OS/MVT.[23] The third operating system, CMS, was a time-sharing monitor that enabled a number of programmers to share a computer simultaneously using teletypes or visual display units rather than decks of punch cards. These three operating systems were the direct ancestors of IBM's current mainframe operating systems.[24]

The original System/360 architecture has undergone numerous improvements and extensions since its introduction in 1964. The major product milestones were designated System/370 in 1970, 370-XA in 1981 (for extended architecture), ESA/370 in 1988,[25] and ESA/390 in 1990. A major constraint on IBM's mainframe architecture has been the need to maintain backward software compatibility, so that in principle—and usually in practice—a program written for System/360 in the 1960s could run without modification on a later IBM mainframe. What has been true for the hardware has also been true for the software. Each of IBM's operating systems has been periodically re-released with major revisions to exploit architectural improvements and support technological advances while maintaining compatibility with earlier releases. For example, when System/370 was introduced, in 1970, the major architectural innovation was virtual storage (VS), a technology that extended the effective memory size of the computer. All three operating systems were re-released—as DOS/VS, OS/VS2, and VM/370.

IBM dragged its feet in unbundling its operating systems, even though by the mid 1970s most other manufacturers had done so. There was considerable debate in the press as to whether IBM's had kept its operating systems bundled in order to maintain its monopoly.[26] The suspicion that IBM used inefficient software to sell more hardware was widespread in the industry, although the evidence was entirely anecdotal. (Surprisingly, this issue and IBM's monopoly in operating systems barely surfaced in the antitrust trial.) IBM eventually unbundled its operating systems in the late 1970s. As new versions of the operating systems were introduced, IBM froze the specifications of the earlier versions (which remained forever free), designating the new versions as paid-for system products.

Whether or not IBM charged for its operating systems, they had to be made available to manufacturers of clone machines for political reasons. (This might have been avoided by technical or legal maneuvers, but it would have been imprudent while IBM was under antitrust scrutiny.) In 1975, Gene Amdahl—one of the architects of System/360—founded the Amdahl Computer Corporation, which introduced the provocatively named 470 series of IBM-compatible processors. Amdahl's use of the best available semiconductor technology made it possible to substitute an Amdahl processor for an equivalent IBM model at a substantial saving. In the wake of Amdahl's success, several other manufacturers of plug-compatible mainframes entered the market, among them National Semiconductor, Fujitsu, Hitachi, and Magnuson.[27] Fujitsu and Hitachi did not use IBM software but instead developed their own highly regarded operating systems; they benefited from their own captive software markets, and by the mid 1980s were among the world's leading software suppliers.[28]

In the world of mainframes, manufacturer-produced operating systems continue to dominate. This is due primarily to a path dependency that originated in the 1960s: Users' software evolved in lockstep with the manufacturer-supplied operating system, and there was never a moment when the dependency could be escaped. The majority of proprietary operating systems for mainframe and minicomputers have pedigrees that go back to the late 1960s or the 1970s. Among the many examples are Unisys's EXEC-8 and MCP, DEC's VMS, and NEC's MODE IV.

Unix and Open Systems

Unix is the only non-proprietary operating system of major significance and longevity. It originated in the early 1970s, and it became popular for use with the minicomputers that were then coming on the market. It is now available for all computer platforms.

The history of Unix is well documented.[29] Like the origins of the Internet (with which it is closely associated), it has entered the folklore of post-mainframe computing. The Unix project was initiated in 1969 by two researchers at Bell Labs, Ken Thomson and Dennis Ritchie. "Unix" was a pun on "MULTICS," the name of a multi-access time-sharing system then being developed by a consortium of Bell Labs, General Electric, and MIT. MULTICS was a classic software disaster. It was path-breaking in concept, but to get it working with tolerable efficiency and reliability took a decade.[30] In reaction to MULTICS, Unix was designed as a simple, reliable system for a single user, initially on a DEC minicomputer. Ritchie and Thomson, who existed in a research culture somewhere between industry and academe, made the code of Unix freely available. Because the system was written in a machine-independent way, it was relatively easy to "port" Unix to different computers by rewriting only 5–10 percent of the code. This quickly caught on in the universities. Although AT&T (which owned Bell Labs) was then prevented by the regulatory authorities from entering the computer market or the software market, it seems unlikely that any rational manager could have anticipated the eventual importance of Unix and capitalized on it. Indeed, it was probably because Unix was virtually free that it proliferated in the cash-poor, manpower-rich university environment.

By the mid 1970s, Unix was in wide use on college campuses. Thus, a generation of computer science graduates grew up in the Unix culture rather than in the IBM ethos of the previous generation. The Unix technology migrated into the workplace with these graduates, both in existing firms and in startups such as Sun Microsystems.[31] By the early 1980s, Unix was available for many platforms, and this happened to coincide with the technological shift from centralized mainframes to "open systems" and networked computers. The universality of Unix greatly facilitated intermachine communication. By the mid 1980s, Unix controlled more computing power than any other operating system, and people began to speak of a Unix industry.[32]

From the beginning, the Unix industry was highly fragmented, with more than 100 suppliers. Most vendors licensed the basic Unix code from Unix Software Laboratories Inc. (a subsidiary of the by-then-deregulated AT&T) and customized it for a particular platform. Early players in the 1980s included SunSoft (a subsidiary of Sun Microsystems), Novell, SCO (Santa Cruz Operation, Inc.), and Microsoft, which supplied operating systems, while hundreds of other firms supplied applications and utilities for the Unix environment. In the mid 1980s, responding to customers' demands, the established computer manufacturers also began to offer Unix, though usually they did so reluctantly in order to preserve revenues from their proprietary operating systems and to keep their systems closed. The manufacturers usually supplied their systems with distinctive branding—DEC had Ultrix, IBM had AIX, Hewlett-Packard had UX.

Database Systems

The term "database" came into use around 1964, initially in a rather loose way to indicate the formal separation of an application program from the systems program that accessed the data.[33] To a degree, early file-management systems such as Informatics' Mark IV had already achieved such separation; however, they were not called database systems; they were batch oriented, and they did not address data integrity.

The problem of data integrity was defined most succinctly by Tom Nies, founder of the database supplier Cincom, as the "my file" problem.[34] This referred to the difficulty of ensuring the consistency of data stored in different files. For example, an organization might maintain one file for its personnel records and another for its payroll. Inevitably the two files would have in common some items—such as employees' names—that could get out of sync. For example, the name of a female worker who married and took her husband's surname might be updated on one file but not on the other. This problem was a legacy of early magnetic-tape storage systems that had fostered the unplanned proliferation of unique files for each application. The central concept of the database was that there would be exactly one copy of any item of data, and that this would guarantee its integrity. In the above example, the personnel and payroll files would simply be different subsets or "views" of a corporate-wide human-resources database.

The arrival of affordable disk stores in the second half of the 1960s provided a one-time-only opportunity for information systems managers to scrap their magnetic-tape files and replace them with a single database. The database fad emerged at about the same time as the concept of a software product, and database packages were supplied both by computer manufacturers and by independent software vendors.[35] In fact, it is a historical accident that the database industry exists at all. The industry is a legacy of the boundary created in the mid 1960s between the operating system supplied by the computer manufacturer and the applications written by the user. The database system came to occupy a middle ground between the operating system and the application program. Had history unfolded differently, the database might have been a fundamental property of the operating system; there would then have been no database industry at all. Instead, the database grew to become the main sector of the independent software industry. Even today, the database industry is perhaps the most vigorous sector of the corporate software products market.

In the late 1960s, when the first database products emerged, the technology had neither matured nor stabilized. There were three basic methods of organizing the data: the so-called hierarchical, networked, and inverted methods. The technical distinctions among these methods are not relevant here; suffice it to say that none was entirely satisfactory. Each required the programmer to be aware of the logical and physical arrangement of data and to "navigate" through it. Indeed, Charles Bachman, a database pioneer at General Electric who was awarded the prestigious Turing Award of the Association of Computing Machinery in 1973, titled his acceptance lecture "The Programmer as Navigator."[36]

Table 5.10 lists the major database products of the late 1970s and their dates of introduction. The leading products were IBM's IMS and Concom's TOTAL. IMS (Integrated Management System) had evolved from BOMP (Bill Of Materials Processor), a program that had been developed in the early 1960s for manufacturing organizations in which products were fabricated from a hierarchy of assemblies, subassemblies, and components. In 1968, BOMP was broadened in scope so that it could be used for a wide range of applications and was renamed IMS. Released in 1969, IMS became a systems product after the 1970 unbundling. It underwent many revisions as IBM worked closely with user associations in specifying additional features.[37] In 1973, DL/1, a less powerful version of IMS intended for IBM's smaller mainframes, was released. Cincom's TOTAL, IMS's chief competitor, was the first true database to be produced by an independent software vendor and enjoyed an enormous first-mover advantage.

Though IBM and Cincom remained the market leaders throughout the decade, there were many new entrants in the early 1970s. By 1975 there were about 50 vendors and more than 80 products.[38] The three most important products to emerge in 1970–71, System 2000, Adabas, and Datacom/DB, all are still major products, though have all undergone great transformations—including the incorporation of "relational" features—to keep up technologically and with users' expectations. All of them required less than one-fourth of a megabyte of computer memory when they were introduced; by the 1990s, no one would have attempted to a operate them on a machine with less than 10 megabytes of memory.

Table 5.10
Leading mainframe database systems, 1979.

Database (date introduced)	Vendors(s)	Technology	Sites	Price	Database revenues
IMS (1969)	IBM (total for IMS and DL/1)	Hierarchical	1,300	$950–$5,000 monthly rental	$42,000,000
DL/1 (1973)	IBM	Hierarchical	750	$300+ monthly rental	
TOTAL (1968)	Cincom	Hybrid	2,700	$10,000–$52,000 purchase $800–$1,300 monthly rental	$30,000,000
IDMS (1973)	Cullinane (1973–1989) CAI (1989–)	Networked (CODASYL compliant)	500	$90,000–$100,000 purchase	$13,900,000
ADABAS (1971)	Software AG	Inverted	375	"Negotiable"	$8,000,000
System 2000 (1970)	MRI Systems (1970–1979) Intel (1979–1985) SAS Institute (1985–)	Hierarchical	280	$12,000 in first year; $8,000 in subsequent years	$10,000,000
DATACOM/DB (c. 1970)	CIM/Insyte (c. 1970–1978) ADR (1978–1986) CAI (1986–)	Inverted	150	$34,000–$70,000 purchase	$1,100,000

Sources: Frost & Sullivan, *Data Base Management Services Software Market*; Palmer, *Database Systems*.

System 2000 and Datacom/DB have survived several changes of ownership, long outlasting the firms that created them. System 2000, developed by MRI Systems, was launched in June 1970. By 1979, when the Intel Corporation acquired MRI, the product had 280 users. Intel intended to use the MRI software in its proprietary database hardware. However, when Intel's microprocessor sales soared in the 1980s, database technology became a non-core activity, and System 2000 was sold to the SAS Institute, a producer of statistical software. Datacom/DB was introduced by Computer Information Management in 1971; shortly thereafter, it was acquired by the Insyte Corporation. It became a major product. In 1978, ADR acquired the package from Insyte. ADR, having languished for most of the 1970s, took advantage of the comeback of investors' confidence in the software industry in the late 1970s by introducing a database product into its aging software portfolio. In 1986, ADR was acquired by Computer Associates, which continues to sell Datacom/DB. Adabas (a contraction of "adaptable database system") has been sold by the German firm Software AG since its inception in 1971. The software was jointly developed by Software AG in Darmstadt and by the nearby AVI Institut. Software AG was the only significant non-US software products firm of the 1970s; its prominence was due in part to the fact that it had a very good product (partly government sponsored) at the right time and in part to the fact that it had a largely autonomous US sales subsidiary.

After the initial rush of firms into the database market, the only major new entrant of the 1970s was Cullinane's IDMS, which made its entry in 1973. Cullinane had been established as a software products firm in 1968. Its founder, John Cullinane, was later to become a major figure in the US software industry and world of business. Cullinane's brief window of opportunity opened in 1971, when the Committee on Data Systems and Languages (CODASYL) introduced a new database standard. Although several computer manufacturers, including General Electric, Honeywell, and Univac introduced CODASYL-compliant packages, Cullinane was the first independent software vendor to introduce a product. The IDMS database had originally been developed by General Electric's computer division. However, when General Electric decided to exit from the computer business, during the 1970–71 computer recession, Cullinane acquired the package. It was rewritten for the IBM System/360 and offered for sale in 1973. The fact that IDMS was CODASYL compliant, unlike the market leaders (IMS and TOTAL), made it attractive to new purchasers of database systems. Cullinane grew meteorically, its annual revenues rising from $2 million in 1975 to $20 million by 1980.

By the end of the 1970s, the database management system had attained a degree of maturity and stability. The top half-dozen products had two-thirds of the market. However, the rise of relational technology would leave all the incumbents shaken.

CICS: IBM's Teleprocessing Monitor

There is no better example of the invisible software infrastructure that runs the modern corporation than CICS, which for more than 30 years has been the world's best-selling computer program. In 1994 the author of a leading textbook on CICS wrote: "For a mere computer program, IBM's CICS has a profound and probably surprising influence on your life. If it vanished tomorrow, never to have existed, modern society and industry would alter radically. It is used world-wide. Yet you are more than likely unaware of its existence, let alone its influence. Many of those around you will never have heard of it. CICS has been immensely influential for two decades, and its importance will grow as we go into the twenty-first century."[39] When this was written, CICS, with gross sales of $4.4 billion, was integral to the business of all but one of the *Fortune* 100.[40] It was estimated that more than 300,000 programmers around the world earned their living developing applications for CICS, and that about 10 billion transactions per day were processed with CICS, on 10 million computer terminals. Whenever an ATM is accessed, a travel reservation is made, or a credit card is used in a retail purchase, the chances are that CICS is involved.

CICS was originally developed by IBM in 1967–68 as a Type II program for utility companies, particularly electrical utilities. SABRE and other airline reservation systems had demonstrated the feasibility of real-time information systems, and utility companies were keen to provide the same kind of instant access for their customers. A few of the major utilities, such as New York's Consolidated Edison, had already implemented real-time systems, but at astronomic cost. CICS (Customer Information Control System) was designed as a general-purpose program to fulfill this need. In the simplest terms, CICS was a piece of software "plumbing" that connected a user's application program to an operating system and a database. It freed application programmers from the logical difficulties of handling many simultaneous transactions. Consisting of about 100,000 lines of code, CICS was not a particularly large program; however, it was logically complex.

CICS was one of the 17 program products that IBM introduced when it unbundled its software, at the beginning of 1970. It was supplied for

System/360 computers running under OS/360 and leased for $600 a month. Sales of CICS rose slowly. Only 50 licenses were issued in its first year, and by the end of 1971 there were still only 111 users. The early CICS system was suitable only for utility companies and would support no more than 50 hard-copy terminals. However, as low-cost visual display units became available and as interest in real-time systems grew, a general-purpose version of CICS suitable for a wider range of applications was introduced. The acronym CICS was preserved even though customer information systems were no longer the dominant application. In 1972–73, about 3,000 licenses were issued. At the same time, IBM encouraged the efforts of third-party CICS application developers and consultants. This eased IBM's problems in supporting CICS as well as consolidating its market position. In 1974 the development of CICS was transferred to IBM's Hursley Laboratories in the United Kingdom, where it became the permanent assignment of a 160-person development team. As medium-size mainframes became capable of real-time teleprocessing, a new version of CICS was produced for IBM's smaller operating systems.

Table 5.11
Numbers of CICS licenses.

Year	Licenses
1971	111
1972	795
1973	2,100
1974	3,330
1975	4,240
1976	5,250
1977	6,100
1978	7,800
1979	9,660
1980	13,150
1981	16,550
1982	19,340
1983	21,350
1984	24,200
1985	26,600
1986	27,800
1987	28,700

Source: Mounce, *CICS,* passim.

This process has continued, versions of CICS being produced for progressively smaller machines—even PCs.

CICS was not without competitors. All the mainframe suppliers supplied teleprocessing monitors for their systems. Several of the independent software vendors also supplied teleprocessing products, the most successful of which were those that were tied in with a database and could be sold as a package—such as Cincom's TOTAL and ENVIRON/1, ADR's Datacom/DB and Datacom/DC, and Software AG's Adabas and Com-plete. In the 1980s, CICS, a revenue producer in its own right, became increasingly critical to IBM's hardware sales. CICS supported only IBM's terminals, which were extremely profitable. IBM marketed CICS so aggressively that by the mid 1980s teleprocessing monitors from the independent software vendors had been "blown away by IBM's CICS" and "the only remaining competitors of note were those tied into DBMS installations."[41] By 1980, 10,000 licenses had been taken out, and it was estimated that running CICS accounted for 16 percent of IBM's mainframe processor cycles.

CICS was continually upgraded to handle new requirements and new platforms. For example, release 1.6 (1982) allowed non-stop operation, so application programs could be modified or replaced without shutting the system down—a major benefit. The software kept growing as new features were added. By 1986 it contained 500,000 lines of code and the supporting documentation amounted to 8,000 pages. The CICS development team in Hursley now had nearly 400 employees. By 1988, CICS had gone through seven major releases without being rewritten and was becoming increasingly brittle from 20 years of ad hoc enhancements. The software was too extensive to be rewritten completely, but it was restructured and new sections were added to make it easier to maintain. IBM's historian of CICS wrote: "One way to describe [CICS version 3] is to say that it is 270,000 lines of new program code on a 500,000-line base; that it has been more formally engineered and exhaustively tested than any previous version; and that to describe all its functions comprehensively involves several million words in 31 publications."[42]

The new release of CICS gave IBM the opportunity to raise prices about 30 percent. Leasing charges now ranged from $3,175 to $7,620 per month, depending on the hardware configuration.[43] Predictably, there were "howls of anguish." Not only had the prices increased; in addition, many systems were brought to a standstill by the new software. Over the years, a generation of programmers had "poked rather too far into the internal code of CICS," and many unofficial work-arounds no longer functioned.[44] As a result, IBM had to continue to support the older versions of CICS for several years while users migrated to the new version.

CICS remains a crucial part of the world's information infrastructure. IBM claims that 470 of the *Fortune* 500 use CICS, and the largest systems can handle a quarter-million simultaneous transactions per second. About 20 billion transactions are processed by CICS every day—three for every man, woman, and child on the planet. In 1999, annual revenues from CICS were reported to be $800 million. If CICS were a software company, it would rank about fifteenth.[45]

Industry-Specific Applications

With several hundred firms by the end of the 1970s, the industry-specific software products industry was so fragmented that it is not possible to study a representative sample. This section, therefore, offers a "horizontal" study of firms that offered software products for the banking sector and a "longitudinal" study of a single firm (ASK Computer Systems) and a single application (Materials Resource Planning).

Banking Software Packages

The supply of banking software was highly fragmented, both in number of vendors and in number of packages. In the 1970s there were between 100 and 200 vendors; by 1984, after some consolidation, one survey estimated there were 150.[46] The range of packages was equally diverse—in 1980 one analyst produced a list with 28 distinct categories (table 5.12). A 1989 survey listed 450 packages from almost as many suppliers.[47]

The large retail banks had computerized their back-office operations in the 1960s, using in-house staff, sometimes augmented by programming contractors or by custom software from programming services firms.[48] In the wake of the early airline reservation systems, the retail banks were the next group of enterprises to go on line. Because their core operations had been computerized before the emergence of the software products industry, packages were developed mainly for non-core activities (e.g., brokerage operations) or for new applications (e.g., supporting ATMs and electronic funds transfer). However, the United States had about 15,000 small and medium-size banking institutions—retail, savings-and-loan, wholesale, commercial, private, and merchant banks. Most of these were too small to afford mainframe computers in the 1960s, but in the mid 1970s, when small-business systems became available, they became a huge market for off-the-shelf banking software pack-

Table 5.12
Banking and financial software packages, 1980.

Department	Application program
Loans	Installment
	Commercial
	Mortgage
	Personal
Trusts	Corporate
	Personal
Retail	Credit card administration
	Telephone bill paying
	Point of sale applications
Deposit systems	Certificates
	Retirement accounts
	Christmas clubs
	Regular savings
General	Electronic funds transfer
	Demand deposit
	Financial control
	Business planning
	Customer information file
	International banking
	Property management
	Proof and transit
	Text and word processing
	Safe deposit box processing
	Automated clearing house
	Commodities transactions
	Personnel systems
	General accounting
	Credit union systems

Source: Business Communications Corp., *Software Packages*, pp. 14–15.

ages. For example, it was estimated that three-fourths of the firms in the wholesale banking sector relied entirely on packaged software.

The most interesting development of the 1980s in this area was the emergence of integrated real-time packages for major banks having international operations. These programs were sophisticated, containing hundreds of thousands of lines of code, and were comparable in complexity to real-time airline reservation systems. Because these software developments were so capital intensive, the two market leaders,

Anacomp and Hogan Systems, were both funded through non-market sources.

Anacomp, established as a programming services operation in 1968, received funding from 30 major banks to develop its CIS (Continuous Integrated System) real-time banking software product. By 1983, Anacomp had pulled ahead of all its competitors and was among the top 10 software products firms, with annual sales of $172 million. However, the company's sales plummeted in 1984 as a result of product-development delays with CIS. After losing $116 million on sales of $132 million, "for all practical purposes, CIS was dead."[49] Anacomp's software was acquired by EDS in 1985, and Anacomp faded from sight.

Hogan Systems had been established in 1977 by Bertie Hogan, a former banking official, with startup capital of $250,000. Hogan Systems reduced its need for working capital for marketing operations by selling its products through IBM. IBM's imprimatur made Hogan Systems' integrated package the market leader and the company the leading supplier of banking software.[50] In 1986, IBM took a 5 percent shareholding in the company. Like Anacomp, however, Hogan Systems experienced problems with product development. For example, a $10 million contract with Britain's Midland Bank eventually had to be canceled in 1987 because of software delays, and Hogan Systems' star began to wane.[51] By 1990 it was clear that integrated software packages for major banking operations were not viable. The world's major banks have continued to rely on custom written software for their core operations and have remained among the largest employers of programmers and analysts outside the software industry.

ASK and MRP Software

Manufacturing was later to computerize than financial services, in part because manufacturing was less information intensive and in part because the average manufacturing firm was much smaller than the average financial services firm.

Since the 1950s, the most popular application for manufacturing operations had been Materials Resource Planning (MRP). An MRP program was used to control production in discrete manufacturing operations. Manufacturing a complex product might require coordinating hundreds or even thousands of manufacturing steps, so it was a natural candidate for computerization. In 1974, no fewer than 20 of the 62 packaged programs developed for the manufacturing industry were MRP packages. By 1980, there were about 80 MRP packages available, from more than 70 suppliers, including all the mainframe makers, a few of the minicomputer makers, several of the big software products firms (e.g., Cincom and Informatics), and dozens of much smaller software firms.[52] Packages ranged in price from $5,000 (for the least expensive minicomputer programs) to $200,000 (for mainframe products). The number of users of a package ranged from two or three (for the lesser-known minicomputer programs) to about 500 (for mature mainframe packages). ASK, which later came to dominate the MRP market, was a mid-size player; its $50,000 MANMAN package had 75 users.

ASK Computer Systems was founded by Sandra Kurtzig, a mathematics graduate of UCLA.[53] After graduation, Kurtzig trained as a programmer. She was subsequently employed as a salesperson by GE Time-Sharing in Palo Alto, where she developed a rudimentary MRP program in response to a customer's request. In 1972 she decided to strike out on her own as a contract programmer specializing in MRP applications. She quickly became aware of the revenue potential of multiple sales of software products, in contrast with one-off sales of custom programs; however, like many others in the early 1970s, she was held back by the capital barriers for marketing. Having developed a simple MRP program called MANMAN, Kurtzig decided to sell it on a per-use basis through the Tymshare service. She was paid a royalty of 20 percent of the cost of the on-line time that customers purchased to use the program. This not only overcame the marketing barrier; it also proved surprisingly lucrative. By July 1974, Kurtzig was able to take on a couple of staff and incorporate her enterprise.

Sandra Kurtzig became aware of the imminent decline of time sharing and the rise of minicomputers in the mid 1970s, earlier than most others in the industry. To find a new distribution channel for MANMAN, she made use of her connections at Hewlett-Packard's Palo Alto facility and rewrote the program for that company's minicomputer. The new version of MANMAN was sold directly by Hewlett-Packard, so ASK did not have to build a sales operation. In 1976, six copies were delivered at $35,000 each. Hewlett-Packard, however, made more profit on each sale than ASK did. Seeing an opportunity to maximize profits, and with ASK now having some significant resources, Kurtzig began to deliver turnkey systems, supplying a computer and software as a package. ASK's turnkey system was a hit, primarily because it could be located on a machine-shop floor rather than in a corporate mainframe suite. In effect it became another machine tool under the direct control of the manufacturing operation. Soaring profits were reinvested in marketing, product refinement, and in complementary products that enhanced the MANMAN package. By 1980, ASK had delivered 75 systems. The following year, it made an initial public offering, valuing the firm at $50 million—of which Sandra Kurtzig had a 61 percent stake. After the IPO, Kurtzig achieved great prominence in the software industry and minor celebrity status outside the industry.

The cash infusion from the IPO set ASK on a program of expansion that continued through the 1980s. New products were developed, and other products (e.g., the Ingres database) were acquired. By 1991, ASK, with sales of $315 million, was the number-10 independent software vendor and the leading producer of manufacturing software. In 1994, however, during a cyclical downturn and profits slump, it was acquired by Computer Associates. MANMAN remains a leading MRP package.

Cross-Industry Applications

As is true of industry-specific applications, it is difficult to convey the full scope of cross-industry applications owing to the multiplicity of suppliers and products. Here, three of the major sectors will be examined: mainframe accounting software, word processing, and computer-aided design. In each case, the sector is examined from the perspective of the leading supplier—MSA, Wang Laboratories, and Computer Vision, respectively.

MSA and Accounting Software
MSA was incorporated in 1963 as Management Science Atlanta to undertake custom programming and consulting for the Atlanta-area textile industry. By 1967 it had 100 employees. Taking advantage of the 1968 boom in software stocks, it made a private placing and obtained $8 million for expansion. Renamed Management Science America, it trebled its staff in the next 2 years, expanding into software products, general consulting, and computer services. In 1970, during the downturn, MSA lost $7 million on sales of $9 million. In 1971, John Imlay, a prominent Atlanta-based businessman and computer consultant and one of the most charismatic and prominent personalities in the industry, was brought in to reorganize the firm under Chapter X bankruptcy.[54]

When John Imlay arrived, MSA, aside from the fact that it was hemorrhaging money at an extraordinary rate, was undifferentiated from a hundred other medium-size software firms. By 1981, it was the biggest software products firm in the world, with annual revenues of $73 million. MSA's rise was due largely to strategic decisions made by Imlay.

Imlay decided to refocus the firm on software products and to close down the loss-making consulting and computer services. This entailed firing about 60 percent of the work force. That software products offered greater profit potential than custom programming and computer services was a commonplace in the industry. Less obvious was the decision to switch from MSA's historical vertical market in the textile industry to cross-industry applications that could be sold across the economy. Another less obvious realization was that software packages were not price sensitive and that the market would bear much higher prices than was widely perceived. MSA's prices were flexible and were based on the value to the user rather than on the need to amortize software development costs over a number of users. MSA's products were targeted at *Fortune* 1,000 mainframe owners with annual revenues exceeding $40 million, and it was therefore able to charge prices much higher than the industry norms.

At a time when the median price for an accounting package was about $10,000, MSA products were priced between $35,000 and $150,000, depending on the program modules supplied. MSA's high profitability enabled it to grow dramatically during the 1970s, even though the software industry was out of favor among investors.

MSA's main product, General Ledger, was introduced in 1972. It had been developed exclusively for IBM System/370 mainframes, which then constituted 75 percent of the computers in the field. The high margins on General Ledger enabled MSA to plow profits into refining its products and creating complementary modules. New modules were developed for purchasing and inventory control, for accounts receivable and payable, for fixed asset management, and so on. MSA spent more than 20 percent of its revenues on product development, a percentage about twice the industry norm and believed to be the highest in the industry. The complementary modules were winners in the market. Perhaps the most common problem experienced by users of software products was the difficulty of integrating packages obtained from different suppliers and having incompatible interfaces and data formats. Users of MSA's General Ledger could now get all their accounting software from a single supplier. In the late 1970s, human-resources modules were added, including one for payroll accounting. By 1981, MSA had enhanced General Ledger into a complete applications suite for accounting and human-resources management (figure 5.3, upper diagram). At that time, MSA had about 4,000 customers around the world—in manufacturing, financial services, health care, and the public sector (figure 5.3, lower diagram). Together, the diagrams in figure 5.3 make up as good a historical snapshot of a cross-industry package as one is likely to find.

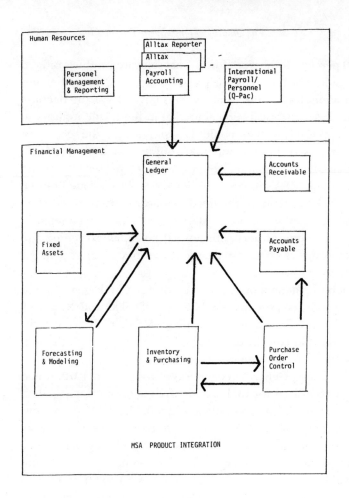

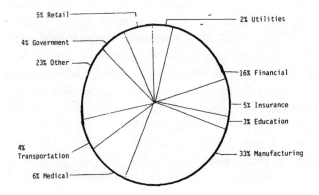

Wang Laboratories and Office Automation

The term "office automation" came into use in the early 1970s to describe the application of computers to the preparation of documents and other office tasks. In the early 1970s there were two methods of automating document preparation. The first entailed the use of a stand-alone word processor. The stand-alone word processor, available since the 1950s, consisted of an electric typewriter and a controller less powerful than a computer. Dedicated to a single user, it was priced between $10,000 and $20,000. The second method enabled multiple users to prepare documents on electric typewriters or visual display units connected to a mainframe computer running a word processing software package. However, sharing a mainframe was rarely satisfactory when a user could be interrupted at any time by the higher priority of corporate data processing. It was also an expensive solution.[55]

Thus, in the mid 1970s there was a vacuum for a low-cost computerized word processing system, and the minicomputer rushed in to fill it. None of the minicomputer manufacturers (DEC, Data General, Prime, Hewlett-Packard) had the expertise to develop word processing software in house, but they all encouraged software houses to develop packages. These packages, of which there were more than 50 by the late 1970s, were priced around $5,000, and the more successful ones sold in the hundreds of units.[56] The most successful word processors, however, were those sold as turnkey systems dedicated to word processing. A turnkey system could be operated in a departmental environment, free from the corporate mainframe, with little or no technical support, and with the high reliability of a dedicated system. By far the most successful entrant into this market was Wang Laboratories Inc.

Wang Laboratories was founded in 1951 by the scientist-entrepreneur An Wang. During the 1960s, the company was highly successful as a manufacturer of electronic calculators.[57] With the decline of the calculator market in the early 1970s, Wang diversified into minicomputers and word processors, the latter essentially as a way of selling the former. Wang's first computerized word processing system, introduced in 1976, was a screen-based system with the visual display units, which were then just coming on the market. It was years ahead of its competitors. (IBM

Figure 5.3
Two 1981 diagrams pertaining to MSA's General Ledger software product. The lower diagram depicts unit sales by industry. Source: Frost & Sullivan, *MSA, The Software Company*, pp. 13, 22.

did not introduced a screen-based product until late 1978.) Three models were offered: the single-user WP-10 ($12,000), the three-user WP-20 ($18,000), and the WP-30, which could accommodate up to fourteen users ($30,000 and up).[58] The Wang system was cheaper and more flexible than stand-alone word processors or mainframe-based systems. By 1978, Wang was the world's largest supplier of word processors, with several hundred installed. Wang's 30-second TV commercial during the 1978 Super Bowl was probably the first instance of a computer firm's appealing to office workers rather than to data processing managers.[59]

By the late 1970s, Wang had been joined by more than a dozen competitors, including Lanier, Jacquard, and Phase-Four Systems. Each of those firms sold hundreds of systems, but Wang remained the market leader, with about 32,000 systems in place by 1982. That, however, was the year in which IBM-compatible PCs began to sell in quantity. In the next 3 years, the market for dedicated word processors collapsed. Wang suffered huge losses in the late 1980s and filed for Chapter X bankruptcy in 1992.

Computer Vision and CAD Systems

What the word processor did for the ordinary office, the computer-aided design system did for the engineering design office. But whereas the turnkey word processor displaced an earlier generation of non-computerized word processors, the CAD system came out of nowhere. CAD was made possible by the minicomputer, the graphical display tube, and the graph-plotter, all which became affordable in the late 1960s.

The first entrant into the CAD turnkey market was Computer Vision, founded in Bedford, Massachusetts, in 1969.[60] Rather than develop a software package to run on a commercially available minicomputer, Computer Vision developed its own processor, which was specially designed for graphics work and was therefore more effective than a general-purpose minicomputer.[61] This initial advantage was short-lived. In the early 1970s, Computer Vision was joined in the CAD market by Intergraph, Calman, Applicon, Auto-trol, and Gerber, all of whose systems used standard minicomputers. IBM, DEC, Prime, and Data General also provided systems for the booming turnkey market.

CAD systems were expensive, costing perhaps 10 times as much per user as word processing systems. A medium-size system with four workstations cost at least $300,000, a large systems at least $1 million. By 1980, CAD systems were a significant sector of the computer industry in their own right, with a total market exceeding $1 billion. Computer Vision's annual sales were nearly $200 million.

Although originally intended for the manufacturing and engineering industries, CAD came into use in architecture, cartography, aerospace, and other industries and professions to which drawing was central. With CAD, drawings could be produced more efficiently, stored permanently, and called up and modified on demand.

In 1981, the *New York Times* explained the CAD system to its readers: "Picture an engineer hovering over a tablet in front of a display screen that looks like a small television set. When the engineer draws on the tablet, the drawing appears on screen. Imagine the engineer has drawn a bolt. With the help of a nearby keyboard, the engineer makes the drawing three-dimensional, rotates it or alters it. Punching a few more buttons, he tells a computer to assess the bolt's strength. The print-out arrives. With a dissatisfied expression, the engineer alters the design."[62] By the time this article appeared, the CAD system, like the dedicated word processor, was being threatened by the personal computer. But whereas stand-alone word processors had just about vanished by the end of the 1970s, dedicated CAD systems continued to benefit from a market willing to pay for ever more powerful systems. They remained a significant sector of the computer market despite competition from personal computers.

Summary

In the 1970s, the positive effects of IBM's unbundling decision on the software industry were overshadowed by the effects of the crash in computer stocks and the computer recession of 1970–71. At the end of the decade, the market for software products was less than $2 billion. Nonetheless, the 1970s was the decade in which the industry was fully shaped and in which its numerous submarkets were defined.

Three major sources of supply of software products emerged in the 1970s: computer manufacturers, independent software vendors, and turnkey suppliers. The computer manufacturers' place in the industry owed less to any proactive entrepreneurial spirit than to the inevitable consequences of unbundling. The term "captive revenues" reflected users' lack of enthusiasm for manufacturers' often lackluster products. In contrast, the independent software vendors were newcomers with reputations to establish, and it was from their ranks that many of today's global software giants developed. Turnkey suppliers owed their existence to the convenience of supplying not simply a software product but a bundle of hardware components integrated by software. Most users of word processors or CAD systems would not have thought of themselves as dealing with a "software" firm.

Two other historically significant forms of software distribution emerged in the 1970s: software brokering and time sharing. Both were unsuccessful in the 1970s, but both have been tried since then. Software brokering failed in the 1970s because productization was in its infancy and brokers could not supply the necessary customer support. When the idea was tried again in the 1980s by the publishers of videogames and personal computer software, it was more successful. But the most successful firms were those that acquired software products outright, then productized and marketed them. Selling software though time-sharing services was a fascinating strategy, but before it could become a significant form of delivery it was overtaken by the collapse of the computer utility in the mid 1970s. The idea has recently resurfaced in connection with the Internet; the suppliers are called "application service providers" (ASPs). Because the concept was never fully tested in the 1970s, history has few lessons to offer to the ASP industry.

Perhaps the most enduring legacy of the 1970s, and the focus of the second half of this chapter, was the classification of software products. It is easy to overlook the significance of this development, but consider trying to find a library book without a catalog. Industry information providers such as ICP and INPUT played an important and largely unsung role in this process. Much as the World Wide Web could not function effectively without search engines, the software industry could not have existed without intermediaries to facilitate information searching.

Now the fastest-growing industry in the world has its own newspaper.

The software industry, already a massive market, will be one of the premier growth industries of the 1980's and it deserves more than just an occasional article or a department or a column. It deserves a publication of its own. Introducing Software News, the monthly computer software newspaper.

Software News is published by Technical Publishing, a company known for its responsiveness to the DP market and its information needs. Technical publishes Datamation, the monthly information source for the EDP professional.

The editorial staff of Software News will save readers time and money by collecting, researching, analyzing, cataloging, and reporting on the products now being offered by more than 3,000 software producers. And they'll bring incisive coverage of application packages, systems software, program development aids, language processors, data bases, productivity enhancements, user ratings and surveys, data and software security, software legal issues, job opportunities and much more.

Software News has a controlled circulation with a guaranteed minimum of 50,000 software buyers and specifiers: qualified subscribers are accepted only on a direct request basis.

For information and a complete media kit, call Jean Gallant (617) 562-9308. Be part of the excitement of the software industry's first newspaper.

Technical Publishing
a company of
The Dun & Bradstreet Corporation

An ad proclaiming the US software products industry the fastest-growing industry in the world.

6

The Maturing of the Corporate Software Products Industry, 1980–1995

It took the software industry nearly a decade to recover from the stock market collapse that followed the go-go years. True, in the 1970s sales of software products grew at an annual growth rate of about 25 percent, but this was far less in real terms in the inflation-ridden 1970s. The turning point for the industry came toward the end of the decade, when, for no discernible reason other than the passage of time and the rebirth of confidence, the packaged software industry began to grow exponentially, sustaining a growth rate exceeding 40 percent per year in real terms. By the mid 1980s, it could be said without hyperbole that "few users write their own software anymore"—and with a reported 8,000 products from more than 3,000 vendors, there were plenty to choose from.[1]

The first beneficiary of the improved climate was Cullinane, whose 1978 initial public offering was the first by a software firm in nearly a decade. Cullinane, a private company created and largely owned by its charismatic founder, had an unblemished 10-year record of steady organic growth. In the 2 years after the IPO, Cullinane's earnings doubled and its share price rose from $20 to $50. In 1982, Cullinane was the first software products company to be listed on the New York Stock Exchange.

After Cullinane's IPO, dozens of software companies, large and medium-size, followed suit, including American Management Systems (in 1979), Pansophic (in 1980), and MSA and Computer Associates (in 1981). In each of these cases, the infusion of capital set the firm on a path of expansion—mainly through acquisition. Public ownership was a two-edged sword, however, since a firm could just as easily be the target of a predator as it could be a predator.

The Ascendancy of the US Software Products Industry

In 1984 the US Department of Commerce published a landmark study titled *A Competitive Assessment of the United States Software Industry*.[2] It was almost self-congratulatory in tone, and with good reason: The United States utterly dominated the world market for software products. Overall, the United States had two-thirds of the world's software market. But because most overseas activity was in custom programming and in computer services, the United States supplied at least 95 percent of the software products market. For example, in the United Kingdom the sales of the entire domestic packaged software industry were less than $200 million, about equal to the sales of any one of the top three US firms.[3] In Japan, only five of the 30 best-selling corporate software products were domestically produced; most of the rest were imported from the United States, a few from Software AG in Germany.[4] In fact, Germany was the only country besides the United States to have a significant player in software products. Software AG had been a major player in the database software industry since the early 1970s; its 1984 revenues of $38 million put it just in the top 20. Later, Software AG was joined by another German firm, SAP. (The reasons for Germany's success in software products are a matter of speculation. The most compelling explanation, perhaps, is that Germany's federal structure made it difficult for custom software firms to become national players because the plethora of federal regulations favored small local players; hence, the only way to grow was by producing generalized software products that were applicable across Germany, and hence throughout the world.)

Although Japan and the European countries have strong indigenous computer services firms, none of them other than Germany has yet produced a top-ten software products firm, and they have only a few firms in the top 100. Anxiety about the United States' domination of the burgeoning software industry was evident in *Software: An Emerging Industry*, a major OECD study published in 1985. In Britain, the Advisory Council for Applied Research and Development (ACARD) conducted its own inquiry, reported in a volume titled *Software: A Vital Key to UK Competitiveness*.[5] Sadly, both of these inquiries were dominated by policy makers unduly influenced by technical experts who saw the problems of the industry as primarily technological. As a result, their recommendations focused on national investments in R&D and scientific manpower training. Meanwhile, the US software industry, whose crude but effective software technologies were viewed with some disdain in Europe, kept growing exponentially without serious competition.

In the period 1977–1982, the number of US independent software products vendors increased from about 600 to 1,800. Although there was some consolidation at the top, so that the gap between the largest and smallest firms widened, the industry retained its "boulders, pebbles, and sand" character. Figure 6.1, which originally appeared in the Department of Commerce's report on the US software industry, illustrates this. In 1982, the 50–60 firms with annual revenues exceeding $10 million had approximately 50 percent of the market, the approximately 430 firms with revenues between $1.1 million and $10 million had 30 percent, and the approximately 1,400 firms with revenues of less than $1 million had the remaining 20 percent. Despite the presence of some very large firms, the industry was remarkably unconcentrated. Even MSA—the largest firm in 1982, with sales of $96 million—had only about 2 percent of the market. The mainframe hardware market was much more concentrated, with fewer than 20 significant firms worldwide and with IBM having about half of the market.

Throughout the history of the US software industry, there has been a vast amount of merger and acquisition (M&A) activity. The history of M&A activity in the computer software industry is well documented, having been tracked by the M&A specialists Broadview Associates since 1970.[6] The number of firms changing ownership has fluctuated with the economic climate, but it has averaged between 100 and 200 deals a year,

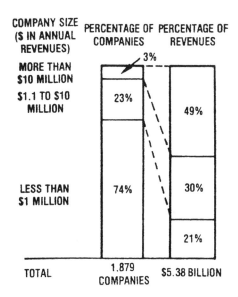

Figure 6.1
The structure of the US software products industry in 1982. Source: US Department of Commerce, *A Comparative Assessment of the United States Software Industry*, p. 6 (attributed to INPUT).

with as many as 500 in peak years. Paradoxically, the high rate of M&A activity has led to neither a much higher concentration nor to a decrease in the number of firms. Concentration has been little affected because consolidation has barely kept pace with the growth of the industry. In 1990, Computer Associates—by far the most acquisitive firm—had annual revenues of about $1.3 billion, representing between 3 and 4 percent of industry sales. The next biggest consolidators, Dun & Bradstreet and Legent, had market shares below 1 percent. The number of firms in the industry has continued to increase, as each year more firms have entered the industry than have exited through acquisition or because they ceased trading.

There is no one best way to understand the growth and development of the corporate software industry. For example, the US Department of Commerce's and the OECD's reports on the software industry of the mid 1980s contain a dizzying amount of detail that does fair justice to all sizes of firm. However, because of their statistical basis and their policy orientation, these reports are tedious to read and difficult to internalize. Most of the readable journalistic accounts focus on individual large firms, treating them in isolation. In this chapter, I opt for a middle course, focusing on the three largest firms that came to prominence in the 1980s: Computer Associates, Oracle, and SAP (all founded in the 1970s). This allows at least some scope for a narrative historical treatment of the individual firm that will not lose sight of the larger picture. Focusing on the largest boulders does leave the pebbles and the sand out of the picture. However, Computer Associates, Oracle, and SAP all started out small, then became medium-size before becoming large. It also happens that the three firms were the most successful exponents of the three phenomena that shaped the 1990 corporate software products industry: consolidation, the rise of the relational database, and Enterprise Resource Planning software.

Table 6.1 lists the leading corporate software products firms for selected years. Of the top 15 firms of 1995, three—Computer Associates, Sterling Software, and Dun & Bradstreet Software—can be characterized as consolidators. Most of the exits in table 6.1 can be accounted for by consolidation. Of the top 15 of 1984, Informatics, Uccel, ADR, Cullinet, ASK, and Pansophic were acquired by Computer Associates, and MSA by Dun & Bradstreet. Of the four top-15 firms of 1984 that were neither acquirers nor acquired and that continue to be major players, three—Cincom, Compuware, and the SAS Institute—were privately held. Private ownership was the only sure guarantee against a takeover.

Consolidators such as Computer Associates and Sterling Software are often portrayed as predators that "hoover up" ailing firms in order to strip them of their software assets. Although there were some such cases, most of the acquisitions were for premeditated strategic reasons. The two main strategic reasons were diversification and economy of scale. Nearly every software products firm started as a single-product company, and for each of them growth slowed as the market saturated. Diversification by the acquisition of new products provided an immediate resumption of growth and was generally cheaper and less risky than developing new products internally. Because most corporate software sales were made through direct marketing, there were great economies to be achieved by deploying a sales force to sell a portfolio of programs rather than a single product. In the early 1980s, the product portfolios of major firms were astonishingly small—for example ADR, one of the more diversified companies, offered only eighteen programs, and the industry leaders Informatics and MSA each offered fewer than ten.

By 1995, after 15 years of consolidation, Computer Associates had 300 products. But Computer Associates' strategy as a consolidator proved difficult to emulate. To integrate its many acquisitions into an apparently seamless product family, Computer Associates evolved a distinctive set of capabilities in market analysis and product searching, in the rejuvenation of sometimes flagging products, and in the rewriting of program interfaces.

The rise of the relational database was a major phenomenon of corporate software in the 1980s and the 1990s. When that technology emerged, Oracle Systems (founded in 1977) was the first to exploit it commercially. Soon it was joined by Informix, Sybase, Ingres, and dozens of producers of complementary products. As the relational database technology was perfected, it began to displace the older database technologies, forcing firms such as Cullinane and ADR close to bankruptcy or into the arms of an acquirer. By 1990, Oracle was the dominant firm in the database industry.

A third force shaping the corporate software industry was Enterprise Resource Planning software, the applications software phenomenon of the 1990s. The term "Enterprise Resource Planning " was introduced early in the decade to differentiate the fully integrated business applications software of SAP and its competitors from the business software of older vendors. Until the arrival of ERP software, there had been two distinct markets for business programs: a cross-industry market and an industry-specific one. Cross-industry software performed business

Table 6.1
Revenues from packaged software of leading independent software vendors, 1985–1995. Revenue is for software products only; consulting, custom software, and computer services income are excluded. An asterisk denotes a privately held company. Unless a footnote indicates otherwise, firms listed were still trading in 2000. Sources: 1985 data from "The ICP 200—Top 45 Software Product Suppliers," *Business Software Review* (June 1986); 1990 data from "Top 100 Independent Software Vendors," *Software Magazine* (June 1991); 1995 data from "Top 100 Independent Software Vendors," *Software Magazine* (July 1996); "Top Software Suppliers," *Datamation*, June 15, 1996.

1985	1990	1995
1 Culline[a]: $183 million	1 Computer Associates: $1.311 billion	1 Computer Associates: $2.461 billion
2 Computer Associates: $170 million	2 Oracle: $677 million	2 Oracle: $1.527 billion
3 MSA[b]: $152 million	3 Dun & Bradstreet: $539 million	3 SAP: $1.349 billion
4 Applied Data Research[c]: $134 million	4 SAP: $286 million	4 Sybase: $786 million
5 Uccel[d]: $118 million	5 Software AG*: $261 million	5 Informix: $536 million
6 Cincom Systems*: $95 million	6 SAS Institute*: $240 million	6 SAS Institute*: $534 million
7 Pansophic Systems: $78 million	7 Ask Computer Systems[f]: $240 million	7 BMC Software: $397 million
8 SAS Institute*: $78 million	8 Pansophic Systems[g]: $215 million	8 Sterling Software[i]: $390 million
9 Information Builders*: $64 million	9 Cincom*: $208 million	9 Compuware*: $388 million
10 Sterling Software: $61 million	10 Legent[h]: $175 million	10 Software AG*: $376 million
11 Candle*: $60 million	11 Information Builders*: $167 million	11 Attachmate*: $347 million
12 Dun & Bradstreet: $60 million	12 Candle*: $151 million	12 Dun & Bradstreet: $285 million

The Corporate Software Products Industry 171

13 Software AG*: $59 million
14 Kirchman: $50 million
15 ISSCO^c: $40 million

13 BMC Software: $125 million
14 Informix: $121 million
15 American Mgt. Systems: $121 million

13 Systems Software Assoc.: $280 million
14 J. D. Edwards: $237 million
15 Platinum Technology^k: $230 million

a. formerly Cullinane; acquired by Computer Associates in 1989
b. acquired by Dun & Bradstreet in 1989
c. acquired by Computer Associates in 1988
d. formerly University Computing; acquired by Computer Associates in 1987
e. acquired by Computer Associates in 1986
f. acquired by Computer Associates in 1994
g. acquired by Computer Associates in 1991
h. acquired by Computer Associates in 1995
i. acquired by Computer Associates 2000
j. acquired by GEAC in 1996
k. acquired by Computer Associates in 1999

processes (e.g., payroll, accounting, and human-resource management) that were applicable to almost every business. The vendors of cross-industry software, such as MSA, American Management Systems, and Dun & Bradstreet, were among the industry's biggest players. However, in the mid 1980s no one firm offered all the cross-industry packages that a typical user needed. This left the customer with the task of integrating packages from several different vendors, a task that was made difficult and expensive by incompatible data formats and interfaces. One user put it succinctly: "If the customer ends up with eight different vendors, none of the software works together."[7]

Industry-specific software vendors overcame the problem of dealing with products from several suppliers by offering a complete software solution for a narrowly defined industry, such as banking, insurance, health care, manufacturing, or retail. This sector of the software products industry included some major players, such as Hogan Systems in banking software and Policy Management Systems in insurance, though most of the players were second- or third-tier firms in niche sectors. Industry-specific software inevitably led to a one-size-fits-all solution that required a business to conform to the dictates of the software. For medium-size firms using small-business systems, this rigidity was acceptable; indeed, without the thousands of industry-specific packages on the market, small computers such as the IBM System/3 and the AS/400 would not have been viable. However, only custom-written software or carefully tailored and integrated cross-industry packages could handle larger firms' historically idiosyncratic accounting systems and diverse overseas operations.

ERP software effectively hit the middle ground. It was a body of generalized, integrated software that could be customized for virtually any large business. The largest player in ERP software was the German firm SAP. It was soon joined by the Dutch company Baan and by a number of American imitators, including Systems Software Associates, PeopleSoft, and J. D. Edwards. SAP's dominance of the ERP sector was due more to luck than to strategy or prescience. In the 1970s and the the 1980s, because of local circumstances and the need to serve the European operations of American multinationals, SAP developed a unique style of tightly integrated generic business software. By the early 1990s, SAP had made the most its first-mover advantage and had become the leading player in what had come to be called ERP software.

Of the top 15 firms of 1995, all but three can be classified as consolidators, database vendors, or ERP software providers. However, this still leaves firms such as BMC Software, Compuware, and Attachmate. These firms can be classified as second-tier operations supplying complementary products to the database industry or specialist software tools to large computer users. In the software industry there were dozens of similar firms with sales between $100 million and a $500 million.

IBM and Other Computer Manufacturers as Suppliers of Software Products

The biggest seller of corporate software products was not an independent software vendor; it was IBM. As table 6.2 shows, in 1986 IBM sold $5.5 billion worth of software products—many times more than its nearest rivals, Unisys and DEC. Only three independent software vendors made it into the *Datamation* top 15: Computer Associates and the personal computer software enfants terribles Lotus and Microsoft. By 1992, the independent software industry had caught up only somewhat, with six firms in the top 15. In 1992, IBM's software sales exceeded the combined

Table 6.2
The *Datamation* top 15 in software, 1986. Firms in italic type are independent software vendors; those in roman are computer manufacturers.

Rank		Software revenues	Software revenues as percentage of total revenues	Total revenues
1	IBM	$5,514,000,000	11.1	$49,591,000,000
2	Unisys	$861,000,000	9.1	$9,431,000,000
3	DEC	$560,000,000	6.7	$8,414,300,000
4	NEC	$507,100,000	8.0	$6,324,600,000
5	Fujitsu	$389,200,000	5.9	$6,575,700,000
6	Siemens	$387,100,000	8.8	$4,387,100,000
7	Hewlett-Packard	$375,000,000	8.3	$4,500,000,000
8	Hitachi	$331,000,000	7.0	$4,728,800,000
9	Nixdorf	$299,500,000	14.4	$2,075,100,000
10	*Lotus Development*	$283,000,000		
11	*Microsoft*	$260,200,000		
12	CGE	$238,100,000	23.2	$1,025,000,000
13	*Computer Associates*	$226,500,000		
14	Olivetti	$225,300,000	5.8	$3,865,200,000
15	Wang	$200,000,000	7.5	$2,668,900,000

total of the next six biggest suppliers. Not until 1998, when Microsoft overtook it, did IBM cease to be the world's largest software supplier.

Although IBM had unbundled in 1970, and all the world's other computer manufacturers in the early 1970s, they did not begin to market their software aggressively until later in the decade (to compensate for falling hardware revenues). In 1979, with the arrival of Japanese clones of its mainframes, IBM cut its hardware prices drastically, causing a recession in the computer industry and precipitating a general restructuring to secure more revenues from software and services. In September 1981, Burroughs acquired the Systems Development Corporation to enhance its software capabilities. Univac increased its spending on software development and soon claimed to be supplying 95 percent of the software for its computers.[8]

The situation was much the same for non-US computer manufacturers. In Japan, NEC, Fujitsu, and Hitachi were all deriving large fractions of

Table 6.3
The *Datamation* top 15 in software, 1992. Firms in italic type are independent software vendors; those in roman are computer manufacturers.

Rank		Software revenues	Software revenues as percentage of total revenues	Total revenues
1	IBM	$11,365,900,000	17.6	$64,520,000,000
2	Fujitsu	$3,524,900,000	17.5	$20,142,200,000
3	*Microsoft*	$2,960,200,000		
4	NEC	$1,840,300,000	12.0	$15,359,000,000
5	*Computer Associates*	$1,770,800,000		
6	Siemens Nixdorf	$1,058,400,000	12.7	$8,345,100,000
7	*Novell*	$968,600,000		
8	Hitachi	$982,500,000	8.7	$11,352,000,000
9	*Lotus Development*	$810,100,000		
10	DEC	$800,000,000	5.6	$14,162,000,000
11	*Oracle*	$782,000,000		
12	Unisys	$712,000,000	9.1	$7,832,000,000
13	Olivetti	$707,800,000	12.3	$5,762,000,000
14	ICL	$692,400,000	16.3	$4,254,800,000
15	*Finsiel*	$633,000,000		

their data processing revenues from software by the mid 1980s, and they had developed impressive capabilities in software development, mainly in systems software and operating systems.[9] The pattern was repeated in Europe, where Britain's ICL, Germany's Siemens and Nixdorf, France's CGE, and Italy's Olivetti all had large software-development operations producing systems software and some business applications.

IBM dominated the market for systems software for its own computers. Although software accounted for only 4 or 5 percent of IBM revenues in the early 1980s, it was estimated to have one-third of the world market for

Table 6.4
IBM's software and services revenues, 1975–1995.

	Data processing revenues	Software and services revenues	Software revenues	Software revenues as percentage of total revenues
1975	$11,116,000,000	$1,112,000,000		
1976	$12,717,000,000	$1,653,000,000		
1977	$14,765,000,000	$1,329,000,000		
1978	$17,072,000,000	$2,000,000,000		
1979	$18,338,000,000	$3,301,000,000		
1980	$21,236,000,000	$5,983,000,000	$800,000,000	3.8
1981	$26,340,000,000	$4,480,000,000	$1,000,000,000	3.8
1982	$31,500,000,000	$5,300,000,000	$1,683,000,000	5.3
1983	$35,603,000,000	$2,302,000,000	$2,200,000,000	6.2
1984	$44,292,000,000	$3,397,000,000	$3,197,000,000	7.2
1985	$48,554,000,000	$4,465,000,000	$4,165,000,000	8.6
1986	$49,591,000,000	$5,814,000,000	$5,514,000,000	11.1
1987	$50,486,000,000	$7,686,000,000	$6,836,000,000	13.5
1988	$55,003,000,000	$8,862,000,000	$7,927,000,000	14.4
1989	$60,805,000,000	Not stated	$8,465,000,000	13.9
1990	$67,090,000,000	$11,452,000,000	$9,952,000,000	14.8
1991	$62,840,000,000	$12,542,000,000	$10,524,000,000	16.7
1992	$64,520,000,000	$17,776,000,000	$11,366,000,000	17.6
1993	$62,716,000,000	$20,664,000,000	$10,953,000,000	17.5
1994	$64,052,000,000	$28,182,000,000	$11,529,000,000	18.0
1995	$71,940,000,000	$33,092,000,000	$12,949,000,000	18.0

Source: *Datamation* 100, 1976–1996.

packaged software. It had a virtual monopoly in operating systems, its CICS package accounted for 90 percent of the market for teleprocessing monitors, and it had half of the market for databases (by far the most competitive systems software market).

In 1982, IBM formed an independent business unit, Information Processing Services, in an effort to improve its position in software by "developing, acquiring, packaging, supporting, and servicing applications, in the areas of industry, cross-industry, user interface, application development, and productivity tools"[10]—in short, every sector of the software industry in which IBM did not already have a strong presence. IBM's efforts in business applications were lackluster, so it focused on acquiring products it could resell through its network of software marketing offices (by now established in twelve US cities). IBM's Information Processing Services unit was a boon to small independent software vendors, who could not have afforded the kind of exposure that IBM's marketing organization provided. By late 1983, IBM was offering about 2,000 application programs.

One of the paradoxes of the software industry was how IBM came to be so dominant despite the mediocrity of its software. Jokes about the quality of IBM software abounded. Here is one that was making the rounds in the 1970s:

Q: What is an elephant?
A: A mouse with an IBM operating system.[11]

Another, from the early 1980s, made a similar point about IBM's aging IMS database:

Q: Name two movie stars and a dog
A: Rin-Tin-Tin, Benji, and IMS.[12]

Although the pot-shots at IBM's software were partly rhetorical and symptomatic of software in general rather than of IBM software in particular, there was an underlying truth to them. For example, two of the most successful products of the 1970s and the 1980s, SyncSort and CA-SORT, provided solutions to IBM's perennial inability to create an efficient sorting program (on the face of it a modest enough goal). On the other hand, where there were big bucks to be made (in operating systems and teleprocessing monitors, for example), IBM controlled the market almost completely and charged prices so inflated that they constituted a price umbrella under which the software industry could shelter. For example, in the mid 1980s IBM's software tariff included its basic MVS/XA operating system (which rented for $4,280 a month) and its COBOL compiler ($1,872 a month). The IMS database "dog" rented for $8,900 a month, plus $1,465 for maintenance.[13] Nonetheless, IMS was the best-selling database of its era, and its price defined an industry norm of purchase prices as high as $300,000 for databases from the independent software vendors. Leasing, rather than one-off sales, gave IBM a secure, constant revenue stream and a way to raise its prices from time to time. When it became evident that software had a life of 10, 15, or even 20 years, versus 5 years for hardware, the advantage of leasing proved even greater. In October 1986, IBM introduced "value-based pricing"—that is, it began to charge software prices such that the users with the most powerful machines paid the most, on the ground that "customers get more value out of software running on a 3090 than from the same software running on a 9370."[14] Under the new pricing regimen, the cost of the same software could vary by a factor of 3.

IBM's main advantage over its competitors, however, was that of what now was called "bundling." In its new sense, this word now meant selling a bundle of software packages that were known to work harmoniously together. "In this technique, IBM defines a high-level functional area of systems software. Once defined, IBM then develops a group of products that work closely together to provide the user with end-to-end functionality not possible with independently developed products. This group of products is marketed and packaged as a unit, making ordering, installation, and maintenance significantly easier."[15] In an era of one-product independent software companies, this gave IBM a strong advantage. For users, the alternative to buying from IBM was integrating products from different suppliers that had not been shown to work together and which might receive random and uncoordinated fixes and upgrades from their individual producers.

IBM's biggest software initiative of the 1980s was its Systems Application Architecture (SAA), announced in March 1987. In an article titled "IBM's Moment of Truth," a perplexed *New York Times* reporter opined: "The solution everyone is waiting for, in fact, is a product that will never be sold separately and almost defies explanation."[16] Indeed the SAA announcement was too arcane for most computer professionals, never mind the public at large. The firm that had charmed the world with the Charlie Chaplin advertisements for its PC had reverted to type when it came to corporate software.

The initiative that led to SAA began in 1983, when IBM began to address the switch from centralized mainframe-based computing to distributed computing in which the three tiers of computing—mainframes, midrange computers, and (increasingly) PCs—were networked. IBM had software products in all three tiers of computing, but programs written for one platform would not run on another and could not be easily converted. This problem also existed for the independent software industry, which needed to develop packages separately for each of IBM's different platforms. The solution was SAA, a set of rules and interfaces that enabled a program written for one platform to be easily "ported" to another. This sounds easy, but in effect it took 3 years of negotiation between IBM's globally distributed software and hardware development centers. The SAA story—"IBM's Moment of Truth"—is, in its way, as compelling as the development of System/360 or that of the PC, but it has never been told well.[17] By 1987, to make its software SAA compliant and to develop new products, IBM had increased its programming staff by 6,000 to 26,000, or 7 percent of its work force.

Beginning in the mid 1980s, IBM increasingly acted as a reseller of applications software for the independents, reportedly making a 50 percent markup in the process. As was commonly observed, IBM had made occasional forays into applications software but had never produced a really good applications package. The usual explanation for this was that IBM's developers were too remote from the real business context. In 1988, in an attempt to remedy this, IBM created "business partners" (such as Hogan Systems in banking software) with the aim of providing single-source solutions for various industries. The following year, IBM made strategic investments of about $100 million in some of the top software firms, including MSA and American Management Systems, thereby gaining board-level strategic and tactical influence over these firms.[18]

Though IBM had become an impressively large software vendor, its software sales fell well short of the 30 percent of total revenues by 1990 it had promised investors at the time of the SAA launch. However, in view of the fragmented nature and the low concentration of the corporate software market, this was perhaps never a realistic goal (barring massive acquisitions, which IBM's fiscal troubles of the early 1990s prevented).

Computer Associates, the Ultimate Consolidator

In February 2000, Computer Associates acquired Sterling Software for $4 billion. The two largest consolidators in the software industry were now combined into an organization that enshrined a veritable history of the software products industry. Computer Associates, whose 1999 revenues

Table 6.5
Financial statistics on Computer Associates.

	Revenues	Annual growth	Employees
1977	$1,242,000		NA
1978	$3,219,000	159%	NA
1979	$12,000,000	265%	NA
1980	$18,000,000	52%	200
1981	$30,000,000	69%	300
1982	$43,000,000	43%	380
1983	$58,000,000	35%	654
1984	$159,000,000	173%	1,000
1985	$228,000,000	44%	NA
1986	$510,000,000	124%	NA
1987	$629,000,000	23%	2,700
1988	$842,000,000	34%	4,450
1989	$1,159,000,000	38%	6,250
1990	$1,244,000,000	7%	6,900
1991	$1,301,000,000	5%	6,700
1992	$1,509,000,000	16%	7,400
1993	$1,841,000,000	22%	7,200
1994	$2,148,000,000	17%	6,900
1995	$2,623,000,000	22%	7,550
1996	$3,505,000,000	34%	8,800
1997	$4,040,000,000	15%	9,850
1998	$4,719,000,000	17%	11,400
1999	$5,253,000,000	11%	14,650
2000	$6,766,000,000	29%	18,000

Sources: revenue data for 1977–2000 from http://www.cai.com/invest/history.htm, accessed July 2001; data on employees from Computer Associates' annual reports and *Hoover's Guide to Computer Companies* (various years).

were $5.3 billion, had been established in 1976. Since then it had absorbed more than 70 firms, ranging from medium-size companies valued at a few million dollars up to Legent (acquired in 1995 at a cost of $1.8 billion). Sterling Software had been founded in 1981 and was itself a result of about 35 acquisitions. Its most controversial acquisition was its hostile takeover of Informatics in 1985. (The enlarged Computer Associates had many more ancestors than the hundred-plus firms acquired by Computer Associates and Sterling Software, for many of those acquired firms were themselves created by merger and acquisition. For example, when Sterling Software acquired Informatics in 1985, Informatics was itself the product of more than 30 acquisitions.)

After the acquisition of Sterling Software, Computer Associates was the number-two firm among independent software vendors, second only to Microsoft. Since the mid 1980s, Computer Associates has always been among the top half-dozen firms, despite having been eclipsed by personal computer software firms such as Microsoft, Lotus, and WordPerfect. Yet Computer Associates has been more consistently profitable than any personal computer software company other than Microsoft, and it would have yielded a much better return on investment. Its image was neatly summarized by the title of a *Business Week* company profile in 1996: "Sexy? No. Profitable? You Bet."[19]

As befits its anonymity to the general public, the literature on Computer Associates is sparse, and most of what has appeared focuses on the management style and idiosyncrasies of its founder, Charles Wang.[20] The firm has often been portrayed as a corporate bully that seeks out ailing software firms and strips them of their software assets while firing most of the employees—an approach once described as a "neutron bomb" acquisition strategy.[21] It has also been portrayed as having a cheap and uncaring culture. Computer Associates was famous, for example, for having continued to use secondhand office furniture and to inhabit shabby offices long after becoming a billion-dollar-a-year operation. Charles Wang has been portrayed as uncompromisingly aggressive, streetwise, and hostile to "kiss-ass MBAs."[22] In the mid 1990s, Computer Associates established a public relations department and adopted more enlightened human-resources policies, and in 1998 *Fortune* ranked it among the 100 best companies to work for. The older image persisted to a degree, but the turnaround was remarkable.

Computer Associates shows the imprint of its founder as much as Microsoft bears the hallmark of Bill Gates or Oracle that of Larry Ellison. Yet, unlike Gates and Ellison, Charles Wang is virtually unknown outside

Table 6.6
Computer Associates' principal acquisitions, 1980–1995.

	Corporation acquired	Activity (cost, where known)
1982	Capex	Mainframe systems and utilities ($22 million)
1983	Information Unlimited	PC applications ($10 million)
1984	Sorcim	PC applications ($27 million)
1984	Johnson Systems	Mainframe utilities ($16 million)
1985	Top Secret Software	Data security tools ($25 million)
1986	Software International	DEC VAX utilities ($24 million)
1986	Integrated System Software Co.	DEC VAX presentation graphics ($67 million)
1987	BPI Systems	PC accounting packages ($11 million)
1987	Uccel	Mainframe system and banking software ($800 million)
1988	DistribuPro	PC software distribution ($12 million)
1988	ADR	Database and systems software ($170 million)
1988	Atrium Information Group	PC resource management
1989	Cricket Software	PC and Apple Macintosh graphics (undisclosed)
1989	Bedford Software	PC and Apple Macintosh accounting software ($13 million)
1989	Cullinet	Mainframe databases ($320 million)
1990	DBMS	Database tools
1991	ManageWare	PC spreadsheet and database
1991	Information Science	Human resources ($7 million)
1991	On-Line Software International	Mainframe systems and databases ($104 million)
1991	Pansophic Systems	Mainframe systems and CAD ($290 million)
1991	Access Technology	DEC VAX applications (undisclosed)
1992	Howard Rubin Associates	PC Project management tools. Three tools sold to CA (undisclosed)
1992	Scancom (Europe) Ltd	PC auditing
1992	Nantucket	PC databases (c. $80 million)
1992	Glockenspiel	Workstation applications (undisclosed)
1994	ASK Group	Mainframe database tools and applications ($315 million for 98%)
1994	Newtrend Group	Banking software (c. $40 million; CA already had a half share of company with EDS)
1995	Legent	Mainframe systems management ($1.8 billion for 98%)

the small world of corporate computing. This may be due less to a lack of media attention than to Wang's inability to communicate a "vision." In 1994, Wang tried to articulate a view of computing in *Techno Vision*, a book aimed primarily at business executives.[23] *Techno Vision* attracted less attention than Bill Gates's *The Road Ahead* (published the following year), and it offers few insights into Computer Associates' strategy. As yet there are no biographies of the Charles Wang, although in 1999 he became America's highest-paid businessman (with a compensation package exceeding $600 million).

Charles Wang (pronounced "Wong"), was born in China in 1944, the younger son of an eminent Chinese lawyer. The family fled to New York in 1952. Charles went on to study mathematics and physics at Queen's College; his elder brother Anthony (later to serve as Computer Associates' general counsel) studied law. After a spell as a programmer and salesman for a software house based in New York, Wang and a colleague named Russell Artzt established a small custom programming and consulting firm. Aware of the limited growth oportunities of a programming services operation, Wang seized the opportunity in 1976 to market a sorting program developed by a Swiss company. Marketed as CA-SORT, in direct competition with SyncSort, the package replaced IBM's standard sorting utility for the System/360 computers. Perhaps uniquely in the 1970s corporate software industry, Wang elected to sell the program without the aid of a direct marketing force, instead offering the $3,000 package by mail order and selling it on its merits after a free trial. In its first year, 200 copies of CA-SORT were sold, producing revenues of $600,000. For the first few years of its existence, Computer Associates followed a strategy of developing mainframe utility packages and selling them through low-cost channels at prices well below the industry norms. By 1980 the firm had a portfolio of twenty products and annual sales of $17.8 million.[24]

In 1981, taking advantage of the boom in software flotations, Computer Associates made an IPO. This netted only $3.2 million, but it enabled Computer Associates to issue publicly tradable stock. In 1982, Computer Associates acquired Capex for $22 million in stock. Capex's mainframe utility programs complemented Computer Associates' own products and immediately added 50 percent to its sales. In the next few years, Computer Associates acquired several other mainframe utility makers, including Johnson Systems, Top Secret Software, Software International, and Integrated Software Systems.

Computer Associates made a dramatic transformation with its sixteenth acquisition, the purchase of Uccel for $800 million in stock in

August 1987.[25] This was 10 times as large as CA's previous largest acquisition. Uccel (formerly UCC, the University Computing Corp.) was a Dallas-based consolidator very much in the mold of Computer Associates and was regarded as its closest competitor, having a portfolio of dozens of mainframe utilities and a significant position in banking software. Of the 1,200 Uccel employees, 550 were fired. This highly public and traumatic acquisition set the tone for several years of adverse reportage of Computer Associates.

By 1987, Computer Associates had a portfolio of 90 products, a customer base of 30,000 installations, and annual sales of $629 million. This made it comfortably the world's largest independent software vendor—well ahead of Microsoft and Lotus, whose sales were $346 million and $396 million respectively.

In the early 1980s, Computer Associates had been one of many corporate software firms that tried unsuccessfully to enter the market for personal computer software. This it did entirely by acquisition, buying Information Unlimited (publisher of the Easy Writer word processor), Sorcim (developer of the SuperCalc spreadsheet and other productivity applications), and BPI Systems (maker of an accounting package for small firms). Later it acquired DistribuPro (a distributor of personal computer software), Cricket Software (a maker of graphics presentation software), and Bedford Software (another publisher of accounting packages). Like nearly every other corporate software vendor, however, Computer Associates failed to capture a significant share of the market for personal computer productivity applications. (It did significantly better with accounting software.) Wang later reflected on the reason for this failure, pinpointing the decision not to integrate the West Coast personal computer software firms into the main East Coast business: "They told me the micro business was different, so I didn't put my own people in it."[26] In fact there was probably no particular reason for the failure. The inability of vendors of corporate software to succeed with software for personal computers was so widespread as to be endemic. Despite the desire Charles Wang expressed in the mid 1980s to have Computer Associates' revenues divided equally among utilities, applications, and personal computer software, utilities (the "software plumbing" of corporate computing) have consistently accounted for more than 70 percent of his company's revenues.[27]

After the Uccel acquisition, Computer Associates made a number of more deliberate acquisitions, usually for cash rather than stock. To a degree, these acquisitions were opportunistic, but there was an underlying strategy of portfolio filling and product integration. This is perhaps best illustrated by the company's entry into the database market. The first foray into that market came in October 1988 with the acquisition of Applied Data Research (ADR), the oldest software products company, whose principal product at the time was the DATACOM/DB database. ADR had recently been acquired by AmeriTech (a Bell subsidiary) for $215 million, but it was losing money, and Computer Associates picked it up for $170 million. After the usual rationalization of ADR (primarily downsizing), the database was marketed as a CA-Datacom/DB, with almost its entire original customer base intact. The software was not allowed to languish; it was enhanced and made to operate with CA's other products. Two years later, CA acquired the troubled Cullinet, and again it managed to turn its IDMS database into a CA product with its customer base largely intact. Other complementary acquisitions were made in the form of database access software by taking over On-Line Software and Pansophic. By 1991, Computer Associates was the number-2 database vendor, with 17 percent of the market—up from zero only 3 years earlier.[28] Although IBM still had 50 percent of the market, Computer Associates' nearest independent rival was Oracle, with 10 percent. Through these strategic and well-managed acquisitions, Computer Associates has evolved a comprehensive portfolio of database systems and utilities, all of which have been incrementally improved and made to work together.

In the wake of IBM's SAA announcement, Computer Associates made its implicit strategy of product integration explicit by announcing its CA90s "architecture." *Business Week* noted: "Last April, Wang and other top executives hit the road to explain their vision. There's nothing radical in the CA90s plan, but that's the beauty of it: It's designed to soothe customers by assuring them that if they stick with CA, it will make sure that more of their incompatible programs work smoothly together."[29] But whereas IBM's SAA had been limited to its own platforms, CA offered a heterogeneous environment. The CA90s architecture, which underlay 250 program products, was particularly well placed to capitalize on the switch from mainframe-based computing to client-server distributed computing in the early 1990s. In effect, Computer Associates had become a one-stop software shop for customers who could now buy their hardware from a commodity-type market.

Since the early 1990s, Computer Associates' strategy has been remarkably consistent: After acquiring companies with low valuations but historically strong products and customer bases, it has integrated the products into its own product line. This culminated in 1995, when, in what was then the largest-ever acquisition in the software industry,

Computer Associates took over Legent for $1.8 billion in cash. Computer Associates' has shown a unique ability to keep a portfolio of 500 products (many of them decades old) constantly rejuvenated and to give them the semblance of a product family.

Oracle and the Maturing of the Database Market

In May 2000, Larry Ellison, the billionaire founder and chief executive officer of Oracle, appeared on the front cover of *Business Week*. Inside, an article titled "Why Oracle Is Cool Again" suggested that the Internet would power the already-successful company to ever greater heights.[30] Embroiled in the *US v. Microsoft* anti-trust action, Microsoft had seen its stock price plunge. Now, Bill Gates had a potential rival as the richest man in the world. Although Gates and Ellison are both strongly associated with the firms that they created, they have very different media personae. Gates is perceived as geek-like and hands-on. The adjectives commonly applied to Ellison are—in purely alphabetical order—aggressive, aloof, cultured, and urbane. Gates is noted for his financial restraint. Ellison flaunts his wealth—he has several multi-million-dollar homes, a 76-foot yacht, and an eclectic collection of fine art. But perhaps the hallmark of Ellison's persona is his arrogance. This was the abiding image of Mike Wilson's 1997 biography *The Difference between God and Larry Ellison*.*[31]

Though comparisons are often drawn between Microsoft and Oracle, they are very different firms with very different business models. Some statistics for the year 2000 make the point: Microsoft had sales of $23 billion and a work force of 39,000; Oracle had sales of $10 billion and a work force of 43,000. A typical Microsoft product sells 10 million copies at $200; Oracle is more likely to earn the same revenue by selling 10,000 copies at $200,000. It is the difference between consumer goods and capital goods.

Oracle (founded in Belmont, California, in 1977) was the most successful of three relational database firms that emerged around that time in Northern California. The other two were Relational Technology Inc., in Alameda, and Relational Database Systems, in Menlo Park. All three were near IBM's San Jose Research Laboratory, where the pioneering research on relational databases had been done in the early 1970s. The technology had diffused from IBM through the University of California at Berkeley to the new firms. As time went on, Northern California fostered other producers of relational database software (including Sybase, Illustra, and Unify) and makers of complementary products (e.g., Gupta Technologies). It remains the center of activity in relational technology.

Table 6.7
Financial statistics for Oracle.

	Revenues	Annual growth	Employees
1984	$13,000,000		NA
1985	$23,000,000	81%	NA
1986	$55,000,000	139%	556
1987	$131,000,000	138%	1,121
1988	$282,000,000	115%	2,207
1989	$571,000,000	102%	4,148
1990	$916,000,000	60%	6,811
1991	$1,028,000,000	12%	7,466
1992	$1,179,000,000	15%	8,160
1993	$1,503,000,000	27%	9,247
1994	$2,001,000,000	33%	12,058
1995	$2,967,000,000	48%	16,882
1996	$4,223,000,000	42%	23,113
1997	$5,684,000,000	35%	27,421
1998	$7,144,000,000	26%	36,802
1999	$8,827,000,000	24%	43,800
2000	$10,130,000,000	15%	41,320

Sources: revenue data for 1984–2000 from http://oracle/com/corporate (accessed July 2001); data on employees from Oracle annual reports and *Hoover's Guide to Computer Companies* (various years).

While the phenomenon of "regional advantage" is well known in the hardware industries,[32] it is much less common in software. Northern California's advantage in regard to relational technology was due largely to fact that the technology was little published and non-obvious (certainly relative to programming languages and word processors, which had spread like wildfire), and so the ideas were slow to diffuse.

The invention of the relational database was a classical example of what some Harvard economists have called a "disruptive technology." Using examples ranging from radial tires to rigid disk drives, the Harvard economists have made numerous case studies of the phenomenon.[33] The common pattern observed is of an emergent technology that is ignored by incumbent manufacturers because it offers no immediate benefit to them or to their customers. In the absence of competition from the incumbents, new firms enter the market to commercialize the new tech-

Table 6.8
Oracle and its competitors in 1995. (Ingres was acquired by the ASK Group in 1991 and by Computer Associates in 1994.)

	Unix database market share (1994)	Revenues	Revenue mix, software:services	Employees	Rank
Oracle	35%	$2,967,000,000	54:46	16,882	3
Sybase	20%	$957,000,000	64:36	5,865	6
Informix	19%	$709,000,000	76:24	3,219	8
Progress	5%	$180,000,000	62:38	1,105	25
Ingres	5%	NA	NA	NA	NA

Data on revenues, revenue mix, and employees from *Hoover's Guide to Computer Companies* (1996); data on 1994 market shares from Salomon Bros., cited in Richard Brandt, "Can Larry Beat Bill?" *Business Week,* May 15, 1995: 38–46; rankings from "Top 100 Independent Software Vendors," *Software Magazine,* July 1996 (Progress Software interpolated).

nology, initially for specialist markets. However, as the technology is perfected it begins to compete with the established technology and eventually usurps it. This is very much the pattern we observe with database technology. By about 1980, the market for databases was relatively mature, with many thousands of installations worldwide and a stable set of leading suppliers (including ADR, Cincom, Cullinet, Software AG, and of course IBM). In the 1970s, when the market for first-generation databases (based on technology developed in the 1960s) was taking shape, the second-generation technology of the relational database was being developed in IBM's San Jose Research Laboratories.

The technical history of the relational database has been well documented.[34] In brief, the underlying mathematical basis for the relational database was developed by Edgar F. Codd, an IBM research scientist, in 1970. The particular advantage the relational database offered was its ability to handle complex queries. Somewhat later, an interrogation language called SQL was developed so that ad hoc queries could be phrased as in the following example.

```
SELECT EMPLOYEES WHERE
    SALARY > 20,000
    AND SEX=FEMALE
```

With first-generation "navigational" databases, such ad hoc queries could be handled (with some difficulty) only by a programmer intimate with the internal structure of the database. In practice this was not a material disadvantage for users of corporate mainframe databases, for whom efficiency and security of the data were paramount. Relational technology was computationally inefficient and not yet robust. IBM's San Jose facility developed an early prototype, known as System/R which was announced to the research community in 1973. In Northern California's university-industry research milieu, the concepts spread to the university at Berkeley, where an academic researcher named Michael Stonebraker developed Ingres, a relational database that was distributed freely in the Unix community. In 1980, Stonebraker left academia to form Relational Technology Inc. (subsequently renamed the Ingres Corporation). Stonebraker was also a mover behind Informix and Illustra.

In 1977, when relational technology was breaking, Larry Ellison and two colleagues had started a custom programming operation known as System Development Laboratories.[35] Ellison, not an academic, had previously been a programmer (first with the Ampex Corporation and subsequently with Precision Instrument, where he had worked on information storage software). System Development Laboratories' first customer was Precision Instrument. The Precision Instrument project had a database requirement. Catching wind of the relational database, Ellison's group developed a program, which they called Oracle. In 1978, Ellison transformed System Development Laboratories in a software products firm and renamed it Relational Systems Inc. Shortly thereafter, Relational Systems shipped Oracle version 2 (there was no version 1). This was the first commercially available SQL-compliant database. It was initially sold at $48,000 for Digital Equipment Corporation (DEC) minicomputers. The company was renamed Oracle Systems in 1982. In 1983, version 3 of Oracle was released. For this version, Oracle was rewritten from the ground to run on most of the popular minicomputers and mainframes. For the first 5 years of its commercial existence, the Oracle database was generally bug ridden, late in delivery, and poorly supported, and many sales were made on the basis of promises not fulfilled completely. Sales were sufficient for the Oracle Corporation's survival but not so high that its reputation was permanently damaged. Meanwhile the software was steadily being perfected. In 1984, Oracle attracted venture funding that enabled rapid expansion, and it made its IPO in the spring of 1986. At that time, Oracle's annual revenues were $55 million, and it claimed to have 3,000 minicomputer and mainframe installations and a similar number of installations sold through turnkey and OEM channels.³⁶

Because relational technology was initially too computationally intensive and too immature to attract the mainstream database suppliers, Oracle and its competitors were able to develop the technology without competition from the big players. By about 1983, however, computer speeds had caught up with relational technology and the software was acceptable in terms of scalability and stability. Relational technology was now a threat looming on the horizon of the established producers, each of which had to decide how and when to switch to the new technology.

IBM was the first of the incumbents to make the break, which it did ambiguously by announcing its "dual database" strategy in 1983.³⁷ By this strategy, IBM would continue to support its legacy database users by maintaining IMS indefinitely and would satisfy the relational market with SQL/DS (derived from its System/R research prototype). In 1985, SQL/DS was rebranded DB2. IMS and DB2 remain IBM's main database products. The second-biggest incumbent, Cincom, made a huge R&D investment (reportedly 21 percent of its revenues between 1979 and 1983) in order to pursue a similar path to IBM's. Cincom's strategy was to continue evolutionary enhancement of its existing TOTAL database while simultaneously developing relational products—SUPRA for the IBM mainframe market, ULTRA for the minicomputer market. The R&D burden significantly reduced Cincom's profitability for several years, but the private privately held company had no shareholders to disappoint.³⁸

ADR and Cullinet were less fortunate. In the case of ADR, the relational database challenge was only one of the factors affecting what was an aging product line; several of the company's big-earning products were nearly 15 years old. In 1985, ADR was acquired by AmeriTech for $215 million.³⁹ This did not work out, and 32 months later ADR was acquired by Computer Associates for $170 million. Cullinet, once hugely successful, was even worse affected. Its IDMS database was second only to IBM's IMS in market share, and the firm was regarded as a bellwether of the industry. Indeed, in 1984 the prominent industry analyst Stephen McClellan of Merrill Lynch was moved to write: "It is obvious that [John] Cullinane has done his job well. Cullinet sales, which were only $2 million in 1975, swelled to $110 million in 1983 and are expanding at 50 percent a year. ... Can anything stop Cullinet? We don't believe so. In this, the golden age of software, Cullinet will be the first billion-dollar software company."⁴⁰ It is salutary to note just how wrong an industry pundit can be.

In the wake of IBM's 1985 commitment to relational technology, Cullinet announced plans to make IDMS fully relational. It was a classic error. The announcement "legitimized the technology for the Cullinet installed base, which later was forced to turn to other vendors, primarily IBM."[41] Sales collapsed, and Cullinet's 20 percent share of the mainframe database market was halved over the next 2 years. In the face of mounting losses, John Cullinane made way for a new CEO to lead a strategy of diversification. This did not work out, and in August 1989 the company was acquired by Computer Associates for $320 million. Computer Associates immediately laid off half of Cullinet's 1,800 employees.

During the 1980s, as the market increasingly switched to relational technology, Oracle grew extremely rapidly—its revenues nearly doubled every year from 1984 to 1990. By 1988 it was the number-4 software vendor overall and the largest database supplier other than IBM. Most of Oracle's 2,000-plus employees were engaged in sales, and their commissions would become legendary in the industry. The company also diversified into consulting operations, taking on 350 consultants (15 percent of its work force) to ease customers into its complex software. Ellison reportedly had boasted that one day Oracle would have 10,000 consultants, "topping even Arthur Andersen and Co.'s 6600 IMS consultants."[42] This was one Ellison boast that eventually worked out. Most of Oracle's sales were made in the newly emerging client-server environment, while its mainframe sales were barely noticeable in comparison with those of IBM, Computer Associates, and the other established vendors.[43]

Oracle's annual growth was finally checked in 1991–92, when the firm came close to financial and technological collapse. The financial disarray was triggered by the US Financial Accounting Standards Board's recent imposition of new criteria for reporting revenues. Oracle, like many other software firms, was in the habit of inflating receivables in its annual accounts by including software sold but not delivered. Oracle, with 48 percent of its revenues in receivables, was among the worst offenders and was obliged to restate its 1990 earnings.[44] To compound Oracle's problems, version 6 of its database (released in 1988) was notoriously unreliable, and the firm was getting a reputation as a den of "thieves, crooks, and bandits."[45] As these problems came to a head, Oracle's "reputation, sales, and stock collapsed," the stock losing 61 percent of its value.[46]

In 1991, Larry Ellison brought in a new management team to take control of his sprawling 7,400-employee empire. With tighter accounting controls, a stable version 7 release of Oracle, and an ever-growing consulting operation, the company soon resumed double-digit annual growth. Though Oracle and the relational newcomers failed to penetrate the mainframe market, which IBM and the old-line database firms continued to dominate, they captured most of the burgeoning Unix market for distributed databases.

SAP and ERP Software

As was noted earlier, the term "ERP software" came into common use in the early 1990s. It was used to differentiate the integrated business applications software of half a dozen manufacturers—of which SAP was the most prominent—from that of other vendors, which tended to be less comprehensive and less well integrated. For example, in 1990 the largest vendor of business application software was Dun & Bradstreet Software, which had acquired MSA and McCormick and Dodge earlier in its history. Dun & Bradstreet could satisfy just about all of a user's business application needs; however, the packages themselves had long pedigrees, and that made their integration a major task. Almost by definition, ERP software was written from the ground up by a single firm in an organic process, rather than being aggregated through acquisition. Typically, an ERP package was the only application software a firm needed.[47]

A common characteristic of ERP software was extreme complexity. Few corporate information systems departments were capable of installing ERP software without the assistance of external consultants. Indeed, it was estimated that the leading ERP package, SAP's R/3, cost 5–10 times as much to install as the software itself. In the 1990s, the installation of ERP software became a huge new business for computer services and consulting firms, and they and the producers of ERP software grew symbiotically. The 1993–94 vogue for business process re-engineering encouraged firms to sweep away years of evolutionary growth in their software systems and replace them in one fell swoop. It happened that ERP software offered a better solution for wholesale re-implementation of a corporate software system than the less-well-integrated products of firms such as Dun & Bradstreet.

The rise of ERP software was something of an unpredictable Darwinian process. Hasso Plattner, one of SAP's founders, was not being entirely disingenuous when he remarked "When people ask how we planned all this, we answer 'We didn't. It just happened.'"[48]

SAP was formed in 1972, in the town of Walldorf, a few miles south of Munich, by five programmers from IBM Germany. SAP (Systems, Applications, and Products in Data Processing[49]) was established as a software

Table 6.9
Financial statistics on SAP. (1 Euro = 1.95583 DM.)

	Revenues		Annual growth	Employees
	DM	Euro		
1972	600,000	300,000		9
1973	1,100,000	500,000	69%	11
1974	1,900,000	1,000,000	78%	13
1975	2,300,000	1,200,000	22%	18
1976	3,800,000	1,900,000	66%	25
1977	6,200,000	3,200,000	64%	38
1978	8,500,000	4,300,000	36%	50
1979	9,900,000	5,100,000	17%	61
1980	13,700,000	7,000,000	38%	77
1981	16,300,000	8,300,000	19%	84
1982	24,200,000	12,400,000	48%	105
1983	40,500,000	20,700,000	67%	125
1984	48,000,000	24,500,000	19%	163
1985	61,200,000	31,300,000	28%	224
1986	106,000,000	54,200,000	73%	290
1987	152,000,000	77,700,000	43%	468
1988	179,700,000	91,900,000	18%	940
1989	366,900,000	187,600,000	104%	1,367
1990	499,100,000	255,200,000	36%	2,138
1991	707,000,000	361,500,000	42%	2,685
1992	831,200,000	425,000,000	18%	3,157
1993	1,101,700,000	563,300,000	33%	3,648
1994	1,831,000,000	936,200,000	66%	5,229
1995	2,963,300,000	1,378,600,000	47%	6,857
1996	3,722,100,000	1,903,100,000	38%	9,202
1997	5,910,100,000	3,021,800,000	59%	12,856
1998	8,440,600,000	4,315,600,000	43%	19,308
1999	9,994,700,000	5,110,200,000	18%	20,975
2000	12,255,200,000	6,266,000,000	23%	24,178

Courtesy of SAP.

contractor, much as custom software firms had been established in the United States a decade earlier. Of the five founders, Hasso Plattner and Dietmar Hopp played the leading roles in the firm's commercial and technical development, respectively. Today they are perhaps the world's least-well-known software billionaires.

SAP's first contract was to develop a financial accounting program for the synthetic fiber works of Imperial Chemical Industries (ICI) at Ostragen. At the time (1972), unbundling of software products was very much in the air in Europe, and Plattner and Hopp got explicit permission to sell the ICI program. The program was named System R, the R standing for "real time." Having a real-time program was a significant advantage for SAP, whose US rival's software had generally evolved in earlier, non-real-time environments. System R took about a year to develop, with SAP operating much like a US programming services firm of the 1950s. For example, SAP owned no computer of its own; it used its client's computer during night shifts and on weekends. Amazingly, SAP continued to operate in this fashion until 1980, when it acquired its first mainframe computer.

Upon completing the ICI contract, SAP immediately obtained two further orders, each of which required some modification of the ICI software. In this case and in subsequent cases, the software was enhanced to incorporate more functions. Thus, it became increasingly universal without compromising its tight integration or fragmenting into different products. SAP continued to operate as a 50-person custom programming outfit rather than a software products firm. The switch to a software products firm came in 1978: SAP decided to rewrite its software as R/2, with the medium-term aim of turning it into a product.

The development of R/2 owed a great deal to the German programming milieu, which has always been more formal and more disciplined than its American counterpart. Hopp had a doctorate in mathematics, and in the early 1980s he recruited dozens of individuals who had just earned such degrees. This created an environment particularly suited to the long-term evolution of a stable product. Later, SAP was able to garner a deserved reputation for "BMW quality" software.[50] R/2, like the earlier System R, was developed over a long period (6–8 years), each new contract being used as an opportunity to add to the program's comprehensiveness. By 1980, SAP was ranked the 17th-largest software firm in Germany and had half of Germany's top 100 corporations as clients. Little known outside Germany, and with sales of less than DM 10 million, it was not among the world's top 100 software enterprises.

R/2 was introduced in 1981 as a program product. SAP, with about 100 developers, increasingly focused on the program, relying on third-party firms to install the software for customers. SAP would often supply additional software components or introduce additional features into the package to support individual clients. By 1982, 250 German corporations were using R/2, almost all the sales having been made through word-of-mouth referral. The isolation of the German market enabled the unique R/2 product to evolve relatively free of competition from US firms, whose products were generally designed for the US market and were rarely well internationalized. There were two brakes on the further development of SAP, however: international sales and the supply of R/2 consultants.

In 1984, SAP established its first international office in Geneva. The Trojan Horse effect of R/2's having been installed in the offices of German subsidiaries of multinational firms was probably a more effective sales device. For example, the US agricultural machinery firm John Deere, having used R/2 in its Mannheim subsidiary, implemented an SAP financial accounting system in its English and African subsidiaries. Likewise, ICI, headquartered in London, was the first UK-based SAP user. As more international sales were made, R/2 was incrementally enhanced to incorporate the many currencies and fiscal regulations that international trade demanded. For example: "In some countries [the software] must allow for three-digit inflation values, fifteen-digit sales amounts, and—as in Brazil—three currencies (the real, an index currency, and the United States dollar). Moreover, SAP must monitor the legal and tax developments in the countries . . . in order to keep the software up to date."[51] Thus, the SAP package was completely generalized, and was tailored to an individual country and firm by setting thousands of parameters and switches. The task of installing the software was extremely complex, requiring years of experience to develop competence. This was the second brake on SAP's development. In the early 1980s, SAP established training courses for consultants. Thousands of students passed through its Walldorf headquarters. In 1987, SAP formally established an international training center. By that time, SAP was a substantial software firm, with more than 750 employees and offices in five European countries; however, it still had no presence in the United States. Its 1988 IPO in Germany provided the capital for entry into the United States, the world's biggest software market.

SAP opened its first US office not in Silicon Valley but in Philadelphia. The apparently unpromising location was chosen for its proximity to

SAP's first US customers, primarily the multinationals whose overseas subsidiaries were already R/2 users. These included ICI America, DuPont, Dow Chemical and other firms in the oil, chemical, and pharmaceutical industries. The American market required much more aggressive marketing than the German. Branch offices were opened in most major US cities, where account representatives earned Oracle-size commissions that left SAP employees back home in Germany open-mouthed with astonishment. Since the US market required a user group, one was established in 1990. It held an annual conference for SAP installers. SAP America established partnerships with the major computer services firms and consultants (such as CSC, EDS, and Arthur Andersen), most of whose overseas subsidiaries already had SAP experience. Other partnerships were formed with computer manufacturers, including IBM, DEC, and Hewlett-Packard.

By 1990, SAP was ranked number 4 among corporate software firms, vying with Dun & Bradstreet as the world's largest supplier of mainframe business applications. However, the US market accounted for less than 5 percent of SAP's sales. As late as 1993, only 70 of the world's 1,800 R/2 installations were in the United States. The US market was finally won over with SAP's successor product, R/3.

Whereas R/2 had been a mainframe product, R/3 had been designed for client-server computing. It was released in Germany in 1992 and in United States the following year. The R/3 product was distributed on two CD-ROM discs, which somehow concealed that the fact that the software was stupendously complex and had reportedly cost $920 million to develop. To a degree, SAP was now rewarded for its early recognition of the switch to client-server computing. However, it also benefited from the fad for business re-engineering that swept America in 1993–94. (The most widely cited catalyst for this fad was the publication of Michael Hammer and J. Champy's *Re-engineering the Corporation*, a classic of the airport-bookstand genre of business literature.[52]) In 1993, SAP went into hyper-growth. Its revenues trebled and its work force doubled in the next 2 years. Much of the growth was due to US sales, which now accounted for nearly half of SAP's revenues.

In the world of client-server computing, perhaps half of the employees of a corporation had computer terminals. This led to the adoption of per-workstation pricing in place of the conventional practice of setting a standard price for a package regardless of how many people would make use of it. SAP charged between $2,700 and $4,000 per workstation. In one well-publicized installation (by no means SAP's largest), Pacific Gas &

Electric installed 5,000 terminals in 300 branch offices, the total software cost exceeding $10 million. In the sedate world of mainframe computing, half a million dollars for a database package was probably the most anyone had ever paid for a single software product. If anything, R/3 was even more formidable to install than R/2, but rather than a disadvantage this came as manna to the consultants. Increasingly, the cost of implementing an SAP system far exceeded the cost of the software itself. In the case of Pacific Gas & Electric, the total cost of the R/3 implementation was $70 million—twice the original estimate, and 7 times the cost of the basic software.[53] In 1995, one-third of SAP's 7,000 employees were offering consulting services. SAP expanded its training facilities and franchised its courses to third-party trainers and academic institutions. Since R/3 had been installed primarily in large firms, a domino effect came into play: The suppliers to those large firms became targets for R/3. By 1998 there were 20,000 R/3 installations.

Although SAP benefited to an unusual extent from its first-mover advantage, several competitors emerged in the 1990s.[54] The most significant of these was Oracle, whose database was at the heart of 80 percent of R/3 installations. Oracle's R/3 competitor, Oracle Applications, was introduced in 1995. Offering tight integration with the Oracle database, it quickly gained 10 percent of the ERP software market. Another

Table 6.10
SAP and its competitors, 1995.

	ERP software market share, 1996	Revenues	Employees	Rank
SAP	33%	$1,875,000,000	6,857	5
Oracle	10%	$2,967,000,000	16,882	3
J. D. Edwards	7%	$341,000,000	2,153	18
PeopleSoft	6%	$228,000,000	1,341	22
Baan	5%	$216,000,000	1,525	23
SSA	5%	$394,000,000	2,000	17
Other	34%			

Sources: for revenues and employees, *Hoover's Guide to Computer Companies* (Hoover's Business Press, 1996); J. D. Edwards, *INPUT Vendor Profile*, 1996; for market shares, Gail Edmondson et al., "Silicon Valley on the Rhine," *Business Week*, November 3, 1997: 38–46, attributed to Advanced Mfg. Resources, Inc.; for rankings, "Top 100 Independent Software Vendors," *Software Magazine*, July 1996); Baan interpolated.

European-based company, Baan (established in the Netherlands in 1978), had a trajectory similar to SAP's, having grown organically in Europe before entering the US market in 1993. Baan made a spectacular entrée by selling its Triton system to Boeing in Seattle, said to be the world's largest data processing operation. Other US-owned competitors to SAP to emerge were J. D. Edwards, PeopleSoft, and System Software Associates, all of which were established software products firms that brought out new products to capitalize on the ERP market. They generally succeeded by selling into the markets where they already had a strong customer base—J. D. Edwards in IBM AS/400 installations, PeopleSoft in human-resources software, and System Software Associates in manufacturing companies. All these firms have cooperated with consultants and systems integrators to install their products; J. D. Edwards and Oracle also have extensive in-house consulting operations.

In many ways, SAP is a more potent monopolistic threat to the United States than Microsoft. In 1997, *Business Week* noted: "... SAP, based in the German town of Walldorf, is a giant. With one-third of the $10 billion enterprise-applications market, it not only dwarfs Oracle's $1 billion applications-software and consulting business but also outstrips the next six contenders combined. Today, SAP's R/3 runs the back offices of half of the world's 500 top companies—scheduling the manufacture of washing machines at General Electric Co. and shipping soda pop on time at Coca-Cola Co. SAP is as much a part of the fabric of global business today as IBM was in the 1980s."[55] An estimated 1.5 million workers around the world used R/3 in their primary job function.

If overnight R/3 were to cease to exist (say, if its licenses were made intolerably expensive), the industrial economy of the Western world would come to a halt, and it would take years for substitutes to close the breach in the networked economy. Were Microsoft's products to vaporize overnight, it would take only days or weeks to find substitutes, and the economic disruption would be modest.

Summary

In the 1980s, to the chagrin of other nations that considered themselves software sophisticates, the United States' lead in software products became seemingly invincible.

One often-forgotten reason for US leadership in software products is that the makers of mainframes were also major vendors of software packages for their own machines. And since US companies dominated the world market for mainframes (except in Japan), they controlled most of the captive software market too. Independent American vendors dominated almost every other non-captive software niche.

The 1980s saw the rise of a small number of truly global independent vendors of software products, with sales in excess of $1 billion. Computer Associates, Oracle, and SAP were the most successful exponents of, respectively, consolidation, relational database systems, and ERP software. These firms all supply forms of software infrastructure, and they are perhaps best viewed in this light.

Consolidation in software products (a remarkable phenomenon little discussed in the software literature) has a strong resonance with integration in other fields, such as transportation and telecommunications, where the rationales are not so much scale as "system building."[56] In transportation or telecommunications, the more dense and interconnected the infrastructure becomes, the more powerful and valuable the system. In software, one can discern strong parallels with other historical episodes of system building, from railways in the nineteenth century to airlines in the twentieth. By seeking out more and more products for their portfolios, consolidators have been able to supply all of an organizations' software. Such portfolio filling may appear opportunistic and unsophisticated at first sight; however, besides seeking out the right products, creating such systems has required major investments in R&D and in re-branding to integrate packages into a product family.

That the database industry exists at all is something of a historical accident. In 1970, when IBM unbundled, database technology was in its infancy, and independent software vendors were able to compete. Had IBM unbundled in 1975, the database would have been a mature product, possibly bundled with the basic operating system, and would never have been considered as other than a captive market, like programming languages. And a firm like Cincom would never have existed or would have been in some other niche. The rise of the relational database reminds us that in software nothing is forever. To compete with this new technology, existing database products had to be adapted and extended. Fortunately, software is infinitely malleable, and incumbent firms were able to retain most of their customer bases. There are not many other fields where this would have been possible—it was like a manufacturer of piston engines suddenly having to switch to turbines.

In the 1990s, for the first time, a little tremor of optimism spread through Europe that there was at least one software niche that the United States did not dominate. ERP software has been the only real commercial threat to the US software industry ever. The reason European firms were able to make this market their own was related to the cycle of infrastructure investment. ERP was not conceivable until the early 1980s, when computers were able to support such complex software and when enough was understood about users' requirements to allow the creation of monolithic, all-encompassing software. At this time, European organizations were undercomputerized relative to American ones, so ERP software from indigenous firms was able to fill the software vacuum when computerization took off in the 1980s. In the United States, widespread computerization had taken place 5 or 10 years earlier than in Europe, and organizations were locked into earlier, discrete software packages that had been painfully integrated by in-house staffs—an infrastructure that is usually graced with the name "legacy software." Traditionally, Europe has always been considered to be at a disadvantage to the United States because of its aging transportation and manufacturing infrastructures, which were laid down decades before those in the United States.[57] It may be that the United States, having been first to create a software infrastructure, will be the first to face the problem of renewing such an infrastructure.

Solve your personal energy crisis. Let VisiCalc™ Software do the work.

With a calculator, pencil and paper you can spend hours planning, projecting, writing, estimating, calculating, revising and recalculating as you work toward a decision.

Or with the Personal Software™ VisiCalc program and your Apple* II you can explore many more options with a fraction of the time and effort you've spent before.

VisiCalc is a new breed of problem-solving software. Unlike prepackaged software that forces you into a computerized straight jacket, VisiCalc adapts itself to any numerical problem you have. You enter numbers, alphabetic titles and formulas on your keyboard. VisiCalc organizes and displays this information on the screen. You don't have to spend your time programming.

Your energy is better spent using the results than getting them.

Say you're a business manager and want to project your annual sales. Using the calculator, pencil and paper method, you'd lay out 12 months across a sheet and fill in lines and columns of figures on products, outlets, salespeople, etc. You'd calculate by hand the subtotals and summary figures. Then you'd start revising, erasing and recalculating. With VisiCalc, you simply fill in the same figures on an electronic "sheet of paper" and let the computer do the work.

Once your first projection is complete, you're ready to use VisiCalc's unique, powerful recalculation feature. It lets you ask "What if?," examining new options and planning for contingencies. "What if" sales drop 20 percent in March? Just type in the sales figure. VisiCalc instantly updates all other figures affected by March sales.

Or say you're an engineer working on a design problem and are wondering "What if that oscillation were damped by another 10 percent?" Or you're working on your family's expenses and wonder "What will happen to our entertainment budget if the heating bill goes up 15 percent this winter?" VisiCalc responds instantly to show you all the consequences of any change.

Once you see VisiCalc in action, you'll think of many more uses for its power. Ask your dealer for a demonstration and discover how VisiCalc can help you in your professional work and personal life.

You might find that VisiCalc alone is reason enough to own a personal computer.

VisiCalc is available now for Apple II computers with versions for other personal computers coming soon. The Apple II version requires a 32k disk system.

For the name and address of your nearest VisiCalc dealer, call (408) 745-7841 or write to Personal Software, Inc., 592 Weddell Dr., Sunnyvale, CA 94086. If your favorite dealer doesn't already carry Personal Software products, ask him to give us a call.

PERSONAL SOFTWARE

VisiCalc was developed exclusively for Personal Software by Software Arts, Inc., Cambridge, Mass.

TM—VisiCalc is a trademark of Personal Software, Inc.

*Apple is a registered trademark of Apple Computer, Inc.

VisiCalc, the original spreadsheet application, was the top-selling personal computer software product of the early 1980s.

7
Early Development of the Personal Computer Software Industry, 1975–1983

The personal computer, which seemed to come from nowhere in 1977, had a surprisingly long gestation.[1] Its critical component, the microprocessor, was introduced by the Intel Corporation in 1971. The microprocessor contained all the essential parts of a processor on a single chip, and, when combined with memory chips and peripheral equipment, was capable of performing as a stand-alone computer. However, because the microprocessor had been developed in the electronics industry, the computer industry did not perceive it as a threat to its products. Ever-increasing chip complexity had already resulted in a stream of consumer goods (electronic games, hand-held calculators, digital watches) and many other less visible devices (numerically controlled machine tools, automotive engine management systems, electronic cash registers). The microprocessor was initially seen as little more than a continuation of this trend, dramatic though it was in itself. For example, a 1975 *Business Week* article titled "Lift Off Time for Microcomputers" stated: "Despite the recession, the microcomputer . . . will nearly double in sales this year. Thousands of new products, ranging from cash registers to controllers for traffic signals and factory production lines, have been designed with the new device. And production orders are momentarily expected to turn into a flood."[2] Nowhere in that article was the personal computer so much as hinted at.

The Origins of the Personal Computer Software Industry

Wherever microprocessors were used, they had to be programmed. This programming activity was far removed from the existing software industry and the traditional corporate data processing department. Programming for the new devices was open to any electronics technician who was interested enough to try it, and there were many—often working in commercial engineering development departments or university laboratories—who did try it. Although some professional software development practices diffused into microprocessor programming, much of the software technology was cobbled together or re-invented, an amateurish legacy that the personal computer software industry took several years to shake off.

The first microprocessor-based computer (or certainly the first influential one) was the Altair 8800, manufactured by Micro Instrumentation Telemetry Systems (MITS). This machine was sold in kit form for assembly by computer hobbyists, and its appearance on the cover of *Popular Electronics* in January 1975 is perhaps the best-known event in the folk history of the personal computer. The cover reads: "Exclusive! Altair 8800. The most powerful minicomputer project ever presented—can be built for under $400."[3] The Altair computer was positioned in the market as a minicomputer. Costing one-tenth as much as the cheapest commercially available model, and targeted at the electronics hobbyist, the Altair 8800 was successful in its niche. Several hundred were sold in the 6 months after its introduction. A number of imitators followed, the most important of which was the IMSAI 8080, produced by Information Management Sciences (IMS), a small company founded by William Millard. Thirteen thousand IMSAI 8080 computers were sold between 1975 and 1978.

The market for kit computers for hobbyists was very limited. With many competitors entering the field, MITS and IMS soon began to lose market share and eventually went out of business. However, a number of individuals who developed software for these machines—including Bill Gates, Paul Allen, and Gary Kildall—were to get a first-mover advantage that would give them early dominance of the personal computer software industry.

The transforming event for the personal computer was the launch of the Apple II in April 1977. The tiny firm of Apple Computer had been formed by the computer hobbyists Steve Jobs and Steve Wozniak in 1976. Their first machine, the Apple, was a raw computer board designed for kit-building hobbyists. The Apple II, however, was an unprecedented leap of imagination and packaging. Looking much like the computer terminals seen on airport reservation desks, it consisted of a keyboard, a CRT display screen, and a central processing unit, all in one package. Though Jobs was not alone in having such a vision of the personal computer, he was by far the most successful at orchestrating the technological and manufacturing resources to bring it to fruition.

During 1977, the Apple II was joined by many imitators from entrepreneurial startups, and by machines from two major electronics manufacturers: the Commodore PET and the TRS-80. The PET was a product of Commodore Business Machines, a major manufacturer of calculators; the TRS-80 was made by the Radio Shack subsidiary of the giant electronics retailer Tandy. Each of these three machines was initially targeted at its manufacturer's natural markets: the Apple II at the hobbyist, domestic, and education markets, the Commodore PET at the engineering calculator and education markets, the TRS-80 at home and videogame users. The launch of these genuine consumer products created a significant consumer market for personal computer software. Software packages were typically priced between $30 and $100.

The role of the personal computer in business was not obvious at this time, since most corporate information processing tasks were already computerized. Word processing was done on dedicated text processing systems of the type sold by Wang and IBM. Technical and financial calculations, when not performed on an electronic calculator, were handled by time-sharing services such as Tymshare and the GE Time-Sharing System. Data processing and decision support tasks were handled by corporate mainframes.

Perhaps the second most celebrated event in the folk history of the personal computer is the fall 1979 release of VisiCalc, a software package that transformed the perception of the personal computer as a business machine. Up to that time few people had even heard of a "spreadsheet," but within a few years the term had entered the lexicon of business and fiddling with spreadsheets had transformed the working lives of middle managers in the Western world. In the wake of the spreadsheet came other so-called productivity applications—word processors, databases, communications software, and so on. These packages, for which corporate America was willing to pay several hundred dollars a program, firmly established three major players in the personal computer software industry: Personal Software (publisher of VisiCalc), MicroPro (publisher of the WordStar word processor), and Ashton-Tate (publisher of the dBase II database program). None of these firms had any direct connection with the existing software industry.

First Movers in Systems Software: Microsoft and Digital Research

The period 1979–1983, when the personal computer software industry was being shaped, can be likened to a gold rush. Hundreds and then thousands of firms set out to make their fortunes. Before there could be a software industry for the microcomputer, however, there had to exist the foundation of an operating system and programming languages. The market for these systems programs was shared by Microsoft and Digital Research. These firms were ahead of the gold rush. They were established before the personal computer was a reality, at a time when the microcomputer was still perceived as either a dedicated controller or a low-cost minicomputer.

Programming Languages
Microsoft got its start developing the first programming languages for microcomputers.[4] The first microprocessor-based computers were so simple that they did not require an operating system, but programmers did need a programming language to develop application programs. Bill Gates and Paul Allen filled this void in 1975 by providing a programming language for the MITS Altair 8800.

Gates was born in 1955 into a well-to-do and socially accomplished Seattle family. He was educated in private schools. In the fall of 1973 he began an undergraduate program at Harvard University, expecting to follow his father in the legal profession. Some years earlier, however, at the age of 13, Gates had become an enthusiastic user of a time-sharing terminal that his school had rented. He had become an accomplished programmer in BASIC, a programming language designed for novices. Gates and his schoolmate Paul Allen had explored the inner software complexities of the time-sharing system and had become real experts.

In the fall of 1971, when Gates was still in high school and Allen was beginning his freshman year as a computer science major at Washington State University, they decided to form a company, to be called Traf-O-Data, that would analyze the data collected by automatic vehicle sensors. (There was a lot of traffic census activity in Seattle at the time.) Having learned of the newly developed Intel 8008 microprocessor, then being used primarily for electronic control equipment, they obtained one and used it to build a rudimentary computer to automatically analyze the traffic data. Traf-O-Data was typical of hundreds of small firms in the United States that were using microprocessors in novel, specialized applications. And like most of them, Traf-O-Data folded for lack of business. Between this experience and the day Gates went off to Harvard, he and Allen were summer interns at the systems integrator TRW, further honing their programming skills.[5]

According to Gates and his many biographers, in his sophomore year at Harvard he saw the January 1975 issue of *Practical Electronics*, which had the Altair 8800 kit on the cover, and saw in a flash the opportunity to become the leading vendor of programming languages for microcomputers.[6] The Altair 8800 used the Intel 8008 microprocessor, with which Gates and Allen were already familiar. Working mainly at night, Gates and Allen wrote a BASIC compiler for the Altair during the next month. Although a considerable achievement, writing a BASIC translator was a task that a first-rate senior undergraduate at any good computer science school could have been expected to accomplish. The manufacturer of the Altair, Albuquerque-based MITS, distributed the software for Gates and Allen, paying them a royalty of about $30 a copy. Allen, meanwhile, had completed his computer science undergraduate program, had briefly worked as programmer for Honeywell's mainframe division, and had then decided to move on to MITS to become Director of Software at that eight-person firm. A few months later, Gates decided to discontinue his studies at Harvard and join Allen in Albuquerque.

In the summer of 1975, Gates and Allen formed Micro-Soft (the hyphen was later dropped) as a company that would supply programming languages to the microcomputer industry. During the next 18 months, as the personal computer became fully defined (culminating in the launch of the Apple II and other consumer machines), Microsoft had a classic first-mover advantage. At the beginning of 1979, Gates and Allen relocated Microsoft, then just 13 people strong, to Seattle. During the first year in Seattle, Microsoft's sales reached $2.5 million, derived almost entirely from software royalties from the BASIC translator. Payment by royalties was effectively a new business model for the software industry. Twenty years earlier, when the Computer Sciences Corporation had set out to develop programming language translators for mainframes, it had sold them for a fixed price, then moved on to the next customer. This was also true of some of the early microcomputer software firms. But the ever-increasing flow of royalties would turn Microsoft into a money-making machine. Payment by royalties was not, however, a Microsoft invention—it was endemic to the videogame and personal computer industry.

After Microsoft's success with BASIC, other programming languages followed, including the industry standard COBOL and FORTRAN. At this time Microsoft was almost exclusively a supplier of programming languages to the microcomputer industry through OEM channels.[7] Although it had competitors, all of them tiny startups, none was nearly as successful in this niche. But Microsoft was little known outside the industry, and it would be several years before the firm was perceived as a maker of consumer software.

Operating Systems

The first vendor of microcomputer operating systems was Digital Research, founded in 1976 by Gary Kildall (1942–1994).[8] In 1972, when the Intel microprocessor first came to his attention, Kildall was a computer science instructor at the Naval Postgraduate School in Monterey, California. Kildall, with a PhD in computer science, got interested in the challenge of programming the new devices and became a consultant to Intel, producing PL/M, a well-regarded programming language for the Intel 8008 microprocessor. During the period 1974–1976, Kildall developed an operating system to enable the Intel chip to make use of floppy disks, which were then coming onto the market. He called the system CP/M (Control Program for Microcomputers).

During the hobby-computer boom of 1975–76, Kildall sold CP/M by mail order for $75. In 1977 he licensed CP/M to the manufacturers of the IMSAI 8080 for $25,000. Up to this time, the most popular personal computers, such as the Apple II and TRS-80, had had proprietary operating systems, either written in house or contracted out. The cost of developing an operating system, perhaps $50,000 or $100,000, was a major barrier to entry to the manufacture of personal computers. But CP/M (essentially the Unix of the microcomputer world) could be adapted at minimal cost to run on any 8008-based machine, and this resulted in dozens of new entrants into the burgeoning personal computer market in 1977 and 1978. Digital Research grew rapidly with the market, creating new versions of CP/M to run on different microprocessors. By enabling any CP/M-compliant application to run on any computer using the CP/M operating system, Kildall liberated the market for personal computer software (much more so than Microsoft's programming languages). By 1980, about 200 makes of computer were running CP/M.

Digital Research and Microsoft had set out at much the same time in complementary system software markets, and their trajectories were very similar. Each effectively monopolized its market. In 1981, each had sales of about $15 million. Each had about 100 employees. The turning point in the relative fortunes of the two companies came with the award to Microsoft of a contract to develop the operating system for IBM's new personal computer, launched in August 1981.

The history of the IBM PC is well known.[9] In mid 1980, IBM decided to create a personal computer. To sidestep the company's slow bureaucratic development processes, it decided to outsource nearly all components and subsystems. Several firms had the good luck to share in creating the

IBM PC, although none of them could have imagined the rewards it would eventually bring. As the two industry leaders in systems software for microcomputers, Microsoft and Digital Research were approached by IBM to develop the programming languages and the operating system for the new computer, respectively. Microsoft quickly sealed a contract with IBM; Digital Research did not, and that mistake ultimately led to Digital Research's being relegated to minor-league status.

How Digital Research came to pass up the opportunity to create the IBM operating system has become one of the most poignant episodes in the folk history of the personal computer. One version of the story has Kildall contemptuously flying his private plane while the IBM "suits" were cooling their heels below; another version has Kildall departing for a two week cruise in the Caribbean without waiting to strike a deal with IBM. These stories are apocryphal, but it is true that Kildall failed to sense IBM's eagerness to reach an agreement. It was a justifiable error: Digital Research was the leader in microcomputer operating systems, and the IBM personal computer held no special significance for Kildall—IBM was simply another of his firm's major corporate clients (Hewlett-Packard and AT&T were others), none of whom were noted for their sense of urgency.

In any case, the window of opportunity closed for Digital Research, and Microsoft was asked to produce an operating system in addition to the programming languages. This Gates and Allen agreed to do. To speed the development, they obtained the rights to a locally developed operating system for a reported $100,000.[10] They quickly re-engineered it into what became known as MS-DOS (MicroSoft-Disk Operating System). The operating system was supplied to IBM on a royalty basis. It was estimated that Microsoft received $10 for each copy sold.[11]

The IBM PC had been designed as an "open" machine, so that non-IBM suppliers could produce add-on hardware and software. As the machine gained acceptance, manufacturers produced IBM-compatible clones in huge numbers. By 1983, nearly a million IBM-compatibles had been sold, 90 percent of them with a copy of MS-DOS.[12] Microsoft, now with sales of $70 million and 383 employees, had steamed past Digital Research to become the leading personal computer software vendor. However, this had not been accomplished on the strength of MS-DOS, which had netted only an estimated $8.5 million in 1983. Microsoft was already becoming the most opportunistic and diversified of the microcomputer software firms, its many products including add-on hardware cards and computer games in addition to programming languages.

The Making and Selling of Microcomputer Software

By the end of 1983 there were estimated to be 35,000 different personal computer software products from about 3,000 vendors, an increase of 50 percent from the previous year.[13] Table 7.1 shows the classification system for the personal computer software industry used at the time by industry analysts and the technical press. The taxonomy is much like that used for the corporate software industry, but modified to emphasize where big money was to be made: primarily in systems programs and productivity applications.

Beyond the taxonomic similarity, however, there was almost no point of contact between the booming microcomputer software industry and the software industry for corporate mainframes and minicomputers. The greatest differences were in volume and price. Corporate software was typically priced between $5,000 and $250,000, and a few hundred sales made

Table 7.1
Taxonomy of personal computer software, 1975–1983. Examples in rightmost column give name of product, publisher, and date published.

Systems		
Operating systems		CP/M (Digital Research, 1976)
		MS-DOS (Microsoft, 1981)
Programming languages		MBASIC (Microsoft, 1975)
		CBASIC (Compiler Systems, c. 1980)
Applications		
Productivity	Spreadsheets	VisiCalc (Personal Software, 1979)
		Lotus 1-2-3 (Lotus Development, 1983)
	Word processors	WordStar (MicroPro, 1979)
		WordPerfect (Satellite Software International, 1982)
	Databases	dBase II (Ashton-Tate, 1980)
		PFS:File (SoftwarePublishing, 1980)
Industry	Cross-Industry	General Ledger (Peachtree Software, 1979)
		General Accounting (BPI Systems, c. 1980)
	Industry specific	Market too fragmented to highlight individual products
Consumer and educational	Games	Zork (Infocom, 1980)
		Choplifter (Broderbund, 1982)
	Home	Home Accountant (Continental, c. 1982)
		Homeword (Sierra On-Line, 1983)
	Education	Typing Tutor (Kriya Systems, c. 1980)
		Science of Learning (Edu-Ware, c. 1981)

for a successful product. Personal computer software was typically priced between $50 and $500, and tens of thousands of units were sold.

Whereas corporate software firms had typically employed a direct sales force, low unit prices and high sales volumes made this impractical for producers of personal computer software. The first application packages of the mid 1970s, such as Electric Pencil and EasyWriter, were sold by mail order, their sole authors having advertised in the electronics hobby press. During the second half of the 1970s, as the personal computer began to take off, a retail infrastructure came into being.[14] The best-known chain of computer stores was ComputerLand, a franchise established in 1976 by William Millard, the founder of IMS. By 1977 there were 24 ComputerLands in operation. The number nearly doubled each year, and by 1983 there were about 500. Other retail chains started in 1976 or 1977 included CompuShop, Computer Factory, Computer Craft, and InaComp. A little later, as the personal computer became a consumer item, hi-fi stores such as Tandy's Radio Shacks (with more than 700 branches) began to sell computers. Tandy also established several hundred dedicated Computer Centers. By 1981, retailing software was a $150 million business, and software-only stores had begun to flourish. The Program Store, which had opened its first outlet in 1979, was a 20-store chain by 1983. Other software-only stores included Programs Unlimited, Softwaire Center International, and Software City. Mail order, which had begun as an in-house operation by the early software producers, became professionalized with the establishment of operations such as 800-SOFT-WARE. One industry source estimated that up to 25 percent of software sales were being made by mail order in the early 1980s.[15]

In the late 1970s, a typical software development firm consisted of one or two programmers with strong technical skills but no manufacturing, marketing, or distribution capabilities. As a result, two forms of intermediary evolved between developers and retailers: the software publisher and the software distributor or wholesaler. The distributor had a somewhat unglamorous role, shipping boxes of shrink-wrapped software from the publishers to the retailers, though the biggest of them moved enormous volumes and became highly profitable. The biggest distributor, SoftSel, established in Inglewood, California, in 1980, had sales of $80 million and 340 employees only 3 years later.[16] It published a catalogue of about 3,000 packages from 200 suppliers, and the SoftSel "hot-list" became the industry's "hit parade."

However, the crucial link in the software supply chain was the software publisher. The role of the publisher has been understated in the history of the industry, most of the glamour going to the "creative" software developers. The publisher took a raw program from a developer and turned it into a product that the market would accept. This involved design of packaging and point-of-sale materials, manufacturing activities (e.g., the duplication of diskettes and printing of manuals), advertising and promotional activities, and order fulfillment. A publisher typically paid a royalty of 15–20 percent to the developer, and in some cases publishers provided advances as high as $50,000.

Two prominent examples of the symbiotic relationship between developer and publisher were VisiCalc and dBase II. VisiCalc, developed by a two-person partnership called Software Arts, was published by Personal Software, which took on all the downstream activities and paid the developers a royalty. The database program dBase II was written by a lone developer, Wayne Ratcliffe, and might have vanished without a trace had Ashton-Tate not published it. In many respects the relationship between a software publisher and the developer was like that between a book publisher and an author. This was particularly true for the publishers of recreational and games software; for example, Broderbund and Sierra On-Line each published dozens of programs, mostly written by moonlighting lone authors.

Figure 7.1 illustrates the cost structure of the typical business package. Based on empirical research undertaken in 1983 (source: Efrem Sigel, a principal of the market-research firm Computer Trends Inc.),[17] this figure was widely reproduced, informing both policy circles and industry analysts. Perhaps the most interesting revelation was that marketing consumed 35 percent of costs. There was much anecdotal evidence for the high level of spending required to produce a hit product. For example, Fred Gibbons, president of the Software Publishing Corporation, was famously reported as saying that the barriers to entry into the personal computer software business were "marketing, marketing, and marketing," and his firm spent 40 percent of its revenues on marketing—15 percent on magazine advertising.[18] Some of the more colorful episodes in the history of the personal computer concern flamboyant and expensive marketing. Ashton-Tate, which spent one-third of its income on marketing activity, achieved notoriety for having a blimp emblazoned "dBase II" fly over a dealer convention in Las Vegas. Lotus Development was reported to have spent $2.5 million of its $3.5 million in startup funds on a blaze of publicity to launch Lotus 1-2-3.[19]

In the period 1981–1983, the receipts of personal computer software publishers grew from $70 million to $486 million.[20] At the same time, the market for business software, particularly productivity applications, quickly became relatively concentrated, nine publishers accounting for 85 percent of sales (table 7.2). A single product often accounted for most of

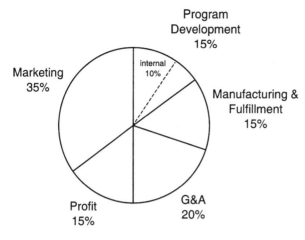

Figure 7.1
A cost breakdown of business and professional personal computer software. Based on data compiled by Efrem Sigel of the market research firm Computer Trends Inc. (Sigel et al., *Business/Professional Microcomputer Software Market, 1984–1984*, p. 69). This figure is reproduced in US Department of Commerce, *A Competitive Assessment of the United States Software Industry and elsewhere.*

Table 7.2
Revenues of top nine business/professional personal computer software publishers, 1981–1983. NS: not significant.

	1981	1982	1983
Microsoft	$16,000,000	$34,000,000	$70,000,000
VisiCorp	$20,000,000	$35,000,000	$60,000,000
Lotus Development	—	—	$48,000,000
MicroPro	$5,000,000	$22,000,000	$45,000,000
Digital Research	$15,000,000	$22,000,000	$38,000,000
Ashton-Tate	$2,000,000	$10,000,000	$30,000,000
Peachtree	$3,000,000	$9,000,000	$24,000,000
Sorcim	NS	$4,000,000	$10,000,000
Software Publishing	NS	$4,000,000	$10,000,000
Total of top nine	$61,000,000	$140,000,000	$335,000,000
Total for US	$70,000,000	$215,000,000	$468,000,000

Based on pp. 18–19 of Sigel et al., *Business/Professional Microcomputer Software Market, 1984–86.*

a firm's revenues. Sorcim derived 80 percent of sales from its SuperCalc spreadsheet, Ashton-Tate 81 percent of its sales from dBase II.[21]

As software publishers became larger, they began to internalize more of their operations. For example, in 1983, Personal Software, publisher of VisiCalc, renamed itself VisiCorp and back-integrated into development, while Ashton-Tate acquired the rights to dBase II from its author and undertook its further development in house. In 1982, the major publishers also began to develop direct sales forces for corporate sales so they would no longer have to bear the hefty markups of the retailers. Lotus Development, for example, sold its 1-2-3 spreadsheet direct to corporate information systems managers, who were by now buying hundreds or thousands of copies for their firms and were keen to negotiate site licenses on special terms at prices well below retail. Other publishers made OEM deals with computer manufacturers who bundled the software packages with their machines. Typically the publisher of an OEM package received only 10 percent of the retail price, but this was still attractive; it eliminated the distribution problem entirely, and it pre-empted purchases of competitive products. In 1983 it was estimated that 30–35 percent of business software was sold through OEM channels.[22] Microsoft, of course, was the pre-eminent OEM supplier long before it was well known in retail outlets.

VisiCalc and the "Killer App" Hypothesis

The invention of VisiCalc is one of the heroic episodes in the history of the personal computer.[23] VisiCalc is often described as the "killer app" that unleashed the corporate personal computer revolution: "Suddenly it became obvious to businessmen that they had to have a personal computer: VisiCalc made it feasible to use one. No prior technical training was needed to use a spreadsheet program. Once, both hardware and software were for hobbyists, the personal computer a mysterious toy, used if anything for playing games. But after VisiCalc the computer was recognized as a crucial tool."[24]

The "killer app" hypothesis argues that a novel application, by enabling an activity that was previously impossible or too expensive, causes a new technology to become widely adopted.[25] Two other commonly cited examples of killer apps are e-mail (credited with popularizing the Internet) and the World Wide Web browser (credited with enabling electronic commerce). Outside the realm of computers, *The Jazz Singer* has been credited with establishing sound cinema in the late 1920s. The killer app is an appealing and intuitive hypothesis that appears to be bound up with the more established economic concept of switching costs. An economist would argue that consumers switch to an alternative technology when the bundle of advantages outweighs the cost of doing so. Although VisiCalc was undoubtedly one of the advantages of personal computers, it was not essential. At some point, the weight of other applications (word processing and personal databases, perhaps) combined with falling hardware prices would have brought the personal computer into business use. In short, the personal computer revolution would have happened with or without VisiCalc. However, it is plausible that the VisiCalc accelerated the process by several months.

The electronic spreadsheet for the personal computer was invented by Dan Bricklin. Bricklin graduated in engineering from MIT in 1973, after which he worked as a programmer for DEC and in the independent software industry. In 1977 he enrolled in the Harvard Business School with the aim of retraining for a career in the financial services industry.

Until the invention of VisiCalc, the spreadsheet was an accounting device primarily used for the financial modeling of business situations. The spreadsheet demonstrated the effect of varying financial parameters, such as raw materials costs or labor charges, in a business model. The problem with the old-fashioned spreadsheet, of course, was the effort needed to recalculate it each time a parameter was altered. At Harvard, Bricklin encountered pencil-and-paper spreadsheets and hit on the idea of implementing a spreadsheet on a personal computer, which would enable the calculation to be automatic and the effect to be seen instantly on the screen.

Spreadsheet-type applications were readily available on mainframes and time-sharing computers (usually in the form of financial modeling packages), but the lack of interactivity limited their usefulness:

Although there have been financial modeling packages and planning programs on mainframes, none had exactly the properties of VisiCalc. . . . It was able to recalculate quickly the rows and columns of a spreadsheet every time a single number was changed. Because the response time was virtually instantaneous, the machine was able to keep up with the thinking speed of the user. Mainframe financial modeling packages were often on-line systems, but they were not real-time. The real-time interactive nature of VisiCalc was not just an improvement on the financial planning software, it was a new paradigm.[26]

After some experimental programming, Bricklin enlisted the help of a friend, Bob Frankston, another MIT engineering alumnus, and together they developed a full-scale program for the Apple II computer. The program contained about 10,000 instructions, which was at the limit of what two programmers could achieve in a few months. In January 1979 they incorporated their partnership as Software Arts and arranged to publish VisiCalc through Personal Software, a small firm established a few months earlier by a Harvard MBA, Dan Fylstra. Bricklin and Frankston—or, rather, Software Arts—agreed to royalties of 36 percent for regular sales and 50 percent for bulk OEM sales.

Fylstra showed early versions of VisiCalc to dealers in the spring and summer of 1979, drawing mixed reactions. The program first went on sale in October 1979 at a retail price of $100. The program sold at the rate of about 500 copies a month—acceptable but not spectacular. After favorable press reviews and word-of-mouth recommendations, sales took off in the second half of 1980, reaching 12,000 copies a month by the end of the year. At the same time, the price of VisiCalc was ratcheted up to $250, a level the market was evidently willing to support. During this period, versions of the program were also developed for other popular machines, including the Tandy TRS-80 and the IBM PC.

By late 1983, an astounding 700,000 copies of VisiCalc had been sold. Almost entirely on the basis of income from VisiCalc, Software Arts and Personal Software had become significant firms. Software Arts had 130 employees and an income of $12 million; Personal Software had 235 employees and sales of $60 million.

In early 1983, Personal Software, having back-integrated into computer services and product development, changed its name to VisiCorp and went on to develop and publish ten complementary products (including VisiWord and VisiFile) collectively known as the VisiSeries. In conjunction with Informatics and other mainstream software firms, it developed programs for mainframe-microcomputer communications. It had a publishing division (VisiPress), an after-sale service division (VisiCare), and a training operation (VisiTraining). Meanwhile, Software Arts, Bricklin and Frankston's development company, had forward-integrated into publishing. It went on to produce several new products, the most successful of which was TK!Solver, an engineering-oriented calculation program.

Productivity Applications: Spreadsheets, Word Processors, and Databases

The term "productivity application" originated around 1982 as a convenient portfolio term for the most commonly used personal computer applications: the spreadsheet, the word processor, and the personal database. Later, graphical drawing packages, scheduling and planning programs, and communications software were added to the portfolio.

Table 7.3
Best-selling applications, fall 1983. (Personal Software Inc. was renamed VisiCorp in 1983. Sorcim was acquired by Computer Associates in 1984. Information Unlimited was acquired by Computer Associates in 1983.)

	Publisher	Units sold (cumulative)	Retail price	Retail value
VisiCalc	VisiCorp	700,000	$250	$175,000,000
WordStar	MicroPro	650,000	$495	$325,000,000
SuperCalc	Sorcim	350,000	$195	$75,000,000
PFS: File	Software Publishing	250,000	$140	$35,000,000
dBase II	Ashton-Tate	150,000	$695	$105,000,000
1-2-3	Lotus Development	100,000	$495	$50,000,000
EasyWriter	Information Unlimited	55,000	$35	$19,000,000

Based on p. 38 of Sigel et al., *Business/Professional Microcomputer Software Market, 1984–86* and on other sources.

Productivity applications, constituting more than 70 percent of the business package market, drove the personal computer software industry. Table 7.3 shows the cumulative sales of the best-selling packages in the fall of 1983. All were productivity applications. The top sellers by value were WordStar, VisiCalc, and dBase II. These products dominated their markets, and they were priced as high as the corporate market would bear. Although the corporate market was not particularly price sensitive, there was considerable price awareness in the education and small-business markets. This created a market segment for medium-price productivity applications, a segment that came to be dominated by Sorcim's SuperCalc spreadsheet, by the PFS:File database, and by medium-price word processors such as EasyWriter and PFS:Write. Below this market segment was another segment for very-low-cost packages.

Spreadsheets
For several years, VisiCalc in particular and spreadsheets in general were by far the best-selling business programs. By 1983, more than twenty competitors to VisiCalc had appeared. Of these the most successful was SuperCalc, produced by Sorcim, a West Coast startup formed in June 1980. SuperCalc was introduced in 1981, initially for CP/M-based computers, a market that VisiCalc had eschewed. With the rising popularity of CP/M machines (about 600 models of which came on the market

between 1980 and 1985), SuperCalc became the dominant CP/M spreadsheet. By the fall of 1983, about 350,000 copies had been sold.

Allowing Sorcim to take the CP/M spreadsheet market was a strategic blunder on the part of VisiCalc's producers. Their failure to capture the market for IBM-compatible PCs was fatal. That market was taken by the Lotus Development Corporation, founded in 1982 by Mitch Kapor.[27]

Kapor, rather like Steve Jobs, was an entrepreneur with a taste for Eastern philosophy. He had studied psychology at Yale in the 1970s and had been employed as a disc jockey, as a transcendental-meditation teacher, and as a computer programmer. As a programmer, he had worked on data analysis software. In the late 1970s, when the personal computer began to take off, Kapor wrote programs for charting statistical data that were marketed by Personal Software as VisiTrend and VisiPlot. When the IBM PC was announced, in August 1981, Kapor recognized the opportunity to develop a spreadsheet for an emerging platform. To raise startup capital, he sold the rights to his programs to Personal Software for $1.5 million. The spreadsheet that Lotus subsequently developed was an integrated program that combined a spreadsheet, a graphical representation module, and a rudimentary database. These three functions suggested the product's name: 1-2-3.

Kapor's new venture coincided with intense interest among investors in the personal computer software industry, and Kapor was readily able to raise a further $4 million from the Sevin-Rosen venture capital fund. This backing enabled Lotus Development to become an integrated operation, controlling both development and publication of the software. However, the bulk of the funds were spent on a spectacular product launch. After its announcement at the fall 1982 COMDEX (COmputer DEalers' EXposition, the industry's major event), 1-2-3 was promoted with full-page advertisements in major newspapers and magazines.

VisiCalc, which had been available on the IBM PC since the time of its launch (well before the launch of Lotus 1-2-3), was rapidly eclipsed. As the market switched to the IBM platform and 1-2-3, the revenues of Software Arts and VisiCorp plummeted. In an acrimonious squabble in late 1983, the two companies sued one another over the rights and obligations of restoring VisiCalc's competitiveness.

Word Processing Software

No market better illustrates the paradigm shift that the personal computer represented than that for word processing software. The best-selling word processing program, WordStar, was priced at $495. IBM's comparable mainframe package, ATMS-III, leased for between $330 and $616 *a month.*[28] In the market for dedicated word processors, the price difference was less marked if one included equipment costs. For example, in 1980 the cost of an Apple II–based word processing system, complete with printer, was estimated to be about $6,000. A dedicated single-person word processing system could be bought for between $5,000 and $8,000; a "shared logic" minicomputer-based word processing system, with perhaps a dozen workstations, offered more advanced features for a similar price per user. The dedicated system would remain the staple of the corporate word processing pool until the late 1980s.

Thus, the personal computer word processor originated not as a replacement for the dedicated word processor but as a useful application for users of home and hobby computers. The first packages, such as Electric Pencil and EasyWriter, were developed by lone authors in 1976 and 1977. The word processor quickly became an essential application for every personal computer on the market. At first, it was a fragmented market, with no dominant product and with each computer having its own word processing package—AppleWriter for the Apple II, Scripsit for the TRS-80, WordPro for Commodore machines. These systems were quite primitive, not least because early personal computers came with screens that were only 40 characters wide, whereas most printers allowed 80 characters on a line. By 1979, however, with the availability of 80-character-wide screens, it became possible to offer facilities comparable to those of a professional word processing system. The most successful package by far was MicroPro's WordStar.

MicroPro was formed in 1978 by Seymour Rubenstein,[29] a seasoned software industry sales professional. Rubenstein had briefly been marketing director of IMS, manufacturer of the IMSAI 8080 microcomputer, and from that vantage point he had observed the growing personal computer software industry and the emergence of retail computer stores as a distribution channel. Having negotiated a CP/M license with Gary Kildall's Digital Research while with IMS, he saw the significance of using the CP/M platform to broaden the potential customer base of a software product.

MicroPro began as a two-person operation, consisting of Rubenstein and a former IMS programmer, Rob Barnaby. Its first product was a rudimentary text processing system called WordMaster. Rubenstein himself marketed WordMaster to retail stores, and this put him in contact with dealers "almost daily." On the basis of knowledge gained through this intimate contact, the program was enhanced and renamed WordStar.

WordStar had most of the features of a professional word processor: "WordStar... provides editing features such as automatic text justification, automatic paragraph intent and undent, simultaneous printing/editing and page break displays; the insertion, deletion, movement and copying of text and features such as boldface type, double striking, underlining, strikeout, subscripts/superscripts, variable character and variable line height, for $495."[30] Most important, WordStar provided a WYSIWYG (what you see is what you get) display; that is, what the user saw on screen closely approximated the printed version. Its price, $495, was remarkably high for a personal computer word processing package.

Launched in mid 1979, WordStar was the first full-feature word processor on the market. Sales were modest at first, but by the end of 1981 about 8,000 copies had been sold. (The only competitor in its class, Magic Wand, had 3,000 sales.) Originally designed to run with the CP/M operating system, WordStar was rewritten for the IBM-compatible PC and quickly became the corporate word processor of choice. MicroPro's fortune soared with that of the IBM PC. By 1984, MicroPro had sales of $45 million and a staff of 425.

WordStar's success inspired many imitators. By mid 1983 there were about 75 word processing packages on the market, including 30 for the IBM PC. The market was highly concentrated, however. WordStar had a 23 percent share; the next most popular package was Sorcim's SuperWriter, with 15 percent. The ten most popular word processing packages had more than 80 percent of the market; most of the remaining packages had less than 1 percent each (table 7.4).

The corporate market for word processing software was insensitive to price. At $495, WordStar was "clearly being purchased by a large number of users who could probably get by with a package for half the price."[31] Hence, the main market for low-cost word processors consisted of the more price-aware home and educational users. Here, Software Publishing's PFS:Write, priced at about $140, was the market leader.[32]

In the corporate word processing market, the main competition was the installed base of dedicated word processors. The leading microcomputer products, such as WordStar and SuperWriter, competed by adding spelling checkers and report generators. For example, MicroPro introduced SpellStar (at $250) to accompany WordStar, licensing the *American Heritage Dictionary* for the purpose. Some software firms worked with third parties to establish training courses and specialized templates and vocabularies for the legal and medical professions. Others attempted to reduce the cost of switching from dedicated word processing systems by

Table 7.4
Leading word processing packages, 1984.

Publisher	Package	Market share
MicroPro	WordStar	23%
Sorcim	SuperWriter	15%
Software Publishing	PFS:Write	9%
MultiMate International	MultiMate	8%
Microsoft	Word	8%
Perfect Software	Perfect Writer	7%
Information Unlimited	Easy Writer	6%
VisiCorp	VisiWord	5%
Peachtree	Peach Text	3%
ARTSCI	Magic Window	2%
Other		14%

Source: Fertig, *The Software Revolution*, p. 164.

producing user interfaces that mimicked those of leading word processing brands such as Wang and Lanier. MultiMate, a word processing package developed by the Connecticut Mutual Life Insurance Company to replace its Wang processors, was eventually spun off as a profitable subsidiary. Leading Edge WP was published by Leading Edge Software, a firm started up by a group of former Wang executives. By focusing so strongly on replicating the interface of traditional word processors, the makers of these packages failed to realize that after 1984 the key to future sales would be increased usability, which would eliminate user training and with it the dedicated word processor.

Databases

In revenues, the leading database product was Ashton-Tate's dBase II, which at $695 was probably the most expensive productivity package on the market. In number of units sold, however, the leading product was Software Publishing's PFS:File, priced around $140. The contrast between the two firms and their products (which coexisted in non-competing niches) is revealing of the rapid maturing and segmentation of the market for shrink-wrapped software in the period 1979–1981.

Ashton-Tate's dBase II was developed on the "old" author-publisher model. The program was developed by Wayne Ratcliffe, a programmer at NASA's Jet Propulsion Laboratory, where he had designed a database system.³³ Like many computer professionals, Ratcliffe bought an IMSAI 8080 computer kit for home assembly. In 1978–79 he developed a remarkably powerful database program that incorporated relational technology. He named the program Vulcan and sold it for $490 by mail order, placing a series of quarter-page advertisements in *Byte* starting in late 1979. Over the next few months, Ratcliffe made about 50 sales, but the process of fulfillment was exhausting: ". . . I really ran out steam. . . . In the summer of 1980, I decided to quit advertising Vulcan and let it drop off to nothing. I would continue to support all the people who had purchased it, but I wasn't going to aggressively go out to find any new buyers."³⁴ Fortunately for Ratcliffe, Vulcan had come to the attention of Software Plus, a software publisher founded in 1980 by George Tate, a former electronics industry salesman. Learning that Vulcan was no longer being actively sold, Tate arranged with Ratcliffe to sell the package on a royalty basis. Tate, a consummate marketer with "a boisterous, Barnumesque instinct for promotion,"³⁵ renamed the program dBase II. (There was no dBase I, but the "II" suggested the program was new and improved.) In January 1981, a full-page ad in *Byte* was captioned "dBase II versus the Bilge Pumps." The product sold well despite its high price. In May 1983 the firm was renamed Ashton-Tate. (There was no Ashton, but it was considered a euphonious and high-sounding name.) About 80 percent of Ashton-Tate's sales were generated by dBase II. In 1983, Ashton-Tate bought the rights to dBase II from Wayne Ratcliffe, and he was appointed vice-president of development and put in charge of 24 programmers. By 1984, more than 200,000 copies of dBase II had been sold. Ashton-Tate had annual sales of $40 million and a staff of 350.

PFS:File, dBase II's main competitor, was a much more consciously planned product. It was developed in 1980 by Fred Gibbons, Janelle Bedke, and John Page, all of whom worked for Hewlett-Packard's minicomputer division in California.³⁶ Collectively, Gibbons, Bedke, and Page had the full spectrum of managerial, marketing, and technical development skills. The PFS:File database system was intended to occupy a perceived gap in the market for a mid-price database that would also exploit John Page's background as the designer of the database software for the HP 3000 minicomputer. The package was developed for the Apple II, with ease of use rather than technical sophistication as its prime selling point. It was priced at $140, barely one-fifth the price of dBase II. Produced and published in house, the package was distributed directly to computer stores. All this was achieved in 1980, when the founders were still full-time employees of Hewlett-Packard. In early 1981, with the success of PFS:File, they secured venture funding and incorporated as Software Publishing. By the fall of 1983, they had sold a quarter-million copies of PFS:File for Apple II, TRS-80, and IBM-compatible machines. Software Publishing continued to exploit the market niche for mid-price software with follow-on products such as PFS:Write and PFS:Graph. It was one of the few market leaders of the early 1980s that was still a market leader at the end of that decade.

Small-Business Software

Although productivity applications accounted for more than 70 percent of business microcomputer software, there was a significant market for conventional data processing applications. The market was similar in structure to the markets for mainframe and minicomputer software, with both cross-industry and industry-specific applications (table 7.5). One might have expected that microcomputer data processing applications would be a natural market for the existing corporate software industry, but even here the dominant firms (including MSA, Computer Associates, and Informatics) were not immediately attracted. Apart from their failure to see the potential of the microcomputer software market, their organizational capabilities were not suited to the low-price, high-volume nature of microcomputer software.

Cross-Industry Applications
Cross-industry applications were targeted at businesses with annual revenues between $250,000 and $10 million and with fewer than 20 employees. The most popular application was the general-purpose accounting or bookkeeping package; then came mailing-list management, payroll, and merchandising. Significant early entrants in this market included Peachtree Software (principal product: Peachtree General Ledger), BPI Systems (BPI General Accounting), and Software Dimensions Inc. (SDI Accounting Plus). All three firms were established by entrepreneurs with backgrounds in the corporate software industry.

Peachtree originated as the Computer System Center, an Atlanta-based Altair dealership formed in 1977 by a group of computer professionals. In 1979, the partners recognized a gap in the market for business software and re-incorporated the firm as Peachtree Software (named for a prestigious street in downtown Atlanta). Its first products, Peachtree Accounting and Peachtree Inventory, were priced at $2,000, a price "considered reasonable because [the principals] had worked previously in the

Table 7.5
The US market for business microcomputer software, 1983.

Total

Application type	Units sold	Average unit price	Retail value	Percentage of total
Productivity	2,049,000	$299	$613,000,000	71
Cross-industry	491,000	$334	$164,000,000	19
Industry-specific	46,000	$1,876	$86,000,000	10
Total	2,586,000		$863,000,000	100

Cross-industry applications

Application type	Units sold	Average unit price	Retail value
Accounting	226,000	$500	$113,000,000
Payroll	65,000	$250	$16,000,000
Merchandising	45,000	$500	$23,000,000
Mailing list	155,000	$75	$12,000,000
Total	491,000		$164,000,000

Industry-specific applications

Application type	Units sold	Average unit price	Retail value
Contractors	10,000	$1,900	$19,000,000
Medical offices	5,000	$2,500	$12,000,000
Wholesale/distribution	2,200	$2,800	$6,000,000
Dental offices	3,000	$2,400	$7,000,000
Pharmacies	1,000	$5,000	$5,000,000
Other	25,800	$1,434	$37,000,000
Total	46,000		$86,000,000

Adapted from p. 104 of International Resource Development Inc., *Microcomputer Software Packages*.

minicomputer and mainframe industry."[37] Peachtree sold its packages by mail order and through computer stores. In June 1981, when Peachtree's sales were about $2 million and its staff numbered 40, it was acquired by Atlanta-based MSA for $4.5 million.[38] MSA invested $10 million over the next 3 years to create a strong Peachtree brand and to acquire new products, including the PeachText word processor and the PeachCalc spreadsheet.[39] By the end of 1983, the Peachtree Division of MSA had sales of $20 million and a staff of 175. However, sales fell by nearly 40 percent during the 1984 shakeout, and MSA sold off Peachtree Software the following year for $1.1 million.

MSA was not the only corporate software firm to get burned entering the microcomputer software industry. In the spring of 1983, ASK acquired Software Dimensions for $7.5 million. In 1984, after heavy losses, ASK disposed of the firm, auctioning the packages separately and netting less than $1 million.[40] Computer Associates' acquisition of the BPI Systems was more successful, although it did not take place until April 1987, when BPI was incurring heavy losses and was priced correspondingly.

Business applications was the one area where there were some significant non-US players. However, most of them were successful only in their own countries, where they were able to take advantage of local knowledge of fiscal regulations and taxation. A few of them became major players. For example, the British firm SAGE plc, established in the north of England in 1981 to develop accounting packages, had sales of $100 million by 1995—in which year it acquired Peachtree Software.

Industry-Specific Applications

Like the market for corporate software, the market for industry-specific microcomputer applications was fragmented. Although five small-business sectors (contracting, medical offices, distribution, dental practices, and pharmacies) accounted for 60 percent of the market, the remaining market was highly specialized, with hundreds of vendors in more than 30 recognized sectors, ranging from advertising to wholesale lumber dealing.

Entrepreneurs often came straight from the ranks of the corporate software industry. Others were physicians, dentists, or insurance brokers who had developed their own packages. Although industry-specific applications were marketed as "packages," they were much more like corporate software products that required some customization and after-sale support. Sales were often attributable to recommendations or to geographical proximity (which made it easier for a vendor to install the software and provide after-sale support). In this respect, these firms were indistinguishable from mainframe and minicomputer application vendors, except with regard to price. But even here, although prices were far lower than was typical in the corporate software industry, the average price (nearly $2,000) was much higher than the $300 that was typical of productivity software and cross-industry applications.

Many of the vendors established in the early 1980s continued to operate on a small scale for many years, with a few employees and with ten or twenty customers. Selling in such a fragmented environment, such vendors had little opportunity to grow into software giants, as happened to some of the corporate software players in banking or medical data processing. In the boulders-pebbles-sand model of the software industry, the microcomputer-industry-specific application vendors were some of the finer grains of sand.

Consumer Software

When the Apple II was launched. in 1977, it was positioned as a "home/personal" computer. The advertising copy reflected that:

The home computer that's ready to work, play and grow with you. . . . You'll be able to organize, index and store data on household finances, income taxes, recipes, your biorhythms, balance your checking account, even control your home environment.[41]

In fact, hardly any of those applications were achievable; the software did not exist. In 1977, the Apple II was too limited and too expensive for home use other than by the most dedicated enthusiast, so it was sold primarily to schools and businesses.

However, during the period 1979–1981 many low-cost machines designed expressly for the domestic market were offered by Atari, Coleco, Commodore, Tandy, Texas Instruments, Timex, and Sinclair. Typically priced between $200 and $600, the most successful home computers sold in vast numbers. The Commodore VIC-20, introduced in 1980, eventually sold 2.5 million units; its successor, the $199 model 64, introduced in late 1982, sold more than 800,000 in its first year.[42] At the bottom of the market, home computers were used primarily by teenage game players.

As personal computer prices dropped, a domestic market for more expensive computers—the Apple II, Tandy's TRS-80 models III and IV, IBM's PC and PC Jr.— developed. Aimed at upper-income families, these computers were promoted in TV commercials featuring Bill Cosby, Dick Cavett, and other celebrities. They promised such boons as home finance management for the head of the household and educational advantages for the offspring.

One of the reasons for the slow development of the home computer market had been the dearth of software titles. To stimulate the market, most of the home computer manufacturers published software (some of it developed in house, some commissioned from third parties) for their machines. By 1982, the manufacturers had about 40 percent of the consumer software market, a situation quite unlike that for business/professional software (table 7.6). The remaining 60 percent of the market was held by independent software companies, the top five producers having 30 percent.

The home software market was divided into three sectors: entertainment, home management, and education. The total market was estimated to be worth $200 million by retail value in 1982. This was less than half the value of the market for business software, but then the unit prices of home software were much lower (typically $30–$50); the number of units shipped was much greater—at least 5 million in 1982.

Entertainment Software

Entertainment software primarily meant computer games. Although there were other creative and quasi-educational programs, such as music applications and drawing packages, games dominated the sector. Indeed,

Table 7.6
Leading US producers of consumer software, 1982. (The total wholesale value of consumer software was $101.85 million. The market was split up as follows: entertainment 57 percent, education 22 percent, home management 18 percent, other 3 percent.)

Rank	Manufacturer	Sales	Market share	Hit products
Computer manufacturers				
1	Tandy	$16,300,000	16%	Superscript
2	Texas Instruments	$10,200,000	10%	
3	Atari	$8,600,000	8%	
4	Apple	$2,800,000	3%	Apple Writer
5	Commodore	$1,900,000	2%	Magic Desk
Total		$39,800,000	39%	
Independent software firms				
1	Sierra On-Line	$10,800,000	11%	Frogger, ScreenWriter II
2	Sirius	$8,000,000	8%	Gorgon, Beer Run
3	Broderbund	$5,250,000	5%	Choplifter, Bank Street Writer
4	Continental	$4,000,000	4%	Home Accountant, Tax Advantage
5	Epyx	$3,000,000	3%	Temple of Asphai
Total		$31,050,000	30%	

Source: Creative Strategies International, *Computer Home Software*, pp. 49, 80, 81, 110, and passim.

home computer games were usually viewed as a subsector of the much larger videogame software market.[43]

Analogies were often made between the personal computer software industry and the recorded-music industry or book publishing. It was with entertainment software that the analogy came closest. The major publishers released many titles, and none was reliant on a single hit product. Sierra On-Line, Sirius, and Broderbund—the three market leaders—each offered between 50 and 60 games. The great majority of these were produced by sole authors on a royalty basis and were ephemeral in the extreme. A few hit products, such as Broderbund's Choplifter and Sierra On-Line's Frogger, stayed on the best-seller lists for more than a year, but the average life of a computer game was about 3 months and the average sales were less than 10,000 units. Few game authors made much money.

Home Management Software

Use of the computer in the home had been projected since at least the mid 1960s. However, home computing only became economically feasible with the microcomputer. Initially there were few compelling applications, and there was great market uncertainty. As late as 1983, an analyst reported: "Home monitoring [is] a newly discovered home software application for the personal computer. Several systems now exist that allow a personal or home computer to run appliances, control security devices, and operate home lighting. However, these must be attached to an interface for analog-digital conversion to be able to function on a personal computer."[44] Not surprisingly, this software genre sunk without trace.

By about 1983, word processing had emerged as the most important use of a home computer. Computers were being used for correspondence and document creation by adults, and for homework assignments by schoolchildren. Because mainstream word processing packages such as WordStar were priced far beyond the reach of the home user, there were many inexpensive packages.[45] Sierra On-Line's Homeword and Broderbund's Bank Street Writer were priced between $50 and $60. Software Publishing's PFS series, priced at $100–$150 each, were favorites with the home-office market. Software Publishing's fortunes rose significantly with the increasing domestic sales of the IBM PC; its PFS programs were approved applications, and IBM later re-packaged them under its own brand as the IBM Assistant series.

After word processing, the most popular domestic application was personal finance management software, typically purchased by the head of an upper-income household. The market leader was Home Accountant, produced by Continental Software. In 1982 it was the best-selling package overall for the Apple II and was said to be the fourth- or fifth-best seller for the IBM PC.

Educational Software
Educational packages, targeted at schools and at an elite of educationally aware users, were typically priced between $50 and $150. This market was fragmented, with thousands of titles.

In addition to entrepreneurial startups such as Edu-Ware and Kriya Systems (the publishers of Science of Learning and Typing Tutor, respectively), many textbook publishers also entered the market. Among them were Addison-Wesley, Hyden, Houghton Mifflin, McGraw-Hill, Milliken, Prentice-Hall, Random House, Readers Digest, Scholastic, and John Wiley. Educational software publishing fitted the business model of the textbook publisher particularly well, and moonlighting schoolteachers appeared to be an inexhaustible source of material.[46] By 1983, Random House had 200 educational titles, some of which (e.g., Galaxy Math Facts and Mechanics of English) sold respectably.

Summary

The personal computer software industry had almost no connection with the existing software industry and therefore established most of its development and marketing practices de novo.

In the early days, when barriers to entry were almost non-existent, there was a gold-rush-style entrepreneurial frenzy. Several thousand companies, offering tens of thousands of products, were started within 2 or 3 years. Only a few years later, most of these firms had failed. Users of software increasingly elected to buy established titles from a select group of publishers, among them VisiCorp, Ashton-Tate, and MicroPro. Seeing the emergence of these highly successful and visible new firms, established vendors of corporate software began to seriously address the new market, but they found it difficult to adapt to the new environment. Hence, most entries were made by acquiring products from, or by taking over, small entrepreneurial personal computer software firms. But even those entries would prove relatively unsuccessful.

The introduction of the IBM PC, in August 1981, simultaneously achieved technical closure and legitimization of the personal computer. IBM made the architecture of its new machine "open," enabling other manufacturers to flood the market with IBM-compatible clones and thus create a de facto standard.[47] IBM's imprimatur transformed the corporate view of the personal computer. Once thought of as an appliance or a videogame machine, it was now perceived as a serious office tool. As a side effect, the arrival of the IBM-compatible standard destabilized the oligopoly of major suppliers of personal computer software, allowing new competitors such as Lotus and WordPerfect to become a major players.

The gold-rush phase of the personal computer software industry was essentially over by the end of 1983. The next chapter describes the maturing of the industry, when significant barriers to entry were established, most sales went to the major players, and newcomers needed significant venture capital.

It's the best thing since 1-2-3.

We asked current 1-2-3® users how to get more out of 1-2-3.

And you told us.

Introducing 1-2-3 Release 2 from Lotus.®

New 1-2-3 is more powerful and a lot more versatile.

You wanted to handle larger jobs with 1-2-3. Now you can. The new 1-2-3 worksheet has been expanded to 8192 rows – 4 times its original size. And your worksheet is actually more flexible because advanced memory management allocates memory more efficiently and allows data to be stored anywhere on the worksheet. When used with new expanded memory boards, new 1-2-3 can address memory beyond 640K.

New 1-2-3 is designed to support the Intel® 8087/80287 math coprocessors so you can now do many calculations faster. We've even added some features that make it possible to do things like regression analysis, string functions and string arithmetic. And new 1-2-3 comes with 40 new macro commands so you can work more efficiently and a lot more productively.

Now you can start 1-2-3 directly off a hard disk without putting a system disk in the floppy disk drive.

But we still kept things simple.

In many respects, new 1-2-3 isn't any different from the original. You wanted us to keep things simple and we did. If you're already familiar with 1-2-3, you're ready to use new 1-2-3.

You don't have to retrain. And new 1-2-3 can read and process existing 1-2-3 files so that virtually all applications already developed can easily be used.

It's even easy to upgrade to new 1-2-3.

If you're a registered 1-2-3 user and want to upgrade to new 1-2-3, you'll find all the details in a mailing from Lotus. If you haven't registered yet, complete and send in your Warranty Registration Card or call 1-800-TRADEUP* so we can send you the mailing.

The cost of the Upgrade product is $150. You are eligible for a free upgrade if you purchased 1-2-3 Release 1A on or after April 24, 1985.

And for everyone who upgrades, there's also a rebate offer of $40 on the Intel Above™ Board, the first expanded memory board certified by Lotus.

We think you'll find new 1-2-3 the best thing since, well, 1-2-3.

*In Canada call 1-800-447-4700.

Suggested retail price of new 1-2-3 is $495. 1-2-3 Release 2 requires 256K of memory. The minimum memory requirement for 1-2-3 Release 1A is 192K.

© 1985, Lotus Development Corporation. Lotus and 1-2-3 are registered trademarks of Lotus Development Corporation. Intel is a registered trademark and Above is a trademark of Intel Corporation.

The top-selling PC software application of the 1980s was the 1-2-3 spreadsheet. The Lotus Development Corporation's revenue stream was sustained by regular upgrades.

8

Not Only Microsoft: The Maturing of the Personal Computer Software Industry, 1983-1995

There is much more literature on Microsoft than on its competitors, and nearly all of it (whether or not Microsoft cooperated with the author) is of the sort that business historians describe as "from the inside looking out." Some books almost have the tone of an imperial history, in which the dominant power defends itself against lesser states. In these histories the lesser states do not have their own histories, but are mere satellites whose paths occasionally cross that of the central subject. As a result, Microsoft is often perceived as big, ugly, and lacking in humanity and in capacity for innovation.

The aim of this chapter is to provide an alternative view of the industry, which is inevitably dominated by Microsoft but in which the lesser players are also seen as creative, autonomous enterprises with their own histories. It is, if you like, a view from the outside looking in.[1]

Microsoft and the Software Industry

One reason for the widespread interest in Microsoft is that its founder and major stockholder, Bill Gates, has become the richest man in the world. Another is that, whereas most computer users never come into personal contact with the products of corporate-software firms of comparable size (Oracle, Computer Associates), they are in intimate daily contact with Microsoft's products when using a word processor, a spreadsheet, or an Internet browser. For many people—perhaps most—Microsoft *is* the software industry.

Microsoft has captured public awareness perhaps more than any other late-twentieth-century company. One measure of the extraordinary interest in Microsoft is the number of books published about the company and its founder. I know of no complete list—books have been published in England, France, and Japan, as well as America—but at the time of writing the Library of Congress lists twenty monographs on Bill Gates and Microsoft. There are more books on Microsoft than on the rest of the software industry.[2]

Microsoft is often perceived as a latter-day IBM, completely dominating the software industry. This is simply not true. IBM, at its peak, in the 1960s, had a three-fourths share of the worldwide computer industry—hardware, software, and services. Microsoft has never even had a 10 percent share of the software market. For example, although by 1990 Microsoft was unquestionably the best-known software firm in the world, its sales ($1.18 billion) constituted only 3 percent of the $35 billion worldwide market for software products, and only one-eighth of IBM's software sales ($9.95 billion). By 1995, though Microsoft's revenues had grown fivefold, to $6.08 billion, it still had less than 10 percent of the worldwide software market, and its sales were still well below IBM's ($12.9 billion). Yet few people outside of the software industry think of IBM as being in the software business at all.

Not until 1998 did Microsoft's software sales exceeded IBM's. In 1999, Microsoft became the most valuable company in the world by stockmarket valuation, but its total revenues ($19.7 billion) were dwarfed by IBM's ($84.4 billion). IBM was the third-most-valuable company. What these figures tell us, of course, is that the stock market perceived Microsoft as the faster-growing and more profitable company. IBM's software is highly profitable too—perhaps as profitable as Microsoft's. (IBM does not disclose profit margins of its individual businesses.) IBM's software sales, however, are overshadowed by much less profitable activities: physical manufacture of computers and peripherals and labor-intensive computer services.

Table 8.1 provides an array of statistics on Microsoft. The upper part of the table shows Microsoft's revenues, revenue growth, and number of employees. Throughout its 25-year history, Microsoft has been the fastest-growing and most impressive of all the personal computer software firms. Its annual revenue growth has been truly spectacular, often exceeding 50 percent a year (even when it was a mature company) and never falling below 25 percent. Its competitors have certainly achieved impressive growth for extended periods of time too, but none has achieved uninterrupted revenue growth for 25 years. Detailed financial data on Microsoft (table 8.1, lower part) has been publicly available only since its 1986 initial public offering. Microsoft is an extraordinarily profitable company, with earnings typically in the range of 30–40 percent of its revenues. Remarkable as this may seem outside the personal

Table 8.1
Microsoft financial statistics.

	Revenues	Annual growth	Employees
1975	$160,000		3
1976	$22,000	38%	7
1977	$382,000	1,636%	9
1978	$1,356,000	255%	13
1979	$2,390,000	76%	28
1980	$8,000,000	235%	38
1981	$16,000,000	100%	130
1982	$24,000,000	53%	220
1983	$50,000,000	104%	476
1884	$97,000,000	95%	778
1985	$140,000,000	44%	1,001
1986	$198,000,000	41%	1,442
1987	$346,000,000	75%	2,258
1988	$591,000,000	71%	2,793
1989	$804,000,000	36%	4,037
1990	$1,186,000,000	48%	5,635
1991	$1,847,000,000	56%	8,226
1992	$2,777,000,000	50%	11,542
1993	$3,786,000,000	36%	14,430
1994	$4,714,000,000	25%	15,017
1995	$6,075,000,000	29%	17,801
1996	$9,050,000,000	49%	20,561
1997	$11,936,000,000	32%	22,232
1998	$15,262,000,000	28%	27,055
1999	$19,747,000,000	29%	31,575
2000	$22,956,000,000	16%	39,170

Revenue breakdown

	Revenues	Systems	Applications	Other
1986	$198,000,000	53%	37%	10%
1987	$46,000,000	49%	38%	13%
1988	$591,000,000	47%	40%	13%
1989	$805,000,000	44%	42%	14%
1990	$1,186,000,000	39%	48%	13%
1991	$1,847,000,000	36%	51%	13%
1992	$2,777,000,000	40%	49%	11%
1993	$3,786,000,000	34%	58%	8%
1994	$4,714,000,000	33%	63%	4%
1995	$6,075,000,000	31%	65%	4%

Data on revenue growth and employees from http://www.microsoft.com/presspass/fastfacts.asp (accessed January 2001); revenue breakdown from p. 5 of Cusumano and Selby, *Microsoft Secrets*.

computer software industry, Microsoft's earnings have not been so much better than its competitors'. Microsoft has come to dominate its sector not through profitability per se but through its ability to gain market share. Since the early 1980s, the personal computer software market has been relatively concentrated, between ten and twenty firms having 80 percent of the market. Microsoft's growth in market share has been by far the greatest. In 1983 it had about one-fifth of the market for PC software; by 1990 it had one-third; by 1995 it had one-half. Thus, within the narrow sector of the PC-software industry, Microsoft does indeed have IBM-like dominance.

In popular histories, Microsoft is usually portrayed as aggressive and predatory, driving competitors out of the business. This is undoubtedly true, but it tells only half the story. At least as many firms were driven out of the business by strategic errors and plain old market forces. This was well in evidence in 1984, the year of the first widely reported shakeout of the personal computer software industry. *Business Week* noted: "No one expected the halcyon days of the personal computer software business to pass so quickly. Industry experts had projected that this market would continue to double annually, and 3,000 hopefuls, as a result, had jumped into the fray. But the glut of suppliers, along with the soaring cost of marketing new products and a flood of me-too programs, is changing the picture dramatically."[3] Perhaps the biggest reason for the hundreds of firm failures was the explosion in the number of competing productivity applications for office workers: "At the last count, there were 200 or more word processors, 150 spreadsheets, 200 data base programs, and 95 integrated packages that offer at least three functions. Moreover, distributors report that of the 20,000 programs on the market, a mere 20 make up as much as half of their total business."[4]

Of the nine largest firms of 1983 (table 8.2), five were in terminal condition by the summer of 1984. VisiCorp and MicroPro, publishers of the leading spreadsheet and the leading word processor, had lost market share to Lotus and WordPerfect. Digital Research had lost sales of operating systems to Microsoft. Peachtree, having been acquired by Management Sciences America in 1981, was losing money and about to be sold off; it would essentially disappear from view for a decade. Sorcim, another victim of the productivity application wars, was bought by the mainframe software maker Computer Associates during that firm's first foray into microcomputer software. Note that only one of these exits was due to Microsoft.

For most of the 1980s, Microsoft grew primarily on the strength of its MS-DOS operating system for IBM-compatible personal computers,

Table 8.2
Revenues of top ten personal computer software companies.

1983	1987	1991	1995
Microsoft: $70,000,000	Lotus: $396,000,000	Microsoft: $1,801,000,000	Microsoft: $7,271,000,000
VisiCorp: $60,000,000	Microsoft: $301,000,000	Lotus: $829,000,000	Novell: $1,900,000,000
Lotus: $48,000,000	Ashton-Tate: $267,000,000	WordPerfect: $603,000,000	Adobe: $762,000,000
MicroPro: $45,000,000	WordPerfect: $100,000,000	Novell: $571,000,000	Autodesk: $544,000,000
Digital Research: $38,000,000	Borland: $56,000,000	Borland: $502,000,000	Symantec: $438,000,000
Ashton-Tate: $30,000,000	Autodesk: $52,000,000	Autodesk: $238,000,000	Intuit: $396,000,000
Peachtree: $24,000,000	MicroPro: $41,000,000	Adobe: $230,000,000	Borland: $208,000,000
Sorcim: $10,000,000	Aldus: $40,000,000	Symantec: $196,000,000	Corel: $196,000,000
Software Publishing: $10,000,000	Software Publishing: $39,000,000	Aldus: $164,000,000	Claris: $184,000,000
	Adobe: $39,000,000	Software Publishing $141,000,000	Santa Cruz Operation: $178,000,000

Data for 1983 from p. 19 of Sigel et al., *Business/Professional Microcomputer Software Market, 1984–86*; data for 1987, 1991, and 1995 from *Software Magazine* (May 1988, June 1992, July 1996). Some missing companies have been interpolated by the author. Inconsistencies between revenues as published in *Software Magazine* and company annual reports have not been corrected.

which probably generated 40–50 percent of its revenues. Though second-tier firms such as Lotus, WordPerfect, Borland, and Adobe did not always grow as rapidly or as consistently as Microsoft, by any normal measure they also expanded remarkably. WordPerfect's sales went from $100 million to more than $400 million between 1986 and 1990. Beginning around 1990, Microsoft achieved much of its growth by publishing applications packages in addition to systems software.

By 1995 the situation in personal computer software was reminiscent of the 1960s computer industry ("IBM and the seven dwarfs"). Microsoft dominated every market in which it operated—operating systems, programming languages, productivity applications. Its competitors survived, and in some cases prospered, by operating in markets in which Microsoft did not participate (yet). For example, Autodesk produced the best-selling computer-aided design (CAD) drafting program, Aldus was successful in desktop publishing. Adobe created the market for laser printer software. Novell and the Santa Cruz Operation (SCO) sold networked and Unix operating systems, respectively. To an extent, each of these firms was a one-product operation and therefore was vulnerable to a competitor, particularly one that turned out to be Microsoft. There was some noticeable concentration through mergers and acquisitions in the 1990s: Lotus was bought by IBM, Ashton-Tate by Borland, WordPerfect by Novell, and Aldus by Adobe Systems. In most cases, the buyers were one-product companies seeking to diversify. A minority of firms made merger-and-acquisition activity their prime strategy for coexisting with Microsoft. Symantec, for example, effectively became a portfolio operator, selling a range of products from many sources, none of which accounted for more than 10 percent of its revenues.

The concentration of the personal computer software market was perhaps its most noticeable feature. It was characterized as a winner-take-all market. Microsoft was the biggest winner, but the second-tier companies were also highly successful at monopolizing their markets. The mainframe software industry was much less concentrated, and in that industry every sector had several competing suppliers of comparable scale; there, a 20 percent market share was the mark of a successful company, and a 50 percent market share was almost unheard of.[5] By contrast, it was quite typical for the major personal computer software firms to dominate their individual sectors, with 60–70 percent of the market. Academic economists described this as the "economics of increasing returns." In the business press, it became known as "Microsoft economics."[6]

The IBM-Compatible PC Standard

Toward the end of the 1970s, academic economists became increasingly interested in the economics of increasing returns, which appeared to explain the market behavior of technological and information firms better than the classical economics of decreasing returns.[7] According to this school of thought, high-tech markets tended to produce natural monopolies in which a single technological "platform" dominated. High-tech goods—such as aircraft, computers, and nuclear power stations—were characterized by high research and development costs and relatively low manufacturing costs. For example, in the 1960s, IBM's manufacturing costs for its System/360 mainframe computers were said to be about 20 percent of the selling price, so each incremental sale of a computer was enormously profitable. Profits were fed back into the development of software (then given away free) and product improvement, which had the effect of making System/360 more attractive to customers; the result was more sales and profits to be invested in further software development and product improvement, and so on. That virtuous circle gave IBM its 75 percent share of the mainframe market.

By the time IBM unbundled its software, System/360 was already a standard platform, and independent software vendors produced software for the IBM platform in order to maximize their sales. These "network effects" further enhanced the desirability of the IBM mainframe. It should be noted that the success of System/360 was largely independent of its original technical merits. It was the mere fact that it was a standard platform that made it desirable. IBM's market dominance could be explained only partially in terms of the economics of increasing returns. IBM was also subject to decreasing returns. For example, selling costs tended to rise at the margin, and IBM's marketing resources and manufacturing facilities were constraints. This left sufficient room in the market for competitors.

Microprocessors and software, however, were essentially information goods. Their marginal manufacturing and distribution costs were nearly negligible. The economics of increasing returns facilitated the creation of a dominant IBM-compatible PC standard that, over a period of years, accounted for more than 80 percent of the market. However, the creation of the IBM PC standard took longer than is commonly supposed. After August 1981, when IBM introduced its personal computer, it was 5 years before IBM-compatible PCs accounted for 50 percent of new purchases. And if one considers the installed base of personal computers, it was not until 1988 that the 50 percent level was reached (table 8.3).

Intel's profits from its 16-bit microprocessor enabled that company to release new microprocessor generations at intervals of about 3 years (table 8.4). With each new product generation, the old microprocessor continued to be sold alongside the new, with careful pricing. In 1984, the original 8088/86 was superseded by the 80286; in 1986, the 80386 superseded the 80286; the 80486 and the 80586 (Pentium) were introduced in 1989 and 1992, respectively.[8] With each new product generation, speed and architectural improvements led to a performance improvement of an order of magnitude. Between product generations, low-cost versions (the 80386SX and the 80486SX) were introduced to exploit the lower end of the market.

The IBM-compatible PC standard consisted of the Intel microprocessor and the Microsoft MS-DOS operating system, both of which were susceptible to imitation. Intel had relatively few imitators because the extreme capital intensity of microprocessor fabrication was a major barrier to entry. A fabrication plant cost about $1 billion before the first microchip rolled off the production line, and this limited competition to a few major

Table 8.3
Shipments of personal computers and installed base, United States, 1981–1992.

	Shipments		Installed base	
	Total	Intel	Total	Intel
1981	780,000	35,000 (4.49%)	1,740,000	35,000 (2.01%)
1982	3,040,000	192,000 (6.32%)	4,780,000	227,000 (4.75%)
1983	5,450,000	698,000 (12.81%)	10,200,000	925,000 (9.07%)
1984	6,660,000	1,942,000 (29.16%)	16,810,000	2,867,000 (17.06%)
1985	5,760,000	2,518,000 (43.72%)	22,270,000	5,385,000 (24.18%)
1986	6,850,000	3,334,000 (48.67%)	28,190,000	8,719,000 (30.93%)
1987	8,320,000	6,081,000 (73.09%)	35,120,000	14,800,000 (42.14%)
1988	8,649,000	6,769,000 (78.26%)	44,988,000	23,538,000 (52.32%)
1989	8,985,000	7,371,000 (82.04%)	52,128,000	30,592,000 (58.69%)
1990	9,337,000	7,835,000 (83.91%)	54,807,000	37,391,000 (68.22%)
1991	9,399,000	7,904,000 (84.09%)	59,303,000	42,792,000 (72.16%)
1992	10,103,000	8,367,000 (82.82%)	63,045,000	48,105,000 (76.30%)

Derived from pp. 210–211 of Steffens, *Newgames*.

players, such as Advanced Micro Devices, Inc. (AMD) and Cyrix. In contrast, Microsoft had numerous competitors because the capital requirements for producing an operating system were relatively modest. Microsoft's original MS-DOS (version 1.0) contained only 4,000 lines of code, had taken less than one programmer-year to develop, and was not by any standard a sophisticated program.[9] It was therefore possible for any two-person operation of reasonable competence to come up with an imitative product, and lots did. The two most serious competitors were Digital Research's CP/M-86 (a 16-bit, general-purpose operating system) and SofTech's USCD p-System. These were both offered by IBM as alternatives to MS-DOS. However, CPM-86 was not available until several months after the launch of the IBM PC, so MS-DOS had an insuperable first-mover advantage. Moreover, Digital Research made a strategic error by pricing CP/M-86 at $240 (4 times the price of MS-DOS). The price was later cut to $60, but it was then too late, and that price was still not sufficiently

Table 8.4
Intel and MS-DOS product improvements.

	Intel microprocessors	MS-DOS versions
1981	8088/86	1.0
1982	—	1.1, 1.25
1983	—	2.0
1984	80286 (1 MIPS)	2.11, 3.0, 3.1
1985	—	—
1986	80386 (5 MIPS)	3.2
1987	—	3.3
1988	80386SX	4.0
1989	80486 (20 MIPS)	—
1990	—	—
1991	80486SX	5.0
1992	—	—
1993	80586 (100 MIPS)	6.0
1994	—	6.22

Data on Intel microprocessors from pp. 210–216 of Steffens, *Newgames*. (Dates given are when the microprocessor model was shipped in volume. This was up to 2 years later than Intel's product announcement.) Data on MS-DOS from appendix A of Ichbiah and Knepper, *The Making of Microsoft*. (Some minor DOS releases have been omitted.)

cheap to tip the market in CP/M-86's favor. Though there are no reliable time series in the public domain for Microsoft's share of the IBM-compatible PC operating system revenues, there is overwhelming anecdotal evidence that MS-DOS accounted for 90 percent of machines.[10] Incidentally, the ease with which Microsoft and many other firms were able to supply an operating system for the IBM PC raises this question: Why didn't IBM write the operating system itself and retain full control over its computer? Some authors have hypothesized that IBM made the decision in order not to provoke the antitrust authorities (IBM's long-running antitrust trial was still ongoing at the time of the PC development), but the best-regarded history of the development suggests nothing more sinister than an aggressive development schedule.[11]

Although Microsoft's operating-system activity came to be associated largely with MS-DOS, the history is more complex. Microsoft's decision to enter the operating-system market predated the launch of the IBM PC. In February 1980, Microsoft had negotiated a license for the industry-standard Unix from AT&T and had begun the development of XENIX, a microcomputer version of Unix, which Microsoft believed would ultimately become the standard for 16-bit microcomputers. Thus, MS-DOS was initially a pragmatic product tailored to the IBM PC. According to Tim Paterson, the developer of MS-DOS, the main design goal was to enable existing software packages to run on the new Intel microprocessor.[12] Thus, MS-DOS enabled prominent software vendors, such as VisiCorp, MicroPro, and Software Publishing, to make their best-selling programs immediately available for the IBM platform. During 1982, minor upgrades were made to MS-DOS, primarily so that it could handle new-style disk drives.

In March 1983, Microsoft released MS-DOS 2.0, a much more sophisticated system that contained 20,000 lines of code and represented an investment of several programmer-years. A major aspect of the design was to provide a smooth migration path to XENIX, Microsoft's ultimate goal. During 1983, Microsoft advertised MS-DOS for ordinary IBM-compatibles and XENIX for high-end machines in the major industry magazines:

If you write and sell 16-bit software, MS-DOS and XENIX give you the largest installed base. In fact, over fifty 16-bit manufacturers offer their microcomputers with MS-DOS or XENIX. IBM, Victor, Altos, Wang, Radio Shack, Zenith and Intel, to name just a few. And the list is growing. That means there's a ready and expanding market for your 16-bit applications software.[13]

There was considerable market uncertainty in the industry, and within Microsoft, as to whether MS-DOS would forever be a product in its own right or would eventually be subsumed into a Unix-style operating system.

At least two dozen operating systems, from about twenty vendors, were competing with MS-DOS (table 8.5). Several of these systems were technically superior to MS-DOS. For example, Digital Research's original CP/M-86 had been superseded by Concurrent CP/M-86, and its "marketing strategy was to hit at MS-DOS's lack of networking facilities."[14] Digital Research's Concurrent DOS offered a rudimentary graphical user

Table 8.5
Principal 16-bit operating systems competing with MS-DOS in 1985.

Vendor	Operating system
Alpha Micro	AMOS
Digital Research	CP/M-86
Digital Research	MP/M-86
Digital Research	Concurrent CP/M-86
Digital Research	Concurrent DOS
G&G Engineering	MP/M-8-16
Hemmenway	MRP
Hunter & Ready	VRTX
Industrial Programming	MTOS
Intel	iRMX
JMI Software	C Executive
Micro Digital	E.86
Motorola	VERSAdos
Phase One	Oasis-16
Pick Systems	Pick
PMS	1-DOS
Ryan McFarland	RM/COS
Silicon Valley Software	Merlin
SofTech	USCD p-System
Software 2000	TURBOdos-16
Systems & Software	REX-80/86
Wicat	MCS
Zenith Data Systems	ZDOS
Zilog	ZERTS

Source: Fertig, *The Software Revolution*, p. 110.

interface (GUI). In the Unix market there were much bigger competitors. IBM had PC/IX, its own version of Unix. The newly deregulated AT&T, Unix's inventor and owner, had gone into PC manufacture and was offering its own version of the operating system.

As a result of this competition and the "fading future hopes of XENIX," Microsoft decided to enhance MS-DOS while simultaneously developing a compatible GUI.[15] But not until 1990 would Windows emerge as a successful product. Meanwhile, MS-DOS was periodically upgraded to accommodate the evolving hardware and software technologies. In August 1984 there was a major upgrade: MS-DOS 3.0, with 40,000 lines of code. This was followed later the same year by MS-DOS 3.1, which had networking capabilities. At that point MS-DOS stabilized for about 2 years, with only minor incremental upgrades. MS-DOS was enormously profitable during this period. Its retail price was $60, though most sales were OEM deals with computer manufacturers, for which Microsoft's royalties were estimated to be $10 a machine.[16] However, the volume of sales made MS-DOS Microsoft's most profitable product by far. In June 1986, Microsoft announced that half of its annual revenues (about $100 million) were from MS-DOS; this number would have corresponded to about 10 million sales worldwide. In mid 1988, another major release, MS-DOS 4.0, allowed the use of a mouse, though it fell short of a complete graphical user interface. As late as 1990, it was reported that MS-DOS still accounted for nearly 20 percent of Microsoft's revenues.[17] Subsequent major releases of MS-DOS came with version 5.0 in mid 1991 and version 6.0 in 1993. By that time, Microsoft was earning far higher revenues from Windows.

Strategic Understanding in the Personal Computer Software Industry: The Case of Autodesk

There were several thousand entrants into the personal computer software industry in its first 10 years. Of these, fewer than 100 became major companies with annual revenues exceeding $50 million. Why did some firms succeed wildly while the great majority either became small businesses employing a dozen or so people or failed entirely?

Brian Arthur has argued that an understanding of the increasing-returns economy was crucial: "What counts to some degree—but only to some degree—is technical expertise, deep pockets, will, and courage. Above all, the rewards go to the players who are first to make sense of the new games looming out of the technological fog, to see their shape, to cognize them. Bill Gates is not so much a wizard of technology as a wizard of precognition, of discerning the shape of the next game."[18] It is clear from *The Road Ahead* that what Gates calls "positive feedback" is an intuitive and informal equivalent to the increasing-returns model of the academic economists.[19] We also know, from a memorandum dated June 1985 and subsequently published in a history of Apple Computer, that Gates was aware of the importance of network effects in establishing an operating-system standard.[20] But until Microsoft opens its archives to independent scholars, we have only some tantalizing hints of the company's strategic thinking and the extent to which Gates was responsible for it.

In the meantime, another firm, Autodesk, has made its early strategic thinking publicly available in a book titled *The Autodesk File*.[21] From 1987 to the present, Autodesk—manufacturer of the leading CAD package—has consistently been among the top ten players in the personal computer software industry. Much more than post hoc recollections, *The Autodesk File* consists of contemporary and unedited strategic position papers produced over 10 years.

The Autodesk File was edited by John Walker, Autodesk's co-founder, and it contains some long and sometimes rambling "Information Letters" in which Walker communicated Autodesk's strategic thinking to his co-workers. In these papers, we can see at work several of the mechanisms identified by increasing-returns economists as enhancing the survival chances of a standardized software product: constant reinvestment of profits in product improvement, the creation of complementary products and a network of vertical application developers, and the creation of training networks to diffuse the standard and lock in users.

Autodesk was founded in January 1982 by John Walker, his colleague Dan Drake, and about a dozen entrepreneurial programmers in the San Francisco area. Walker and Drake were then the principals of Marin Systems, a failing computer hardware company that had been founded in 1977. Marin Systems was failing because the hardware business was both highly competitive and capital intensive. They therefore decided to switch from hardware to software. Walker believed that he and his colleagues stood at a unique moment in time when the door was closing on the period when it was possible for two or three programmers to create a viable software product in a few months and was opening on a period in which software would require significant venture capital for development and promotion. In a memo dated January 12, 1982, he wrote:

Products like Wordstar are selling in the $10–20 million per year range today. Bear in mind—this is a product that any of us could write in about two months. We should consider ourselves extremely lucky to be in this business at this time in history. It's a rare piece of luck to have the field you've chosen as your career explode into the hottest growing entrepreneurial arena just as you hit your prime, and we're now at the point that if we want a chance to get involved we have to act immediately. The game has changed and the pace is accelerating very rapidly. This business is getting very big and very professional, and within one year the chances of success of a tiny, heavily technically oriented company will be nil. If we move now, if we move fast, and if we react extremely rapidly and work ourselves to the bone, we can grab a chunk of this business before it slips away.[22]

Aware that technical competence was no guarantee of a hit product, Walker proposed a cooperative partnership that would publish several products, determine the market acceptance of each, then aggressively develop and promote the one or two that seemed most promising.

Autodesk began operations with a portfolio of actual and potential products created by the partners, including a CAD package called Interact, a personal database called Selector, translators for various programming languages, a sort program, and several other application programs and utilities. All the partners worked for Autodesk part-time at this stage, most in addition to their full-time jobs. Seven products were introduced at the fall 1982 COMDEX. In the fashion of VisiCorp's "Visi-" prefix, each package was given an "Auto-" prefix. The CAD system was renamed AutoCAD, the personal database AutoDesk, a screen editor AutoScreen, and so on. Though there had been high hopes for AutoDesk, the product from which the company had taken its name, it was AutoCAD that grabbed the most attention.

After the COMDEX debut, nearly all Autodesk's resources went into promoting AutoCAD. That product had originally been envisaged as a modest "word processor for drawings," but now Autodesk found itself thrust into the CAD sector, a major sector of the corporate software world occupied by ComputerVision, Intergraph, Calman, Applicon, and several other suppliers. The partners, with little experience of the new world in which they found themselves, had to learn fast. Autodesk—still a tiny firm with only five full-time employees—took a booth at the January 1983 CAD industry fair known as CADCON. There they discovered that corporate CAD and personal CAD were two different worlds, far apart in culture and as markets. Autodesk achieved far greater sales and visibility at COMDEX and other personal computer fairs. Autodesk did not return to CADCON for several years, and it was able to evolve out of sight of the big players in CAD.

AutoCAD differed from most personal computer software in that it was priced high and marketed primarily through specialist dealers, who often bundled it with a computer system. The AutoCAD software package retailed at about $4,000, from which Autodesk derived an average return of $2,000 on each copy sold. When combined with a high-end personal computer, a complete AutoCAD drawing system cost about $10,000—perhaps one-twentieth the cost of a system from ComputerVision. AutoCAD dealers provided after-sale support and training for what was a highly complex software system relative to a word processor or a spreadsheet. Autodesk produced dealer training manuals and later organized dealer training courses, which had the side benefit of locking dealers into AutoCAD and thus making it less likely that they would switch to distributing an alternative package.

There turned out to be a huge market for a low-cost engineering drawing package. Walker liked to quote the statistic that the United States had more than 600,000 manufacturing enterprises, 85 percent of which had ten or fewer employees and did all their drawing manually.[23] Again, there was a vast number of architectural practices and design consultancies that were potential AutoCAD users. AutoCAD's two-page color advertisement in the September 1984 *Scientific American* was its first major promotion. The IBM-compatible PC standard was not yet fully established, so Autodesk aimed to supply AutoCAD on every significant platform in order to become the dominant CAD standard. By spring of 1984, the program ran on 31 different desktop systems. This maintenance burden would diminish as the IBM PC standard began to dominate.

Autodesk's sales went from $1 million in 1985 to $10 million in 1986 (table 8.6). AutoCAD had been regularly upgraded. With 200,000 lines of code, it now represented an investment of 76 programmer-years. The AutoCAD standard was consciously promoted by building a network of complementary products, dealers, and training agencies. Besides Autodesk's own add-ons, dozens of third-party software developers for vertical markets were promoted in the *AutoCAD Applications Catalog*, which listed more than 150 programs. There were 1,300 authorized dealers, all required to attend training courses to obtain and retain their dealerships. There were 43 authorized training centers to introduce new users to the complexities of AutoCAD. More than 600 educational institutions taught engineering drawing using subsidized AutoCAD software, thus locking in the rising generation of engineering graduates.[24] During the next few years, Autodesk continued to perfect AutoCAD. By 1990 the program had grown to a million lines of code. With no major competitors in its PC-CAD software niche, Autodesk had annual revenues of $179 million, was a global player in the CAD market, and was probably more profitable than any of its mainstream competitors.

Table 8.6
Autodesk financial statistics.

	Revenues	Annual growth	Employees
1983	$14,000		26
1984	$1,200,000	8,471%	104
1985	$8,500,000	608%	190
1986	$29,500,000	247%	313
1987	$ 52,300,000	77%	399
1988	$ 79,200,000	51%	414
1989	$117,000,000	48%	576
1990	$178,600,000	53%	905
1991	$237,800,000	33%	1,100
1992	$285,000,000	20%	1,310
1993	$353,000,000	24%	1,510
1994	$405,600,000	15%	1,788
1995	$454,600,000	12%	1,788
1996	$534,200,000	18%	1,894
1997	$496,700,000	−7%	2,044
1998	$617,100,000	24%	2,470
1999	$740,200,000	20%	2,712
2000	—	—	3,024

Data for 1984–1993 from J. Richardson, "A Decade of CAD," *CAD User,* March 1998: 20ff.; data for 1994–2000 from Autodesk annual reports). Owing to a change in financial reporting, revenues for 2000 are not comparable with those for previous years.

Paradigm Shift: The Graphical User Interface

By 1982, the personal computer paradigm had reached technological "closure" with the IBM-compatible PC equipped with an Intel 8086 or 8088 microprocessor and the MS-DOS operating system. In the classic way in which technologies are shaped, however, no sooner had this technical closure been achieved than a new "critical problem" came into view.[25]

The most commonly perceived problem with the personal computer was the lack of "multitasking"—the ability for a user to work simultaneously with two or more applications. The scenario most commonly envisaged was of a manager wanting to process data using a spreadsheet, express the results visually in a pie chart, then incorporate the pie chart into a word-processed document. To do this, the user had to fire up the spreadsheet, extract data from a file, process the data in the spreadsheet, save the results in a file, and then close the spreadsheet application; a similar sequence was then repeated with the graphics drawing package and again with the word processor. Lotus 1-2-3 was one of the first products to address the problem of switching between programs by integrating three applications (a spreadsheet, a graphics package, and a simple database) in a single program. However, the preferred approach would be to allow the user to have several applications active simultaneously and to be able to switch rapidly from one to the other, thus permitting data to be shared between them. This was multitasking.

In 1982 and 1983, a consensus emerged that the best way to achieve multitasking was by means of a windowing system. A windowing system allowed several applications to coexist, each in a separate "window" on the computer screen. One application would command the user's attention at any moment. By choosing to focus on a different window, the user could effortlessly switch to another task, then back again.

The concept of a windows-based operating system had originated in the 1970s at Xerox's Palo Alto Research Center (PARC),[26] where most of the ideas now standard in a graphical user interface, including overlapping windows, pull-down menus, and point-and-click task selection by means of a mouse, originated. The work at Xerox PARC had led to the Xerox Star, announced in May 1981—a failure in the market, primarily because its price ($40,000) was much too high for a personal computer.

The concept of the graphical user interface was also adopted by Apple Computer for its Lisa computer, launched in May 1983. Though universally regarded as a path-breaking product, the Lisa also failed in the market because of a high price ($16,995). Apple Computer's second attempt—the $2,500 Macintosh, launched in January 1984—was much more successful. The Macintosh's unique selling point was its user-friendly interface.[27] It succeeded in capturing 5–10 percent of the personal computer market for the next decade. But because it was a proprietary system, it never attracted as many software and hardware suppliers as the IBM-compatible PC.

As a result of the rise of the graphical user interface, multitasking became conflated with the secondary issue of user friendliness. In 1983, several software firms were developing windowing systems for the IBM-compatible PC, and their distinguishing characteristics tended to relate to ease of use—for example, whether the system allowed pointing and clicking by means of a mouse or required the user to navigate by means of keyboard function keys or obscure keystrokes.[28] Table 8.7 lists the principal windowing systems produced in the period 1983–1985. In all cases, the development of a windowing system was highly speculative both in terms of market uncertainty and in terms of the challenge of creating an unfamiliar technology.

Most of the media attention focused on three companies: VisiCorp, Microsoft, and Digital Research. VisiCorp, at the November 1982 COMDEX, was the first to announce a windowing system. VisiCorp was then the leading personal computer software company in terms of revenues, and so its windowing product, VisiOn, attracted intense interest. VisiOn was designed as an environment that would sit between the MS-DOS operating system and ordinary applications. The project turned out to be far larger than originally envisaged. Development began in early 1981. By the time VisiOn was shipped, 3 years later, it was said to have been rewritten from the ground up three times and to have cost $10 million.[29] It was necessary for software publishers to rewrite their applications to run under VisiOn. Despite VisiCorp's encouragement, none chose to make the investment. When VisiOn was released (in January 1984), the only products available for it were VisiCorp's own productivity applications, such as VisiOnCalc and VisiOnWord. VisiOn was priced at $495; the applications averaged $400 each. Lukewarm reviews and the absence of

Table 8.7
Windowing systems for IBM-compatible PCs, 1984–1985.

Publisher	Product	Price	Announced	Released	Notes
VisiCorp	VisiOn	$495	November 1982	January 1984	Price reduced to $95
Digital Research	GEM	$399	November 1983	September 1984	Price included Concurrent DOS
Microsoft	Windows 1.0	$95	November 1983	November 1985	
IBM	TopView	$149	August 1984	February 1985	
Quarterdeck	DESQ	$399	Spring 1983	May 1984	

Sources: John Markoff, "Five Window Managers for the IBM PC," *Byte Guide to the IBM PC* (Fall 1984): 65–7, 71–6, 78, 82, 84, 87; Irene Fuerst, "Broken Windows," *Datamation* (March 1, 1985): 46, 51–2; Allen G. Taylor, "It's Gem vs. Topview as IBM, DRI Square Off," *Software News* (August 1985): 71–3; Ken Polsson, *History of Microcomputers: Chronology of Events,* http://www.maxframe.com/hiszcomp.htm (accessed December 2000).

any applications except VisiCorp's own resulted in slow adoption, despite a cut in VisiOn's price from $495 to $95 within a month.

In November 1983, Microsoft announced its intention to develop a windowing system. Like VisiOn, Microsoft Windows was to be a software layer between MS-DOS and ordinary applications. The announcement stated that Windows would be available by the spring of 1984, with the expectation that it would run on 90 percent of MS-DOS computers by the end of the year. But developing Windows turned out to be much more complex than was originally expected. The release date was postponed to May 1984, then to August, then to January 1985. Soon Windows was Microsoft's largest development project, with two dozen developers working on it and a dozen more people working on documentation. Windows finally arrived in November 1985. Priced at $95, it was launched with the biggest publicity campaign in Microsoft's short history, complete with full-color, eight-page inserts in leading computer magazines.

Though Digital Research had been eclipsed by Microsoft in 1983, it remained a significant and growing company. Its annual revenues, exceeding $50 million, were derived mainly from its eight-bit control program for microprocessors, the CP/M operating system. In late 1983, Digital Research announced a Graphics Environment Manager (GEM) for the Macintosh-like Atari ST. GEM was never able to penetrate the IBM PC market significantly.

Another well-publicized windowing system was IBM's TopView. Yet another was DESQ, produced by a Santa Monica startup called Quarterdeck. These too were market failures.

The failure of all the windowing systems came as a surprise. The business and computer press had anticipated a "fierce battle" for this new territory, and 1984 had been "heralded as the year of the window."[30] In fact, there never was much of a battle. The reasons were rather mundane: Most of the products had limped into the market after long development delays, and none performed acceptably. With the low-powered 8088/86 microprocessors, the systems were "unbearably slow" and left reviewers "begging for faster hardware."[31] This was true even with the new Intel 80286 microprocessors. The disappointing results illustrated how immature the PC software industry was, revealing its inability to estimate either development times or software performance realistically. Some of the vendors of windowing systems paid a high price for their inexperience.

In the summer of 1984, the failure of VisiOn obliged VisiCorp to lay off half of its 110 workers and then to submit to a takeover by a small company, Paladin Software, a step it viewed as preferable to the "ignominy of

Chapter 11 bankruptcy."[32] The rights to VisiOn were sold off to the Control Data Corporation, a mainframe maker. It was an extraordinary about-face for a firm that had been number 2 in the industry in 1983.

Digital Research was unable to restore its fortunes with GEM or with any of its other operating-system developments. Its 1985 income was down $20 million from the 1984 peak of $56 million, and it cut its 600-member work force by half. Founder Gary Kildall resigned in mid 1985.

Only Microsoft and IBM had the resources to persist with a graphical user interface, which would have to wait for the next generation of Intel microprocessors. In any case, a simple GUI enhancement of MS-DOS was not their only project. In early 1985, Microsoft and IBM had begun joint development of a new operating system intended to be the long-term replacement for MS-DOS.

Meanwhile, inside Microsoft, development of Windows continued under its own momentum. In late 1987, Windows 2.0 was released to modest acclaim. The interface had been polished considerably, and its main visual elements were almost indistinguishable from those of the Macintosh. Microsoft had obtained a license from Apple Computer for Windows 1.0 but had not renogotiated it for the new release. Version 2.0 so closely emulated the "look and feel" of the Macintosh that Apple sued for copyright infringement in March 1988. The Apple-vs.-Microsoft lawsuit consumed many column-inches of reportage and rattled on for 3 years before a settlement in Microsoft's favor was reached in 1991.[33] So far as can be ascertained, the lawsuit was something of a sideshow that had little bearing on Microsoft's or any other company's technical or marketing strategy.

Within a few months of the introduction of Windows 2.0, Microsoft and IBM announced the fruits of their joint development of a new operating system, OS/2. With its high price ($325), its incompatibility with existing applications, and its marginal advantages over MS-DOS, OS/2 was yet another failure in the market.[34] At this point, Microsoft decided to cut its losses, withdraw from the joint development with IBM, and pursue Windows. It had not, in any case, been a happy experience. The cultural differences between the two firms had been insurmountable. Though Windows 2.0 could not begin to match MS-DOS (now selling 5 million copies a year) in sales, it was doing much better than OS/2, thus confirming the advantages of a migration path that maintained user lock-in by augmenting MS-DOS rather than replacing it. In 1988 and 1989, several mainstream application publishers converted their products to run under Windows. By early 1989, 2 million copies of Windows 2.0 had been sold.

With the positive response to Windows, more resources were poured into development. A new version was launched in May 1990, with a reported $10 million spent on a worldwide publicity splash. Windows 3.0 received unequivocal market acceptance, and the paradigm shift had finally occurred. A writer for *PC Computing* caught the moment well:

> When the annals of the PC are written, May 22,1990, will mark the first day of the second era of IBM-compatible PCs. On that day, Microsoft released Windows 3.0. And on that day, the IBM-compatible PC, a machine hobbled by an outmoded, character-based operating system and seventies style programs, was transformed into a computer that could soar in a decade of multitasking graphical operating environments and powerful new applications. Windows 3.0 gets right what its predecessors—VisiOn, GEM, the earlier versions of Windows, and OS/2 Presentation Manager—got wrong. It delivers adequate performance, it accommodates existing DOS applications, and it makes you believe that it belongs on a PC.[35]

In hindsight, it seems extraordinary that Microsoft's development processes and technical competence were such that it simply kept improving and re-launching Windows until the product was finally in harmony with the technology and the market. Yet that was the essential truth. Microsoft was a young company with surprisingly little technological depth. It was a "learning organization"—an organization that fumbled its way to success by making mistakes, learning from them, then making fewer mistakes.[36] Microsoft's three attempts to produce Windows were mirrored by an equal number of attempts to produce application software.

Microsoft and Productivity Applications

Microsoft's domination of the software market is best observed in the sector of productivity applications, a sector in which it had virtually no presence in 1983 but was dominant by 1995.

Microsoft's strategy has been to use the revenues from its successful systems products to develop applications software, without regard to short-term profitability. Invariably, these packages have succeeded only after a third or subsequent product launch. A good product was never sufficient in itself to dislodge any of the incumbents in productivity applications (Lotus in spreadsheets, WordPerfect in word processors, Ashton-Tate in database systems). However, at some point each of those firms temporarily lost its hold on the market, opening a window of opportunity for Microsoft. For example, in the late 1980s each of those firms had experienced a development debacle that had resulted in a product that was either late or unreliable. They had also misjudged the impact of Microsoft Windows, betting instead that OS/2 would become the dominant platform.

In short, Microsoft played a waiting game and had excellent products ready to replace an incumbent when the opportunity came. Had history unrolled differently (for example, had OS/2 and not Windows become the dominant platform), the story may have unfolded differently, but the outcome likely would have been much the same.

Lotus 1-2-3 and Excel

The Lotus Development Corporation was perhaps the premier example of a firm dominating its sector. Lotus launched its 1-2-3 spreadsheet in January 1983, when the IBM-compatible PC was emerging as a standard. By the end of 1984, Lotus was the leading personal computer software company, with revenues of $156 million. It was ahead even of Microsoft, whose sales were $97 million (table 8.8). All this was due to a single product. Lotus 1-2-3 completely eclipsed VisiCalc. VisiCorp, unable to regain its competitive position with VisiOn, was taken over by Paladin Software in 1985. VisiCalc's developer, Software Arts, was acquired by Lotus for $800,000, and Dan Bricklin and Bob Frankston became employees of Lotus. In mid 1985, VisiCalc was withdrawn from the market and its users were offered a half-price upgrade to Lotus 1-2-3.[37]

With Release 2 of 1-2-3 (September 1985), Lotus's fortunes continued to soar. That package dominated the software charts for the next 2 years.

Table 8.8
Lotus financial statistics.

	Revenues	Annual growth	Employees
1983	$53,000,000		291
1984	$156,000,000	194%	750
1985	$226,000,000	45%	1,050
1986	$283,000,000	25%	1,400
1987	$396,000,000	40%	2,100
1988	$469,000,000	18%	2,500
1989	$556,000,000	19%	2,800
1990	$685,000,000	23%	3,500
1991	$829,000,000	21%	4,300
1992	$900,000,000	9%	4,400
1993	$981,000,000	9%	4,738
1994	$971,000,000	−1%	5,522
1995	$1,150,000,000	18%	6,000

Source: *Hoover's Guide to Computer Companies,* various years.

In 1987 it had a 70 percent share of a spreadsheet market estimated to be worth $500 million. A total of 3 million copies of Lotus 1-2-3 had been sold, and its lock on the market was consolidated by a range of complementary products from other vendors, such as Funk Software's Allways and Personics' Look&Link.[38]

Lotus's first major setback came with Release 3 of 1-2-3, "the spreadsheet that nearly wore Lotus out."[39] For this release, Lotus decided to rewrite the program from the ground up, not reusing any of the code from the previous releases. Lotus was not accustomed to development on such a scale. An IBM development manager was hired to establish bureaucratic processes for controlling the project and its 35 developers. Originally scheduled for mid 1988, the program's release date slipped three times before it finally appeared in June 1989. The new release of Lotus 1-2-3 contained 400,000 lines of code; the first version had contained 20,000. Lotus had staked its future on OS/2 as the successor to MS-DOS, and 1-2-3 could run with either. Though Lotus maintained its sales volume during the Release 3 debacle, it was forced to cut the price in order to maintain competitiveness against Microsoft's Excel and Borland's Quattro. This dramatically affected Lotus's profits, causing its share price to fall nearly 60 percent in 1988.

Microsoft's spreadsheet, eventually to become Lotus 1-2-3's main competitor, had started life in 1980, when Bill Gates and Paul Allen decided to diversify into applications to reduce their dependence on systems software. Microsoft's first spreadsheet, then known as MultiPlan, was released in the second half of 1982, with versions for the Apple II and the IBM PC. Though the product got excellent reviews (including a "software of the year" award), it made little headway against 1-2-3. Lotus was making more money on its single program than Microsoft was making on its entire product line. Accepting the impossibility of competing with Lotus 1-2-3 head on, Microsoft decided to develop a GUI-based spreadsheet for the Macintosh. In effect, this strategy would shield it from Lotus's competition and would allow Microsoft to perfect its spreadsheet and interface technology. The new Macintosh spreadsheet, now called Excel, was released in September 1985 and quickly secured 90 percent of Macintosh spreadsheet sales.

Windows 2.0 was released in October 1987. Excel for Windows was released at the same time, and in the absence of competitors it became the preferred spreadsheet for Windows. However, to most observers, including Lotus, OS/2 seemed the more likely successor to MS-DOS. When Windows took off, with version 3.0 in 1990, Lotus and many other software makers were caught without Windows versions of their products.

Excel filled the spreadsheet vacuum created by Windows. When Lotus 1-2-3 for Windows was finally released, in 1991, Excel had the first-mover advantage on what was rapidly becoming the dominant platform. Lotus 1-2-3 never recovered its former market share: by 1995, Excel had more than 70 percent of the world-market revenues for spreadsheets, while Lotus had less than 20 percent.[40]

WordPerfect and Microsoft Word
In 1984, WordPerfect had less than 1 percent of the market for personal computer word processing software; MicroPro's WordStar had 23 percent. Two years later, WordPerfect's share was 30 percent and rising; WordStar's was half that and falling. By 1986, MicroPro's revenues had fallen from a peak of $67 million in 1984 to $38 million, and it was experiencing a $1.2 million loss. Satellite Software International, the manufacturer of WordPerfect, was renamed the WordPerfect Corporation in 1986. By 1987, WordPerfect was the best-selling personal computer software package by volume, ahead even of Lotus 1-2-3 and dBase III.

The WordPerfect Corporation was one of the few personal computer software publishers to have originated in the world of corporate computing. The firm was incorporated as Satellite Software International (SSI) in 1979 by Alan Ashton, a computer science professor at Brigham Young University, and his graduate student Bruce Bastian.[41] Ashton had developed a word processing program in 1977, essentially as an academic exercise during his summer vacation. This was subsequently enhanced by Bastian and packaged for the Data General minicomputer for a client in Orem, Utah. In March 1980, the package, named SSI*WP, was made available to Data General resellers, who sold two or three copies a month at a retail price of $5,500. By the end of 1981, SSI was a modest success, with annual sales of $850,000. As such, it was similar to at least fifty small firms that were developing word processing software for minicomputers and competing with such major players as Wang and Lanier.[42] SSI—young, very small, and with few institutional rigidities—was eventually able to move into PC software, unlike most other corporate software firms.

In 1982, SSI acquired its first IBM PC and decided to convert SSI*WP for use on it, not least because some of its Data General–owning clients were acquiring IBM PCs and wanted a compatible word processor. The word processor for the IBM PC, essentially identical to the one designed for Data General, was renamed WordPerfect and launched in October. The product had little impact at first, mainly because SSI only had experience of direct selling to the corporate computer market, possessing neither the capability nor the advertising budget to address the retail sector. However, as SSI built up experience in selling through ComputerLand and other PC retailers, sales began to rise. Press reviews of WordPerfect were highly favorable, singling out for special praise the toll-free telephone lines that offered technical support and the fact that the software supported more than 200 different printers. The product quickly gained a word-of-mouth reputation for solid dependability,

The turning point for WordPerfect came in late 1984, when MicroPro replaced its aging WordStar with WordStar 2000. MicroPro made a classic error. WordStar 2000 had an entirely new interface and was bigger and slower than the previous version. Because the new interface required so much relearning, users switched to other products and WordStar lost its lock on the market. Many of the switchers chose WordPerfect 4.0, a major new release that had fortuitously appeared just a few weeks earlier and was probably the best available option to WordStar on the market.

Since its initial release, in 1982, WordPerfect had been revised continually, major or minor upgrades being announced each year at the fall COMDEX. By 1987, however, WordPerfect was in need of a major rewrite because of competitive threats coming from two directions. The first was from Microsoft Word, which at that point was taking 10 percent of word processing unit sales.[43] The second came from desktop publishing packages such as Aldus Pagemaker and Adobe Illustrator, which offered integration of text and pictures and which supported the new generation of laser printers. Version 5.0 was the WordPerfect Corporation's biggest undertaking so far. WordPerfect had little experience with major development projects, and the schedule slipped month by month. Although development was well behind schedule, WordPerfect 5.0 was announced at the November 1987 COMDEX in the hope of dissuading users from switching to Microsoft Word. It was eventually released in May 1988 with so many bugs that, in a much-reported story, "the company got so many calls that the 800 lines into Utah were jammed, cutting off Delta Airlines and the Church of Jesus Christ of Latter-Day Saints. AT&T eventually added more 800 lines."[44] Users were surprisingly tolerant of unreliable software. Bugs and all, WordPerfect remained the best-selling word processor, with sales up from the previous year by an astonishing 75 percent. By 1990, WordPerfect had 80 percent of word processor revenues.

WordPerfect's growth stalled in 1991 because of competition from Microsoft's Word for Windows. WordPerfect, like Lotus, had bet on OS/2 and therefore had no product ready when Windows 3.0 was released in

May 1990. In that same month, WordPerfect, quickly sensing the market's acceptance of Windows, announced the postponement of WordPerfect for OS/2 in favor of a Windows version. Intended for release in February 1991, WordPerfect for Windows was not shipped until November, and because of its weak integration with the Windows interface it was poorly received. Meanwhile, as it had done with Excel, Microsoft had perfected Word over a period of years. After its 1983 launch, it had secured a respectable but modest 10 percent market share. Again, as with Excel, a version for the Macintosh enabled Microsoft to perfect its graphical user interface and its laser printer technology out of sight of the mainstream IBM PC market. Released in January 1985, Word was consistently the best-selling application for the Macintosh. In December 1989, Word for Windows was released for the IBM platform. This sophisticated product embodied all of Microsoft's Macintosh know-how and contained 250,000 lines of code (3 times as many as its predecessor). Though its first release had many bugs, Word for Windows 2.0 was a highly regarded product that was able fully to exploit the mass acceptance of the Windows platform.

The WordPerfect Corporation's revenue growth flattened as the market rapidly switched to Microsoft Word, precipitating management restructuring in 1992.[45] WordPerfect never regained its former prominence. By the mid 1990s, its sales were essentially vestigial; Microsoft Word had 90 percent of the market.

Ashton-Tate
In 1983, Ashton-Tate's dBase II was the third-best-selling software product, with cumulative sales of 100,000 copies at a price of $695. Though Ashton-Tate owed much of its success to an early start, it was particularly adept at fostering complementary products. By 1984, about 1,800 companies had signed up for Ashton-Tate's support program for developing dBase II templates for vertical markets, and the *Application Junction* catalog for 1985 listed more than 1,700 complementary products.[46] Ashton-Tate's market dominance was consolidated by the release of dBase III in 1985. By then, Ashton-Tate was enjoying an extraordinary 68 percent share of a market for personal computer databases estimated at $150 million.[47] The rest of the market was shared by more than sixty competitors. Of these, the most important were Ansa's Paradox, Microrim's R:base, Fox Software's Foxbase, and Information Builders' PC/Focus. Besides these full-feature databases, there were myriad personal filing systems, exemplified by Software Publishing's PFS:File.

Though Ashton-Tate had begun as a software publisher rather than as a developer, it established its own software-writing capability during 1984 and thus was able to produce dBase III in house. In 1986, development work started on dBase IV, an ambitious product that was to result in a program containing about 450,000 lines of code produced by 75 developers. Like Lotus and WordPerfect, Ashton-Tate was overwhelmed by the unfamiliar scale of development. Ashton-Tate's vice president of programming, Wayne Ratcliffe (developer of the original dBase II) resigned amid the chaos, acrimoniously claiming that Ashton-Tate "did not understand the software development process."[48] After considerable slippage in the schedule, dBase IV was released in October 1988 as a gargantuan product that took up fourteen floppy disks.

Alas, dBase IV was highly unreliable. Of the 105 bugs reported, Ashton-Tate conceded to 44.[49] Bug-ridden products were by no means unusual in the maturing personal computer software industry. Microsoft, Lotus, and WordPerfect all had shipped products that had had to be withdrawn or remedied with free upgrades. However, a word processor or a spreadsheet that froze up could be tolerated; loss of computer data could not. The risk of data loss was probably exaggerated, but users stopped buying dBase IV. Fortunately, Ashton-Tate's revenues declined less sharply than might have been predicted, largely because the sub-industry of complementary product makers enabled Ashton-Tate's customers to sustain their businesses with the aging dBase III. However, by July 1990, when a reliable version of dBase IV was released, many users had switched to competing products, and Ashton-Tate was running up losses of $20 million per quarter. In a 1991, Ashton-Tate was acquired by Borland (a much smaller company) in a stock swap for $439 million. Although dBase IV continued to sell acceptably on the basis of its historic customer base, it had lost its allure, and it never recovered.

It should be noted that the demise of Ashton-Tate cannot be traced to Microsoft. Even best-selling products, such as dBase IV, VisiCalc, and WordStar, could fall from grace because of obsolescence or unreliability.

Product Integration and Office Suites

Though Lotus, WordPerfect, and Ashton-Tate could be fairly described as one-product companies, in that 80 percent or more of their sales came from a single program, this was not by choice. Having succeeded on one front, they all sought to diversify into other productivity applications, partly to achieve corporate growth but also out of awareness that the market was inherently unstable and that today's hit could easily become tomorrow's also-ran. Early on, each of the two market leaders, VisiCorp and MicroPro, had tried to create a brand image and a portfolio of products to complement its best-selling package: VisiWord and VisiFile complemented VisiCalc, and CalcStar and DataStar enhanced WordStar. But even attractive pricing was not enough to persuade the market to switch from a known and trusted product.

In 1984, both Lotus and Ashton-Tate attempted to augment their hit products with integrated packages that would offer the three main productivity applications in a single program. Lotus 1-2-3 was, of course, itself an integrated package to the extent that it contained basic charting and database capabilities. However, Lotus's new product, Symphony, would incorporate a spreadsheet, a full-strength word processor, a database, and a communications program. Lotus reportedly spent $14 million on the program, including $8 million for an advertising blitz that "included a [TV] spot during the Summer Olympics."[50] At the same time, Ashton-Tate was working on Framework, an integrated package that aggressively targeted Symphony, even matching its price of $685 and its launch date of July 1984.[51] Ashton-Tate encouraged its hundreds of suppliers of complementary products to create vertical applications. Ashton-Tate reportedly spent $10 million on Framework, much of it on publicity. However, Symphony, Framework, and dozens of imitators were failures in the market. Within a year, the fad for integrated products evaporated. "With about 70 integrated software packages on the market," *Datamation* noted, "software companies have been failing with the regularity of Philadelphia commuter trains and new Italian restaurants."[52] Although WordPerfect had begun to develop complementary spreadsheet and drawing programs (PlanPerfect and DrawPerfect), neither of them had reached the market.

For the next 5 years, stand-alone applications dominated the market. In 1990, however, Microsoft introduced a devastating marketing strategy. In a single shrink-wrapped box called "Office," it bundled all its productivity applications for Windows at a price of $750—not much more than the cost of one of the individual programs. Microsoft had tested this approach in 1988 with Office for the Macintosh, bundling Excel, Word, and the new PowerPoint presentation graphics program; the results had been excellent.[53] In 1990, Microsoft Office for Windows, with the same three productivity applications, was introduced. With the GUI environment of Windows, Office provided a degree of integration that had not been possible under MS-DOS. Lotus was obliged to follow suit; in 1991 it introduced SmartSuite, which consisted of Lotus 1-2-3, the AmiPro word processor (acquired from Samna in 1990), Freelance presentation software, and a personal database. In 1992 Microsoft acquired Fox Software (maker of the Foxbase database program) and subsequently added a database program, Access, to the high-price "professional" edition of Office. In 1993, Borland—which owned the Quattro spreadsheet and two leading database packages, dBase IV and Paradox, but no word processor—cooperated with the WordPerfect Corporation to publish Borland Office, consisting of WordPerfect, the Quattro Pro spreadsheet, and the Paradox database.[54]

Microsoft has never explicitly described the strategy behind Office, but analysts have speculated that the package enabled the company to push loss-making applications, such as PowerPoint, onto people's desktops and to gain market share from its one-product competitors while incurring relatively little revenue loss. At all events, by the fall of 1993 Office accounted for more than half of Microsoft's productivity application sales and was increasingly positioned as Microsoft's "primary application" rather than as "simply a way of marketing a group of applications."[55] With intense competition among the three office suites, prices fell to the $300 level during 1993–94. Microsoft Office far outsold the others, gaining an estimated 90 percent market share; Lotus SmartSuite had 8 percent and Borland Office had 2 percent.[56] WordPerfect, its profits collapsing, was acquired by Novell for $800 million in 1994. In 1995 IBM acquired Lotus for $3.5 billion, not to get its spreadsheet but to get its know-how in personal computer software and some interesting products then under development.[57] Perhaps the biggest casualty was the Software Publishing Corporation, whose low-cost PFS programs lost the price advantage they had had for a decade. In 1993 the company laid off 140 employees, more than 20 percent of its work force. Soon it was lost from sight as a significant software company.

Competing with Microsoft

Between 1983 and 1995, Microsoft came to dominate the personal computer software industry to an extent that has no parallel in the corporate software industry. However, though Microsoft bestrode its world like a colossus, other firms were able to compete successfully. In 1995, while Microsoft had approximately 50 percent of the market for personal computer software, the top half-dozen firms after Microsoft had shares of between 2 percent and 10 percent. These were successful firms by any standards. And beneath them were hundreds of lesser but moderately successful firms. Four distinct strategies have allowed these firms to coexist or compete with Microsoft.

First, some firms have published products complementary to Microsoft's best-selling packages. Perhaps the best-known example of this strategy is Peter Norton Computing, established in 1982 with the idea of enabling MS-DOS users to restore files they had accidentally deleted. In 1987, it was reported, "close to a million of the fumble-fingered, fatigued, or forgetful now reach for the $100 Norton Utilities as if it were aspirin," and Peter Norton Computing was a fifty-employee firm with annual sales of $10 million.[58]

From Microsoft's perspective, Norton was little more than a speck of dust. At the other end of the spectrum, Novell—which complemented MS-DOS with networking software—became the number-2 firm in the industry. Originally a hardware firm, Novell introduced its NetWare software product in 1987. Once it had succeeded in integrating isolated desktop computers into corporate networks in the late 1980s and the early 1990s, the company grew rapidly.[59]

Complementers, however, always run the risk that Microsoft will incorporate the functions contained in their software into its own products, either by internal development or by acquiring the technology in a takeover. Merger talks between Novell and Microsoft in 1990 fell through. Microsoft subsequently introduced networking capabilities into its operating systems in the early 1990s, thereby entering into intense competition with Novell. On the other hand, in the case of Norton Utilities, Microsoft has shown the tolerance of an elephant for the tikka bird on its back, allowing Peter Norton Computing "deep into the innards of the operating system" and fostering "tremendous personal relationships between their development teams."[60] This is probably because Norton Utilities complement Microsoft's operating systems, adding to their value—by providing anti-virus facilities, for example—in a way that Microsoft's relatively bureaucratic development processes would find difficult or uneconomical.

A second way of coexisting with Microsoft has been to occupy niche markets where Microsoft has no presence. A classic example of this was Borland, which introduced Turbo Pascal in 1982. It happens that developers prefer particular programming languages for reasons that are more religious than rational. Pascal was an elegant programming language favored by academics and by idiosyncratic firms like Apple Computer, whereas Microsoft and most software developers preferred the more prosaic BASIC or C programming languages. As a result, Borland experienced no competition from Microsoft, whose forte was programming languages; had Microsoft ever produced a Pascal system, it would surely have eclipsed Turbo Pascal. There were other successful niche players among the top ten firms: Autodesk, specializing in CAD software; Adobe Systems, in printing software; Aldus, in desktop publishing; Corel, with its Draw! package; Intuit, in personal finance software; SCO, in Unix-on-Intel operating systems. Each of these firms has been vulnerable to Microsoft's ability to enter its niche with huge development and marketing resources. This has already happened to Adobe Systems and Intuit.

Adobe Systems was formed by John Warnock and Charles Geschke, who pioneered laser printing technology at Xerox PARC in the late 1970s. In 1982, when Xerox had failed to market the technology, Warnock and Geschke started their own company.[61] Adobe grew rapidly, supplying a software technology known as Postscript for manufacturers of laser printers and for the Apple Macintosh. That the Macintosh was subsequently able to dominate the high-end desktop publishing market was due largely to Adobe's technology. By 1984, half of Adobe's income came from Apple royalties. By the late 1990s, however, Adobe's Postscript technology was no longer unique; both Apple and Microsoft had developed their own systems. Recognizing that its niche in printing software was evaporating, Adobe made a number of strategic acquisitions in order to diversify into desktop publishing and electronic document distribution.

Intuit was established in 1983 by Scott Cook, a former Procter & Gamble brand manager. Intuit's Quicken personal finance software pulled ahead of many similar packages, as much because of its heavily promoted brand image as much as because of its intrinsic merits. By 1990 it dominated the personal finance niche, with annual sales of $33 million. The next year, Microsoft introduced its Money package. Despite Microsoft's multiple product relaunches, Quicken continued to outsell Money seven to one. In October 1994, Microsoft made a takeover bid, and Intuit accepted. However, in April 1995, before the deal was consummated, the US Department of Justice brought an action to prevent the merger on the ground that it would diminish competition in the emerging markets of personal finance and home banking. Microsoft withdrew its offer.[62]

Microsoft has invested in niche firms without actually introducing a product of its own, essentially as a way of gaining market intelligence and potential access to technology. One example is Microsoft's involvement with the Santa Cruz Operation. Formed in 1979, SCO specializes in Unix-on-Intel operating systems. In 1982, SCO licensed Microsoft's XENIX operating system as a reseller. After 1983, as Microsoft increasingly focused on the high-volume MS-DOS operating system, it chose to eschew a market it had pioneered, leaving SCO free to develop that market. In 1989, Microsoft took an 11 percent equity holding in SCO while continuing to pursue its non-Unix Windows and MS-DOS operating systems. The investment in SCO has enabled Microsoft to gain access to Unix technologies should that ever become necessary. In a similar way, Microsoft's investment in the corporate database provider Sybase subsequently informed Microsoft's move into database technology.[63]

A third way of competing with Microsoft has been to develop packages that run on a range of operating systems and platforms, typically the big three: IBM-compatible PCs running MS-DOS or Windows, the Apple Macintosh, and Unix. Since the late 1980s, Microsoft has elected to concentrate almost exclusively on the IBM-compatible PC platform, making only a vestigial commitment to the Macintosh and virtually none to Unix. This strategy has enabled Microsoft to dominate the market for IBM-compatibles. However, many corporations operate heterogeneous environments in which PCs, Macintoshes, and Unix workstations coexist and must run common applications software. This was increasingly the case after the networking of isolated desktops in the early 1990s and the emergence of the Internet in the mid 1990s. Developing efficient cross-platform software is technologically demanding. Microsoft has never mastered the skill, and its few attempts at simultaneous development for the IBM PC and the Macintosh have been poorly received.

In 1992, Adobe Systems became one of the first firms to exploit the trend toward heterogeneous environments when it launched Carousel, a product that created electronic documents in a portable document format (PDF) that could be used on any of the common platforms. Carousel was initially unsuccessful owing to a flawed marketing strategy that required all users to purchase a viewing package. When the product was relaunched as Acrobat in 1994, it was much more successful because a viewer program was distributed free of charge; only users wishing to publish documents had to pay for the development tools.[64] Another vendor with a strong cross-platform strategy was Netscape Communications, established in 1994. Netscape developed its Navigator Internet browser software simultaneously for the IBM-compatible PC, the Macintosh, and seven versions of Unix. Although Microsoft introduced its own browser (called Explorer) in 1995 and quickly colonized the PC platform, Netscape's cross-platform capability enabled it to maintain a high market share in heterogeneous environments.[65]

The fourth way for firms to maintain market share relative to Microsoft has been growth through acquisition. By the mid 1980s, it was clear that acquiring firms or products was a much more successful growth strategy than internal development. One reason was that unfamiliar technologies have proved remarkably difficult to develop *ab initio*. Often it took two or three attempts to get a product's features right. Indeed, so often did Microsoft need three tries that in 1990 "even Gates mused aloud that if the company did not change, customers would simply skip the first two versions."[66] And even a technically sound new product was not guaranteed success in the market. As is evident from the fate of the myriad me-too spreadsheets and word processors, success for any new product in an existing niche was positively unlikely. Acquiring an already developed product could eliminate either or both of these uncertainties. Microsoft was well aware of this, though it generally bought for the technology rather than the brand; its own corporate identity was stronger than almost any individual product name. Microsoft's Access database and its Internet Explorer were acquisitions, the former achieved by taking over Foxbase and the latter by licensing technology from Spyglass. Most of the bigger personal computer software companies have made acquisitions, either for technology or to obtain a successful brand. Adobe, for example, became the market leader in desktop publishing by acquisition. It made a strategic decision to diversify into desktop publishing applications after its 1986 IPO. Its internally developed Illustrator program was moderately successful, but in 1994–95 it bought a dramatically larger market share by acquiring Aldus (the publisher of the best-selling PageMaker package) and the rights to FrameMaker.

Acquisitions have been most successful when there has been a good strategic fit. Novell was much less successful when it tried to grow though poorly judged acquisitions. Novell first tried to move into productivity applications in 1990 by attempting (without success) to take over Lotus. In 1994, when its fortunes were at a low ebb, Novell acquired the WordPerfect Corporation. That acquisition was a spectacular mistake, and WordPerfect was sold to Corel 2 years later for $124 million, one-sixth what Novell had paid.[67] Similarly, Borland achieved a growth spurt by acquiring Ansa (the publisher of the Paradox database) in 1987 and Ashton-Tate (the troubled maker of dBase) in 1991. This briefly made Borland the number-3 company in 1992, but several years of losses and downsizing followed, culminating in CEO Phillipe Kahn's resignation in 1995.[68]

Perhaps the leading exponent of the successful takeover in personal computer software has been Symantec, whose president, Gordon Eubanks, effectively invented the idea of a portfolio of software brands. Eubanks, a former graduate student of Digital Research founder Gary Kildall, was one of the pioneers of the personal computer software industry.[69] In 1983, Eubanks, who had briefly been a vice president of Digital Research, bought Symantec, then a startup of no consequence, and became its CEO. Realizing that he could not compete head on with Lotus in productivity applications or with Microsoft in systems programs, Eubanks sought to build up a portfolio of successful utilities in which the products and development teams retained their identities: "There isn't really one Symantec. It's a bunch of people from a bunch of companies. . . . When we acquire we take the core product team and keep them together."[70] Whereas Microsoft bought primarily for the technology, Symantec bought as much for the brand. After its 1990 IPO, Symantec made its biggest acquisition to date: Peter Norton Computing. The famous Norton Utilities continued to be marketed as an individual brand, providing some marketing efficiencies and, more important, long-term product refinement to maintain its competitiveness. By the mid 1990s, Symantec had bought twenty companies, giving it a host of products within the niche of utilities for IBM-compatible PCs. It was a Computer Associates writ small.

Summary

Microsoft has truly dominated the personal computer software industry since the early 1980s. However, as this book has been at pains to point out, Microsoft still constitutes only about one-tenth of this extremely fragmented industry. Nonetheless, Microsoft's amazing success demands historical analysis. Was Bill Gates smart, lucky, or ruthless? A little of each, perhaps.

Without doubt, Gates has "smarts" (a word popularized by his own company). Aside from his considerable though unexceptional technical capabilities, his success has been attributed to an ability to understand and exploit the economics of increasing returns. Steve Lohr, the author of a fine history of computer programming, has written: "Gates, most of all, is someone with a deep understanding of software, and the foremost applied economist of the past half-century."[71] Describing the development of increasing-returns economics, Lohr continues: "One of the seminal papers in this new branch of economics research was written by Michael Katz and Carl Shapiro, 'Network Externalities, Competition and Compatibility,' published in 1985. While they were working on their paper, Shapiro recalled Katz saying there was a guy who had been at Harvard when Katz was an undergraduate, who was doing precisely what they were writing about at a software company outside Seattle. He was speaking of Bill Gates of course."[72]

"The foremost applied economist of the past half-century" is a resounding phrase that perhaps fails to capture the naiveté of Gates's understanding. Autodesk's founder, John Walker, applied essentially the same principles as Microsoft and ended up with a dominant share of the market for desktop CAD software. Walker's understanding was plainly naive, but it was no less effective for that. In each case, an intuitive understanding of increasing-returns economics enabled a firm to dominate its market. Though this hypothesis remains to be proved, it is probable that a majority of the successful firms had a similar vision.

Popular writing on success in the personal computer industry often appeals to the idea that luck—being in the right place at the right time with the right skills—was a major factor. This is a theme suggested by the title of Robert Cringely's excellent and entertaining book *Accidental Empires*, for example.[73] Luck did indeed play a big part. To a degree, understanding increasing-returns economics was luck—the practitioners who invented or stumbled across this way of doing business prospered, whereas others who held to some different belief (say, that high-quality software or low-cost packages would lead to market success) bet on the wrong theory and were weeded out in the Darwinian evolution of the industry. However, Microsoft had more luck than just picking the right theory. The rise of the Windows operating system led to Microsoft's ultimate domination of the most lucrative sector of the personal computer software industry: office productivity applications. Microsoft's principle competitors, Lotus and WordPerfect, bet their futures on the OS/2 operating system, while Microsoft bet on a horse in its own stable: Windows. When Windows 3.0 took off, Microsoft's Word and Excel were the only products for the new platform. This was a turning point for Microsoft. As Word and Excel came to dominate their markets, Microsoft became simply unstoppable. Some writers have suggested that Microsoft somehow conspired to make this happen. The evidence suggests that, if it did conspire, it did not do a very good job: Windows succeeded only after its third release, and Microsoft's word processor and spreadsheet had been gestating for a decade. More likely, persistence paid off.

Microsoft has certainly been ruthless and aggressive in its business practices, and the details of this make up most of the narratives of the many popular histories of the company. However, this book has shown that Microsoft's naked ambition has not differed markedly from that of other successful software firms, such as Computer Associates and Oracle.

The story of Microsoft in the second half of the 1990s is dominated by the antitrust suit filed by the US Department of Justice. Exactly why the action was brought remains a mystery. Microsoft's monopolies and abuses do not seem any worse than some of the others described in this book. For example, IBM's CICS has a monopoly in transaction processing comparable to that of Microsoft's Windows in desktop operating systems. But if CICS were to vanish, corporate America would grind to a halt. If Microsoft Windows were to vanish, one of the available substitutes would fill the vacuum within weeks, perhaps within days. Perhaps Microsoft's real offenses are its aggressive, arrogant, brattish behavior and the extreme wealth of its founder. What was permissible, even admirable in a 5-year-old startup was not acceptable in a company that was, by stock-market valuation, the world's largest. Microsoft had failed, as a mature company, to moderate the external signs of the very behavior that led to its success. It may be that a social historian of the future, rather than an economic historian, will explain the context of the case against Microsoft.

Sonic the Hedgehog was introduced to American gamers in 1989.

9
Home and Recreational Software

Home and recreational software is generally understood to mean products that are produced directly for the home environment, primarily for entertainment and lifestyle. Products are typically sold in shrink-wrapped boxes or CD "jewel cases" through retail stores or by mail order. The biggest classes are videogames, CD-ROM multimedia, and personal finance software.

The Software Publishers Association (SPA), the leading trade association for shrink-wrapped software in the United States, recognizes about 30 software categories, running from recreational through word processors and spreadsheets to personal information managers (table 9.1). Some of these categories clearly fall into the domestic (i.e., home) class; others are sold in both domestic and corporate markets. In general, software analysts do not disaggregate corporate and domestic sales. Thus, a sale of Microsoft Word (say) counts for one unit sold, whether it is used in a corporation or in a home. On the other hand, personal finance software is unequivocally domestic and so is ignored by most software industry analysts. (Its principal producer—Intuit, with nearly 5,000 employees and with year 2000 revenues of $1.1 billion—does not feature in the software industry rankings.)

This chapter traces the histories of three of the most important categories of home and recreational software: videogames, CD-ROM encyclopedias, and personal finance software.

Videogames and Computer Games: Early Development and Classification

If one were to list American cultural phenomena of the 1970s, alongside CB radio, hot pants, and *Charlie's Angels* one would have to include the videogame Pong, a game of electronic table tennis that briefly mesmerized America and much of the developed world. It is not an overstatement

Table 9.1
The Software Publishers Association's 1996 data on shrink-wrapped software products.

	Domestic	World	New units	Upgrade units
Recreation	$168,300,000	$230,600,000	14,916,000	NA
Home creativity	$93,100,000	$101,400,000	6,092,000	NA
Home education	$104,500,000	$123,500,000	7,503,000	NA
Reference	$116,500,000	$193,400,000	18,040,000	818,000
Education sold to schools	$91,700,000	$96,200,000	415,000	48,000
Personal finance	$87,200,000	NA	NA	NA
Tax programs	NA	NA	NA	NA
Accounting	$100,800,000	NA	NA	NA
Personal/business productivity	$39,800,000	$41,400,000	1,946,000	89,000
Word processors	$757,100,000	$1,821,000,000	24,047,000	9,715,000
Spreadsheets	$683,800,000	$1,723,100,000	21,902,000	9,300,000
Databases	$310,900,000	$684,700,000	15,347,000	3,868,000
Internet access and tools	$60,100,000	$82,600,000	2,841,000	NA
System utilities	$316,700,000	$393,700,000	14,282,000	1,151,000
Applications utilities	$2,600,000	$11,800,000	549,000	38,000
Electronic mail	$274,400,000	$482,400,000	54,662,000	12,976,000

Communications	$146,700,000	$184,500,000	4,286,000	302,000
Integrated	$91,500,000	$198,600,000	16,467,000	682,000
Presentation graphics	$285,300,000	$615,100,000	23,845,000	9,759,000
Professional drawing and painting	$99,100,000	$247,800,000	2,787,000	237,000
Clips	$14,800,000	$15,700,000	1,297,000	0
Fonts	$8,600,000	$16,400,000	329,000	NA
Desktop publishing	NA	NA	NA	NA
Forms	$36,100,000	$41,100,000	1,605,000	142,000
Programming languages	$136,700,000	$412,700,000	2,663,000	179,000
Programming tools	$81,500,000	$161,800,000	677,000	105,000
Project management	$129,000,000	$222,200,000	649,000	195,000
Time managers, PIMs	$86,500,000	$148,000,000	23,975,000	3,917,000
Training	$83,000,000	$117,100,000	816,000	NA
General business	$43,700,000	$69,600,000	1,154,000	383,000
Total	$4,687,900,000	$8,947,800,000	282,543,000	56,267,000

Source: SPA Worldwide Data Program.

to say that Pong, produced by Atari, was the springboard for today's vast computer entertainment industry.

By the early 1980s, a multi-billion-dollar videogame industry was worthy of the attention of government and industry analysts.[1] Analysts divided the industry into three sectors, distinguishing among games played on arcade machines, those played on home videogame consoles, and those played on home computers. Coin-operated video arcade machines appeared in 1972, displacing traditional pinball and slot machines in "penny arcades" and in bars. Home videogame consoles, which appeared in 1975, offered essentially the same features as arcade machines, but at much lower cost. By the early 1980s, rare was the American home with a teenage son that did not own a videogame console. Home computers were first sold in 1977, and most of them were used to play games.

Unlike the early corporate software artifacts discussed in this book, most of which have evaporated into the mists of time, early videogames have been well preserved through the curatorial efforts of private citizens and preservation societies. Most of these custodians, now in their thirties and their forties, were teenagers in the 1970s and the 1980s. Enthusiasts trawl garage sales for video consoles and games, and some have impressive collections. On the Internet, one can obtain a simulator and software that will enable the owner of a present-day personal computer to experience just about every game ever produced for a popular videogame system.[2]

Arcade Games

Most of the videogame's pre-history took place in university computer laboratories in the 1960s, when enthusiasts developed games that ran on mainframes and minicomputers equipped with graphical displays. Some of these games, including Space War or Lunar Lander, spread like wildfire and were played in hundreds of university computer science departments and computer-equipped research laboratories.[3]

It was at the University of Utah, in 1965, that Nolan Bushnell, who would go on to found Atari, first encountered a computer game. By his own account, this sowed the seed for an arcade game. Upon graduating, Bushnell got a job as a computer engineer in California.[4] By 1971, the cost of electronics had fallen to a level where it was practical for Bushnell to develop an arcade-style game, which he called Computer Space. The entrepreneurial and charismatic Bushnell sold the idea to Nutting Associates, a major producer of old-style arcade games. Though about 1,500 units were manufactured and installed, Computer Space was less successful from a commercial viewpoint than the average pinball machine. Bushnell concluded that its lukewarm reception was due to the complexity of the game. Developing even modest expertise required a lot of exploratory play

In 1972, Bushnell and a partner developed Pong, a simple and intuitive table-tennis-like game whose only instructions to players were "AVOID MISSING BALL FOR HIGH SCORE." A prototype machine was installed in a bar called Andy Capp's in Sunnyvale, California. What followed has become one of Silicon Valley's most enduring legends. Within 48 hours, the machine broke down and required a service call:

> [Atari's engineer] opened the front panel of the machine with his key and threw the credits switch on the coin mechanism, which allowed him to play without paying. The game worked perfectly. He deduced immediately that the problem was not in the circuitry but that the coin mechanism was broken. He released the latch on the coin mechanism. Inside, the maze the coin traveled down looked like an ant colony clogged with quarters. The game wasn't broken; the sawed-off plastic milk container they were using as a coin box had to be emptied.[5]

Despite this hint of Pong's ultimate revenue potential, Bushnell was unable to obtain committed enthusiasm (or, more to the point, generous royalties) from Nutting Associates or any other arcade machine producer. He therefore decided to manufacture the machine directly. He secured venture capital, and volume production was begun in 1973. Pong was a huge novelty in arcades and bars throughout the United States. Ten thousand machines were eventually delivered—three times the normal production run for a pinball machine. The technology of Pong was easy to imitate, and Bally Midway and Allied Leisure (two major producers of arcade machines) quickly introduced competing games. By 1974 there were 100,000 Pong-type machines around the world.[6] They largely displaced pinball machines, diverting the flow of coins from an old technology to newer one without much increasing the overall take.

In 1978, however, the Japanese arcade game manufacturer Taito introduced Space Invaders, a new videogame that had the effect of dramatically expanding the industry: "With Space Invaders, videogames evolved from a friendly game of Ping-Pong into a violent life-and-death struggle. This was when kids started blowing their allowances and lunch money in the arcades; when they started missing classes, skipping their homework, and learning to kill things dispassionately. Kids stopped listening to music, reading books, and playing sports. Space Invaders was a new form of entertainment that appealed to kids who had not been exposed to the old forms of entertainment."[7] The devotees of Space Invaders were 90 percent male and 80 percent teenage,[8] and the new aggressive form of play attracted the attention of Sherry Turkle and other social theorists.

By the early 1980s, the arcade game industry, a backwater since its heyday in the 1930s, quadrupled in size to become a global business dominated by American and Japanese firms. Annual sales of arcade machines peaked at about $2 billion in 1982. There is no corresponding measure for the income generated by the operation of videogame machines in arcades, although there is anecdotal evidence that a typical machine costing $5,000 paid for itself in about 6 months. Some of the early videogame machines are still in use.

Videogame Consoles

Like the electronic calculator and the digital watch, the home videogame console was made possible by the unprecedented reduction in the cost and the physical size of electronics that occurred in the 1970s.

Though Atari was not the first entrant into the market for home videogames, it was overwhelmingly the most successful entrant, and it shaped the industry. In 1974, Atari designed a home version of Pong. Intended to retail for about $100, it used an ordinary TV receiver to display the action. How to market Pong to the American public through retailers, which tended to be conservative, was not obvious. In a canny piece of negotiating, Nolan Bushnell persuaded Sears to distribute Pong on an exclusive basis through its 900 retail stores and its mail-order operation. In 1975, Sears sold 150,000 units.

In 1976, the semiconductor firm General Instrument introduced a single chip with four tennis-like games and two shooting games programmed into it. That $5 chip enabled any toy manufacturer, experienced in electronics or not, to produce a Pong clone. By the end of 1976, there were 75 Pong-type videogames on the market, and they were being produced in the millions for a few dollars apiece. Innovators in search of fresh profits rapidly turned out new games. Within months, clones of the latest videogames would appear and prices would plummet.

By 1978 the fad was on the wane. Every videogame ultimately lost its novelty and found its way to the back of the toy closet. The solution to this problem was a general-purpose videogame console for which games could be purchased separately. The difference was similar to that between the music box and the phonograph. As envisioned in the mid 1970s, the videogame console would be a general-purpose microcomputer dedicated to the task of playing games. Games would be supplied as plug-in memory modules. Such a system would be quite distinct from the personal computer, which was emerging at about the same time. The personal computer would be user programmable and would be equipped with a keyboard, a cassette or disk storage device, and the ability to control a printer. In contrast, the videogame console would be essentially a black box to which nothing more than a joystick, a TV monitor, and a videogame cartridge could be attached. In 1976 and 1977, besides Atari, several videogame, toy, and consumer electronics firms, including Coleco, Mattel, Magnavox, RCA, and Fairchild, developed consoles. Although Atari was not the first to market, its VCS 2600 videogame console was the overwhelmingly market leader, eventually achieving a 75 percent share.

Recognizing that developing the VCS 2600 would require a tremendous injection of capital, Bushnell sold Atari to Warner Communications in 1976 for $28 million. The new owners injected $100 million into product development. The VCS 2600 was launched in October 1977. The Atari console, like its competitors, sold for about $200 retail, as something of a loss leader. Profits came from the sale of game cartridges, which sold for $30 while costing only $10 to manufacture. Atari announced nine titles when the VCS 2600 was introduced.

At first, games were produced only by the console manufacturers. There were three barriers to entry for third-party software developers. First, the programming codes of the early consoles were proprietary knowledge; they were not disclosed to competitors. Second, unlike home computers, which could be programmed by users, videogame consoles required a costly development system to write and test software. Third, the cartridges required significant manufacturing investment and production runs of several thousand units. Small developers of entertainment software therefore gravitated to the home computer.

One of the few exceptions was Activision, founded in March 1980 by a group of developers from the Atari Division of Warner Communications who had become disaffected with the corporate management style and wanted a more congenial and rewarding working environment. Activision had proprietary knowledge of the Atari console and was able to attract $700,000 in venture capital. In its first year, Activision published four hit titles and achieved sales of $65.9 million.[9] The sales of individual hit games could be stratospheric—Activision's Pitfall! sold 3.5 million copies, for example. Activision was quite exceptional, however. Much more typical of the third-party cartridge developers were established media firms and toy manufacturers such as CBS, Disney, Lucas Films, Parker Brothers, Imagic, and Mattel.

Videogame software cartridges, largely ignored by software industry analysts, became a huge business. It was estimated that five games were sold each year for every one of the millions of videogame consoles in existence. In 1982, when the videogame market peaked, nearly 8 million consoles and 60 million cartridges were sold in the United States (table 9.2). The retail value of the videogame cartridges sold in 1982 was $1.5 billion, several times the retail value of the personal computer software sold that year.

Games for Home Computers

In the late 1970s, personal computers were much more expensive than videogame consoles, and less well defined. As late as 1981, videogame consoles outsold home computers by more than 10 to 1.

Whereas videogame consoles were bought for a single purpose, home computers were bought for a variety of uses: for home offices, for small businesses, for educational purposes, and so on. But in practice most home computers ended up with game playing as their principal use. Games accounted for about 60 percent of home computer software sales.

In the early 1980s, as the price of home computers fell to as little as $200, there were signs of a convergence with videogame consoles. Thus, home computers such as the Tandy TRS 80 Color Computer, the Commodore Vic-20, and the Texas Instruments 99/4 were increasingly aimed at the home entertainment market, while videogame console manufacturers began to produce inexpensive home computers (e.g., Atari's 800 and Coleco's Adam). The most successful home computers

Table 9.2
Retail sales of video game consoles and home computers, United States, 1981–1983.

		1981	1982	1983
Videogame consoles	Units	4,620,000	7,950,000	5,700,000
	Value	$577,000,000	$1,320,000,000	$540,000,000
Videogame cartridges	Units	34,500,000	60,000,000	75,000,000
	Value	$800,000,000	$1,500,000,000	$1,350,000,000
Home computers	Units	360,000	2,261,000	5,027,000
	Value	$393,000,000	$1,400,000,000	$1,900,000,000
Entertainment software	Value	$18,000,000	$157,000,000	$405,000,000

Derived from pp. 3, 7, and 45 of Yankee Group, *Software Strategies*.

sold in the millions, spawning huge user communities, mostly young male hobbyists. There was a vibrant newsstand literature, with titles such as *Power/Play* (for Commodore users), *99 Home Computer Magazine* (for users of the Texas Instruments 99/4), *Antic* (for Atari 800 users), and *Hot CoCo* (for users of the TRS 80 Color Computer). These magazines typically sold 50,000–150,000 copies a month.[10]

As was noted earlier, a crucial differences between a videogame console and a home computer were that the latter could be programmed by the user and that it came equipped with a keyboard and secondary storage. Thus, barriers to entry into software development for home computers were almost non-existent. No additional software development system was needed, there were no proprietary trade secrets to unlock, and programs could be duplicated on the computer itself, with no need for access to a third-party manufacturing plant. The lack of significant barriers to entry led to the phenomenon of the "bedroom coder." Thousands of would-be software tycoons began to write games in their spare time, selling their programs through small ads in computer magazines. The typical game cost $15 and consisted of a smudgy, photocopied sheet of instructions and a tape cassette or a floppy disk in a plastic bag. Most of the programs were disappointing to their purchasers.

From these thousands of entrants, about twenty major players emerged, including Adventure International, Infocom, Broderbund, and Sierra On-Line (table 9.3). Most of them had remarkably similar trajectories. Typically, the founder—usually a man in his late twenties or early thirties—acquired a home computer and had the skill and luck to write a hit game that sold a few thousand copies by mail order. The entrepreneur then sustained the initial success by investing in packaging, advertising, and distribution. Parlaying early success into a long-term business required intuition, entrepreneurial talent, and access to venture capital. Those who succeeded were in no sense "accidental millionaires." Of these success stories, that of Broderbund is perhaps the best documented.[11]

Broderbund was founded in 1980 by Douglas Carlston, a 30-year-old self-employed lawyer in Maine. Carlston had bought a Radio Shack TRS-80 computer for his legal practice in 1978. Instead of automating his office, he got sidetracked into developing a game he called Galactic Empire. Carlston sold the program through a videogame distributor for $15 retail, on a 20 percent royalty basis. Few entertainment titles were available at the time, and Galactic Empire sold well, netting Carlston several thousand dollars. In 1980 he decided to abandon his legal career for that of a software entrepreneur.

Table 9.3
Leading producers of entertainment software for home computers, 1983.

Firm; location; founders	Date founded	Number of employees	Revenues	Best-known games
Adventure International; Orlando, Fla.; Scott Adams (b. 1952)	1978	50	c. $10,000,000	Adventure Pirates Adventure Commbat Preppie I and II
Infocom; Cambridge, Mass.; Joel Berez (b. 1954)	1979	50	$6,000,000	Zork Trilogy
Broderbund; San Rafael, Calif.; Doug Carlston (b. 1947)	1980	60	$12,000,000	Choplifter Lode Runner
Datamost; Chatsworth, Calif.; David Gordon (b. 1943)	1981	65	c. $12,000,000	Snack Attack Cosmic Tunnels Super Bunny Market Mogul Ankh
Sierra Online; Coarsegold, Calif.; Ken Williams (b. 1953) and Roberta Williams (b. 1954)	1980	93	$10,000,000	Wizard and the Princess Frogger Dark Crystal
Synapse; Richmond, Calif.; Ihor Wolosenko (b. 1948) and Ken Grant (b. 1947)	1980	100	$6,000,000	Blue Max Shamus Zaxxon

Source: Levering et al., *The Computer Entrepreneurs*, passim.

Thus far, Carlston's story was far from unique. He recognized early, however, that to sustain a videogame business he would need a constant stream of titles for all the popular platforms and an effective distribution network. In the summer of 1981 therefore moved his operation to San Rafael, California, to be close to America's best concentration of programming talent, software distribution channels, and venture capital. He quickly secured funding of $1 million, which he used to pay royalty advances to software authors, to purchase rights to Japanese arcade games, and to invest in software development systems, cartridge and disk duplicating equipment, and packaging machinery. By 1983, Broderbund was a broad-based publisher of videogames for home computers and consoles, with annual sales of $12 million.

The Crash of 1983

By about 1982, there was every indication that home computers and videogame consoles were converging, and that the home computer, augmented with a joystick or paddle controllers, would supplant the videogame console. This would likely have happened had the entire videogame industry not gone into recession in 1983.

At its peak, in 1982, the size of the videogame industry in United States was estimated at between $3 billion and $5 billion. The videogame crash began during the 1982 Christmas season, though it did not become apparent to analysts until early 1983. There were three main reasons for the crash. First, the market for home systems was approaching saturation. The installed base of videogame consoles and cheap home computers in United States was in excess of 15 million units (3 times the number of personal computers at the same date). Second, there had been an explosion in the number of computer games on the market—for example, whereas the Atari VCS 2600 had started out with just nine games, there were now several hundred. Most of the games were derivative or plain unexciting, and ordinary buyers, unable to distinguish the good from the dross, stopped buying games altogether. Third, retailers who had overstocked in anticipation of big Christmas sales cut their prices by one-half or even two-thirds in the new year to move the unsold stock. This had the unintended consequence of undermining the market for new, full-price videogames. Consumers, faced with choosing between a new $30 videogame about which they knew nothing and a six-month-old $10 game about which they also knew nothing, opted for the latter.

Sales of videogame consoles fell from 7.95 million in 1982 to 5.7 million in 1983, while the average price fell from $166 to $95. Sales of videogame cartridges fell less precipitously, from about five per machine per year to a little under three. Atari, the industry leader, was the firm damaged worst by the crash. Sales of its VCS 2600 fell from 5.1 million units in 1982 to just 3 million the next year. Atari's new VCS 5200 console and 800 home computer fared no better, selling in the hundreds of thousands rather than in the millions. In March 1983, Warner Communications decided to cut its losses and sell Atari. The video console and home computer division was sold to Commodore Business Machines, the arcade videogame division to the Japanese firm Namco.

Arcade machines went out of favor at the same time. In 1978 there had been four major producers in the United States; now there were 20 American and Japanese players, often with decidedly lackluster offerings. The declining quality of games caused the teenage denizens of amusement arcades to look elsewhere for entertainment, and the arcades' proprietors had to find new customers. As J. C. Herz explained in *Joystick Nation*, dingy, matte-black-painted amusement arcades were swiftly transformed into brightly painted "family entertainment centers," and the gangs of scowling adolescents "lured to glowing videogames in darkened rooms like mosquitoes to a bug zapper" were replaced by families playing Skee-Ball, an electro-mechanical game harking back to before World War II.[12] Arcade videogames continued to be manufactured, but their days as a cultural phenomenon and a license to print money were gone forever.

Videogames Redux, 1983–1995

The year 1983 was a low point for the videogame industry. Once-thriving magazines such as *Videogaming Illustrated, Computer Fun,* and *Electronic Games,* which had boasted circulations as high as 250,000, folded.[13] The demise of Atari, once the industry's biggest player, was described in depressing detail in a rush-released 1984 book titled *Zap!* According to one source, only 5 or 6 of the 135 significant videogame software companies survived the downturn.[14] The videogame industry recovered, but it took about 5 years to get back to its 1982 size.

It was almost universally agreed that the principal cause of the videogame crash was the flood of low-quality games that saturated the market in late 1982.[15] During the subsequent recovery, the likelihood of this happening again was lessened by two developments. First, the attrition in the number of videogame software firms stemmed the flow of second-rate games. Second, a new generation of videogame consoles incorporated technical mechanisms that gave their manufacturers control over game production and game quality. This development, introduced by Nintendo, was subsequently adopted by Sega and Sony. From the mid 1980s on, though Japanese manufacturers dominated videogame console production, videogame software remained highly internationalized, with producers in the United States, in Japan, and in Europe.

Managing Risk in the Videogame Software Industry
The product life cycle of videogames differed greatly from that of conventional personal computer software. Even a hit game usually had a market life of less than a year, whereas a word processor or a spreadsheet program had an indefinitely long life and provided a constant source of upgrade fees. Hence, the appropriate business model for a producer of videogames was closer to that of the recorded-music industry than to that of the personal computer software industry. As in the music industry, there was a need for a constant stream of titles, most of which needed to at least break even financially, while the occasional hit provided spectacular but unpredictable profits. Relative to personal computer software, there were few lock-in effects to prevent a good game from becoming successful. Thus, while the personal computer market could bear no more than a few word processors or spreadsheet programs, the teenage videogame market could support an indefinite number of programs in any genre. In this respect, videogames were, again, more like recorded music or books than like corporate software, and they were promoted by similar means: reviews in specialist magazines and word of mouth.

Videogame software was a high-risk business. Probably the first conscious risk-management strategy in the industry was genre publishing. From their earliest days, videogames had been classified as "racin', fightin', or shootin'." With time, more complex classification systems emerged in the industry and in gaming magazines (table 9.4). No two classification systems were quite alike, and many games were not easy to categorize. (Was the ubiquitous Tetris a puzzle, or a strategy game? Was SimCity a simulation, or a strategy game?) In any case, precise classification was an academic problem. The point of genre publishing was that certain categories of game appealed to large groups of users and therefore had less need to overcome market inertia.

The most successful games became subgenres, and transferring hit games to all the available gaming platforms was a major source of revenue. Games were also rewritten for "next-generation" consoles, which provided a richer gaming experience through improved technology (e.g., enhanced 3-D displays). The most popular videogame characters and

Table 9.4
J. C. Herz's phylogeny of video games. (Publication dates represent when games were introduced in the US. All games listed are currently in production—not necessarily with the original publisher, and usually with a minor title change to indicate the latest versions—e.g., SimCity 2000, Tiger Woods PGA Golf 2001.)

Genre	Classic examples (publisher, year introduced)
Action "The largest phylum in videogamedom, comprising most of the home console universe and virtually all arcade games; these games are also known as 'twitch' games and 'thumb candy'."	Mario Bros. (Nintendo, 1985) Sonic the Hedgehog (Sega, 1985) Prince of Persia (Mindscape, 1989)
Adventure "Adventure games are about accumulating an inventory of items that are then used to solve puzzles."	Zork (Infocom, 1980) Legend of Zelda (Nintendo, 1986) Myst (Cyan, 1993)
Fighting "Fighting games are basically comic books that move, pitting steroid-enhanced combatants against each other in hyperkinetic two-player death matches."	Street Fighter (Capcom, 1987) Virtua Fighter (Sega, 1993) Tekken (Nameco, 1995)
Puzzle "Logic exercises... usually in the form of colorful, two-dimensional, grown-up building blocks that stack in deceptively simple ways."	Tetris (Spectrum Holobyte, 1986)
Role playing "Role-playing games [RPGs] unfold as long, drawn-out epic narratives involving a band of distinct characters that travel, fight, and plunder as a team."	Ultima (Origin Systems, 1980) Wizardry (Sir-Tech, 1981) Final Fantasy (Square, 1987)
Simulation "Most simulations are first-person, pseudo-VR vehicle sims—planes, tanks, helicopters, nuclear submarines, manned space missions."	Flight Simulator (SubLOGIC, 1978) SimCity (Maxis, 1989) Gran Turismo (Sony Computer Entertainment, 1998)
Sports "A combination of action and simulation, sports games are recognizably distinct enough to be classed as a separate category."	PGA Tour Golf (Intellivision, 1983) NHL Hockey (Intellivision, 1983)
Strategy "In strategy games... the game play and possibly even the goals of the game are abstract ...just like in a card game or a board game."	Populous (Electronic Arts, 1989) Railroad Tycoon (Microprose, 1990) Civilization (Microprose, 1991)

Quotations and classification from pp. 24–31 of Herz, *Joystick Nation*; data on classic games from various web sites.

themes, such as Mario Bros. and PGA Tour Golf, provided opportunities to repackage the same basic themes. There are several dozen games that have been though many such iterations—for example, the role-playing game Ultima, first released in 1980, is now in its eight incarnation.

As with movies and recorded music, one way of increasing the likelihood of a game's becoming a hit was to improve the "creative inputs," particularly by employing "star" programmers.[16] One of the first firms to explicitly invest in creative inputs was Electronic Arts, formed by the entrepreneur Trip Hawkins in 1982.[17] Hawkins consciously adopted what was subsequently called the Hollywood model of game production: "The business was more than simply treating programmers as artists—as creative people. It would be more accurate to say that we brought a methodology for managing a creative process to what had traditionally been an engineering methodology. This translated into a certain style of recruiting, managing, and rewarding creative people. It also translated into a production process methodology that more consistently, like a cookie cutter, cranked out good titles and products."[18]

In the Hollywood model, a "producer" was responsible for managing the creative process from concept to product. Besides programmers, the creative team included video layout artists, sound and music directors, and script editors. The new creative process was much more capital intensive than the old. A side benefit of the increased capital intensity was to erect barriers to entry that reduced the number of players in the industry and marked the end of the bedroom coder. This trend accelerated in the 1990s, with as many as 50 "creatives" working on a single game, development times on the order of 12 months, and budgets of several million dollars.

Within genre publishing, another way of reducing risk was by means of licensing—the videogame equivalent of Hollywood's "bankable star" system. Sports and movie tie-ins were the most common form of licensing. The Professional Golf Association,[19] the National Baseball League, FIFA, and the Olympics Committee all endorsed videogames. Movie tie-ins were a rich source of potential hits. *Star Wars, Teenage Mutant Ninja Turtles, E.T., The Hunt for Red October, Robocop,* and *Batman* all had videogame derivatives, as did many Disney productions.

Licensing of arcade games was also successful. Costing upwards of $10,000, arcade machines used more advanced technology than home consoles. However, as soon as home consoles caught up with the technology—usually within 2 or 3 years—there was a ready market of teenagers who, having experienced the arcade version of a game, were keen to purchase the home edition. In this way, video arcades were a proving ground for the home videogame industry—hit games had an afterlife of many years, while the failures faded away. Some popular examples from the early 1980s are Wizard of War and Gorf (both licensed by CBS Entertainment from Bally Midway). Later, the Mario Bros. and Sonic the Hedgehog made the transition from arcade machines to home consoles.

A major problem for the big videogame producers was maintaining a steady flow of good games, since in-house creative teams would sometimes lose their touch for no discernible reason. Electronic Arts became a publisher and distributor of games from third parties in order to fill out its catalog and maximize the utilization of its distribution network. In the 1970s videogame publishers had used conventional toy wholesalers and distributors, but Electronic Arts built up it own sales force and distribution network, dealing directly with major retailers such as Toys R Us and Wal-Mart and with gaming chains such as Electronic Boutique. By controlling the entire supply chain and delivering above-average products, Electronic Arts quickly gained an edge over wholesalers who merely shipped unappraised products. This trend accelerated in the 1990s with the emergence of a two-tier industry, one tier consisting of small and often anonymous "code shops" and the other of publishers and game manufacturers. There has been relatively little acquisition. Electronic Arts and Activision, the two biggest players, have acquired very few firms in 20 years. In contrast, the leading firms in the corporate and personal computer software have typically made a dozen or more acquisitions.

The Rise of the Japanese Videogame Console

After the shakeout of 1983, there were only three US producers of home consoles (Atari, Coleco, and Intellivision) and perhaps two dozen significant videogame software firms. Because profit margins had become so small, there was little incentive to innovate in either hardware or software. Japan did not experience a comparable downturn and continued to innovate.

In the 1980s, the biggest name in the Japanese videogame industry was Nintendo, an arcade machine manufacturer that introduced its first videogame console in late 1983.[20] Known as the Famicom (for "family computer"), Nintendo's machine was a major advance over any previous console. The advance was achieved by the use of custom chips (originally designed for Nintendo's arcade machines) instead of off-the-shelf components. For example, the Nintendo machine used a "picture processing unit" that enabled it to display moving images of far higher quality than previously. The Famicom was a best-seller in Japan during 1984, with Nintendo reportedly shipping 2.5 million consoles and 15 million cartridges.[21]

The success of the Famicom was due in part to its excellent performance and in part to the fact that the arcade game Mario Bros. was bundled with it. The brothers, Mario and Luigi, were cartoon plumbers who had attained great popularity in a Nintendo arcade game during 1982 and 1983. A sequel, Super Mario Bros., was published in 1985 for both arcade machines and the Famicom. Because the Famicom and the arcade machine used similar hardware, the player experience was essentially the same on either system (an unprecedented achievement), and many arcade gamers bought a Famicom simply to be able to play Super Mario Bros. at home.

In 1985, Nintendo decided to launch the Famicom in the United States and Europe under the name Nintendo Entertainment System (NES). A copy of Super Mario Bros. came bundled with the NES, and it was a sensation among teen and subteen males everywhere. In 1986, Super Mario Bros. 3 reportedly grossed $500 million, more than any Hollywood film apart from Spielberg's *E.T.* Had Nintendo (the creator and publisher of the Mario Bros.) netted 50 percent of sales, it would have ranked in the top ten software firms by revenue on the basis of that game alone.

It is fair to say that Nintendo was solely responsible for the renaissance of the worldwide videogame industry. However, this had less to do with Mario Bros. as a "killer app" than with Nintendo's control of the software market for the NES. When the console was being designed (around 1983, at the height of the videogame crash), it was recognized that it was necessary to control the supply of third-party software. This was achieved by incorporating inside each game cartridge a chip that acted like a key, unlocking the NES's circuits. Only Nintendo-manufactured cartridges had the chip. Although the chip was protected by worldwide patents, the innovation was doubly effective because it would not have been economically feasible for a third party to clone the chip, legitimately or not.

Many third-party developers were encouraged to produce games for the NES, but Nintendo was the sole publisher of games for its console. This controversial practice lay somewhere between benign intervention and restraint of trade. On the benign side, the practice did have the effect of eliminating poor-quality games, and the "Nintendo Seal of Quality" on each cartridge was a guarantee that all NES games met a high standard. Nintendo artificially restricted the supply of new games, permitting no more than five titles a year from any one publisher; it also determined the selling price and took a 20 percent royalty. Most restrictively, Nintendo required games to be "exclusive" and not available on any other platform.

In February 1986, when the Nintendo console was launched in the United States at a price of $249,[22] it was widely presumed to be a loss leader. Nintendo cartridges, which cost perhaps $10 to manufacture, retailed at $58. In 1987, Nintendo reportedly sold 3 million consoles and 15 million cartridges worldwide; only 6 million of the cartridges were for games developed by third-party suppliers.

A second Japanese player, Sega, entered the US home video console market in 1986. Sega, already established worldwide as an arcade game manufacturer, had produced a home entertainment system that competed successfully with Nintendo's Famicom in the Japanese market. The console, launched in the United States in 1986 as the Sega Master System, was not particularly successful, perhaps because of its late entry and perhaps for lack of an attraction to compete with Mario Bros. While Nintendo sold 1.1 million consoles in the United States in 1987, Sega sold only 125,000 (and Atari only 100,000). By the late 1980s, it was estimated that Nintendo had 85–90 percent of the videogame market worldwide. In 1989, Nintendo launched a $100 hand-held videogame console, the Game Boy, which reportedly achieved sales of 5 million consoles and 20 million cartridges in its first year. In its peak year, 1992, Nintendo produced 170 million cartridges at an average retail price of $40,[23] and NES consoles and cartridges were the best-selling children's toys, reportedly accounting for 19 percent of sales in Toys R Us stores.

By the early 1990s, there was a growing concern in the videogame industry about Nintendo's lock on the market. In response to a number of lawsuits, the company relaxed its exclusivity clauses for third-party developers and various other discriminatory practices. In any case, Nintendo's time in the sun was coming to an end with the arrival of 16-bit home consoles and Sega's secret weapon: Sonic the Hedgehog.

Sega introduced its $200 16-bit Genesis console in 1989. Besides the anticipated improvement in computational speed and visual experience, the Genesis came with Sonic the Hedgehog, Sega's answer to Mario Bros. Visually, Sonic was a blue-furred amalgam of Felix the Cat and Superman. Sonic was a worldwide sensation, eclipsing even Mario Bros. and bringing in his wake animated cartoons, comics, T shirts, and lunchboxes.

Though Nintendo introduced its 16-bit Super NES within a few months of the introduction of Sega's Genesis, it lost its dominant market share as the new 16-bit consoles replaced the old 8-bit models over the next 2 or 3 years. Sega achieved an 80-plus-percent share of new sales. Even the fact that the Super NES was compatible with the original NES, so cartridges for the original machine would continue to function, had little influence on users' purchasing decisions. Videogamers were less interested than personal computer users in backward software compatibility or in protecting their investment in cartridges. Games tended to become "played out" within a few weeks or months and to be consigned to a closet or a garage sale.

By the mid 1990s, 80-plus-percent dominance by a single platform—characteristic of the personal computer market—was also apparent in the videogame industry. But whereas it had taken about 7 years for the personal computer market to switch to a dominant installed base of IBM-compatible PCs, in the videogame world a similar switch occurred in about 2 years. Game consoles were much cheaper than personal computers, and there was little residual value in the software. Another change of dominant platform occurred in 1995 with the launch of the Sony PlayStation, the first 32-bit console.

More than any previous console, the Sony PlayStation was strategically positioned in the market.[24] Sony was, of course, a hugely experienced and innovative consumer electronics giant. It had introduced the ubiquitous Walkman in the 1970s.[25] After failing to make the Betamax VCR the market leader in the 1980s,[26] Sony recognized the need to support any new platform with content. Hence, before introducing its audio CD players in the 1980s, it created Sony Music, which then bought up several worldwide music publishers, including CBS Records.

The 32-bit technology of the PlayStation was a step-function improvement over the 16-bit consoles. This was, however, recognized within Sony as insufficient to ensure success against the Nintendo and Sega 32-bit consoles that were also under development. Three things were perceived as crucial to the PlayStation's success: fostering third-party software developers, reducing the manufacturing cost of videogames, and changing the demographics of videogamers.

Unlike Nintendo and Sega, Sony had no in-house experience in game development. It therefore needed to encourage third-party developers to produce games, which it would then publish. Sony provided unprecedented support in the form of development systems and software libraries, thus allowing developers to focus more on the game experience and less on the writing of code. Rather than cartridges, Sony decided to use compact discs, which would enable far more video, sound, and game content to be incorporated in a game. The use of CDs gave Sony a major advantage over its competitors. They were cheaper to manufacture than cartridges, and their short lead times permitted almost instant reproduction in response to market demand. Cartridges, besides being much more expensive to manufacture and having long lead times, had fixed production volumes. Marketers of cartridge-based games thus had to make forecasts on the sales of games. The forecasts were notoriously unreliable, often resulting either in huge inventories of unsold games or in the inability to exploit a hit game in a timely way. PlayStation games could be priced as low as half the price of cartridge-based games while yielding comparable profits.

The Sony PlayStation has dominated the videogame market since the mid 1990s. In the late 1990s, it was commonly reported that Sony's entire corporate profits came from the PlayStation Division. According to Sony, 40 million PlayStations had been sold worldwide by September 1998, only 3 years after the product's launch.[27]

Interestingly, for the PlayStation there was no "killer app" comparable to Mario Bros. or Sonic the Hedgehog. Though games such as Virtua Fighter and Tomb Raider (Lara Croft) sold in the millions for the PlayStation, they were also available for other platforms. The PlayStation succeeded because Sony recognized the changing demographics of videogamers. The videogame enthusiasts of the 1980s—the 6-to-14-year-old males who had grown up with Ataris, Colecos, and Nintendos—were now in their mid twenties, with spending power far greater than an adolescent's. Today videogaming is an exploding cultural phenomenon with particular appeal to people under 30. Some authors believe that its long-term significance will one day rival that of movies or recorded music.[28]

CD-ROM Encyclopedias: Content vs. Code

Measuring the recreational software economy is made difficult by the blurred boundaries between intellectual content and programming code. The CD-ROM encyclopedia points up the difficulty particularly well.

A CD-ROM encyclopedia consists of intellectual content (the text and images of a traditional encyclopedia) and program code (which provides the user interface—the ability to search the encyclopedia and to display images and video clips). Some CD-ROM encyclopedias (e.g., Microsoft's *Encarta*) are produced by software companies; others (e.g., the *Encyclopaedia Britannica*) are produced by traditional publishers and sold as "new media." *Encarta* constitutes a small part of Microsoft's annual revenues and contributes to its ranking in the software industry, whereas *Britannica* on CD-ROM goes largely unnoticed by analysts of the software industry. For analysts of the media industry, the reverse is true. Beneath the somewhat theoretical debate over what constitutes a software good lies an important phenomenon: the convergence of content and program code, of which the CD-ROM encyclopedia is one of the most visible examples.

The CD-ROM (Compact Disk–Read Only Memory) was developed jointly by Sony and Philips in the early 1980s and was first marketed in 1984. Based on the technology of the audio CD, it offered a storage capacity of more than 500 megabytes—several hundred times that of a floppy disk. Although CD-ROM drives were expensive ($1,000 or more) for several years, their capacity gave them the potential to create an entirely new market for computer-enabled content—a market that computer networks would not be able to satisfy for another 15 years.

While CD-ROM drives remained expensive, their use was limited to corporations and libraries, for which a significant professional market for CD-ROM content developed in the mid 1980s. Content providers tended to be established media companies; for example, the magazine *The Economist* launched a CD-ROM version in 1987. However, these productions were more an evolutionary development of microfilm than a new publishing paradigm. Lotus took a more innovative approach in 1987 by creating One Source, an organization for supplying business information that could be manipulated by means of the Lotus 1-2-3 spreadsheet program.

Gary Kildall and Bill Gates played major roles in establishing a consumer market for CD-ROM media. Kildall was the inventor of the CP/M operating system and the founder of Digital Research Inc., once the leading microcomputer software firm. Gates, who needs no further introduction, turned out to have the deeper pockets and better timing. Both Kildall and Gates hit on the idea of the CD-ROM encyclopedia as the means to establish the home market for CD-ROMs. In an interview published in 1989, Kildall said: "We want to develop applications that will give people a definite economic advantage if they buy them. That's why we went for an encyclopedia as the first CD-ROM application. Everyone knows encyclopedias usually cost about $1,000. Someone can rationalize buying a computer that has the encyclopedia, if it's in the same price range as the printed encyclopedia."[29] In 1985, Kildall, still chairman of Digital Research, established and financed a firm called Activenture (subsequently renamed Knowledge-Set) to create CD-ROM content. Kildall secured the CD-ROM rights to the *Grolier Encyclopedia*, a junior encyclopedia established shortly after World War I. The *Grolier Encyclopedia* on CD-ROM was released in 1985, well ahead of any similar product. Priced at several hundred dollars, the Grolier CD-ROM was worthy but dull. It sold well to schools, but it did not induce consumers to invest in CD-ROM equipment.

The failure of the computer-based encyclopedia to start a CD-ROM revolution was largely a matter of marketing and culture. It was a truism of traditional encyclopedia marketing that encyclopedias were sold rather than bought. The successful publishers of encyclopedias, such as *Britannica*, employed large staffs of a door-to-door salesmen who pressured American families into purchasing $1,000-plus encyclopedias to benefit their children's education. An important selling point of the traditional encyclopedia was its heavy imitation-leather binding. A 4-inch CD-ROM, weighing a fraction of an ounce, lacked this attraction, and consumers were considered unlikely to "fork over $1,500 for something that looked like a music album."[30]

Microsoft probably tried harder than any other firm to establish a consumer market for CD-ROMs.[31] Consumers would not purchase CD-ROM drives until there was a rich choice of content—hundreds if not thousands of titles. At a time when so few consumers owned CD-ROM-equipped personal computers, it was not economically feasible for individual publishers to invest in CD-ROM production. Progress was further handicapped by the lack of a common CD-ROM standard.

Though Microsoft probably never did anything without a degree of self-interest, it is a testament to its growing maturity and strategic vision that in March 1986 it held an industry-wide conference for CD-ROM producers and publishers. Gary Kildall was invited as the keynote speaker—perhaps out of genuine appreciation of his pioneering work, perhaps in an attempt to demonstrate Microsoft's disinterest in controlling the agenda. Several hundred delegates attended. Further conferences were held in 1987 and 1988, each as well attended as the first. It is not clear that the Microsoft conferences had any effect on the development of the CD-ROM market. There was certainly no instant acceptance of Microsoft's CD-ROM standards, and it was to be 5 years before the consumer market took off in the early 1990s. The most that can be said of the Microsoft conferences is that they probably helped to foster business relationships between media firms and software producers.

A CD-ROM publishing operation was established within Microsoft in 1986. With no in-house capability in literary works (apart from a modest publishing activity in product-related literature), Microsoft sought to acquire content. Its first acquisition was Cytation Inc., a tiny startup that had obtained the CD-ROM rights to the *American Heritage Dictionary*, *Roget's Thesaurus*, the *Chicago Manual of Style*, *Bartlett's Familiar Quotations*, and the *World Almanac*. This set of titles resulted in Microsoft's first CD-ROM product, launched in 1987 as *Bookshelf*. Priced at $295, *Bookshelf* was targeted at corporate and professional users rather than consumers. When it came out, "it and Gary Kildall's encyclopedia constituted the entire catalogue of available titles" for home computers.[32] Microsoft did not break into the consumer market for CD-ROM encyclopedias until 1993, with the introduction of *Encarta*.

Encarta is a remarkable example of Microsoft's ability to dominate a market with a thoroughly lackluster product. Microsoft's decision to produce a CD-ROM encyclopedia was no doubt primarily motivated by the desire to own the killer app for a market that would one day be huge, but it also held an emotional appeal for Gates. He waxed somewhat lyrical, if a little artless, in *The Road Ahead*: "When I was young I loved my family's 1960 *World Book Encyclopedia*. Its heavy bound volumes contained just text and pictures. They showed what Edison's phonograph look like, but didn't let me listen to its scratchy sound. The encyclopedia had photographs of a fuzzy caterpillar changing into a butterfly, but there was no video to bring the transformation to life.... I had a great time reading the encyclopedia anyway and kept at it for 5 years until I reached the Ps. Then I discovered *Encyclopaedia Britannica*, with its greater sophistication and detail."[33]

To establish its encyclopedia, Microsoft needed, as with *Bookshelf*, to obtain the rights to one of the leading American encyclopedias, primarily for content but also for the brand. First, in 1985, Microsoft pursued the rights to the *Encyclopaedia Britannica*, the leader in both content and brand. At this time, Encyclopaedia Britannica Inc., a charitable foundation whose profits accrued to the University of Chicago, operated in the classical mode of direct door-to-door selling to consumers. Indeed, the company regarded its sales force as its principal asset, ahead of the literary property. Of course, the former could not exist without the latter, and so Britannica rejected Microsoft's overture. This was to prove a disastrous decision for Britannica—a decision sometimes compared to IBM's failure to take the threat of the personal computer seriously in the mid 1970s. In fact, in both cases the truth was more complex.

Encyclopaedia Britannica Inc. fully recognized the threat of the CD-ROM and, wary of cheapening *Britannica*'s reputation, made its first foray with its lesser-known *Compton's Encyclopedia*. In 1989, about 4 years ahead of Microsoft's *Encarta*, this became one of the first multimedia encyclopedias on the market. *Compton's Encyclopedia* on CD-ROM was highly successful. Indeed, at its peak it was said to own the market, though sales were modest relative those of its printed works.

Spurned by Britannica, Microsoft turned next to the number-two encyclopedia, *World Book*. Microsoft's proposal was rejected by World Book Inc. for much the same reasons as it had been rejected by Britannica. Moreover, like Britannica, World Book was not oblivious to the CD-ROM revolution; it introduced its own product in 1990 (3 years before *Encarta*). The *World Book* CD-ROM was another dull, text-only production. (A multimedia version was released in 1995.)

According to the many histories of Microsoft, the company gradually worked its way down the pantheon of American encyclopedias, finally obtaining the rights to *Funk and Wagnall's New Encyclopedia* in 1989. *Funk and Wagnall's* had an unglamorous but profitable niche as a $3-per-volume impulse purchase in supermarkets. Microsoft's decision to acquire the rights to *Funk and Wagnall's* has often been derided as pandering to the educationally challenged working class. Though this criticism was more than a little tinged by academic snobbery, *Funk and Wagnall's* was indeed a junior encyclopedia of negligible authority. In its characteristic fashion, however, Microsoft proceeded to sculpt this unpromising material into *Encarta*, adding video and sound clips and other multimedia elements. Planned to be launched in time for Christmas 1992, like so many Microsoft products it missed its schedule; it was released in early 1993. Priced at $395, in line with its competitors, it sold modestly.

In fact, the CD-ROM encyclopedia did not turn out to be the killer app that would start the CD-ROM revolution. Quite independently, the price of CD-ROM drives had been falling for several years. By 1992, a CD-ROM drive could be bought for about $200, comparable to the price of a printer or a rigid disk drive. Now there was the possibility of a mass market for CD-ROM media. In 1992 the recreational software publisher Broderbund released, at a price of $20, *Grandma and Me*, which became an international best-seller and probably the best-known CD-ROM title of all time. Thousands of other titles, from hundreds of publishers, soon followed. Publishers of videogames began to release CD-ROM titles with multimedia effects. Prominent examples were Broderbund's *Myst* and Sierra's *Outpost*. Microsoft contributed to this boom, having bought a minority stake in Dorling Kindersley, a UK publisher of children's books, in 1990. In 1992, Microsoft published a film guide titled *Cinemania*.

As with boxed software, price was a significant brake on consumer demand—a fact seemingly little appreciated by CD-ROM publishers until the early 1990s. For the 1993 Christmas season, Microsoft released *Encarta '94* at $99.95, one-third the price of previous editions and well below its competitors' prices. *Encarta* sold extremely well at the lower price, inducing Microsoft to lower the price of *Bookshelf* to $69.99. These titles were bundled with other Microsoft software in consumer PCs, leading to Microsoft's rapid domination of the market for popular consumer reference works. Even more than its conventional software, Microsoft's reference works merited annual upgrades, which produced a long-term revenue stream. In December 1993, Microsoft acknowledged the increasing importance of its consumer products by establishing a separate brand called Microsoft Home.

In the mid 1990s, the "CD-ROM hit parade" featured ephemeral productions (such as videogames and multimedia entertainment) and perpetual best-seller reference works (such as *Encarta* and *Cinemania*).[34] The popular reference works established a price point of $100 or less, which all the encyclopedias eventually matched (table 9.5).

The availability of low-cost CD-ROM encyclopedias had a devastating effect on the market for traditional hard-bound encyclopedias. World Books Inc. saw its sales plunge from a reported 330,000 sets in 1988 to perhaps fewer than 150,000 only 4 years later.[35] Encyclopaedia Britannica

Table 9.5
Principal CD-ROM encyclopedias.

Title and publisher	Date introduced	Launch price	Price in 2000	Age range
Grolier Multimedia Encyclopedia (Knowledge-Set)	1985	—	$59.99	All
Compton's Encyclopedia (Encyclopedia Britannica)	1989	$895	$39.99	All
World Book Multimedia Encyclopedia (World Book)	1990	$700	$89.95	Younger
Encarta (Microsoft)	1993	$375	$99	Younger
Encyclopedia Britannica (Britannica)	1996	$995	$49.95	Adult

Sources of prices in 2000: Tom Ham, "CD-ROM Encyclopedias: A Visual Feast," *USA Today*, January 21, 2000; various other web and print sources.

Inc., rapidly losing sales, reported a $12 million loss for 1991. To ease its cash situation, the company sold off the *Compton's* division to the Chicago's Tribune Company for $57 million and halved its 2,300-person sales force.[36]

Early in the 1996, Britannica finally introduced a multimedia CD-ROM version of its flagship encyclopedia. It had multimedia enhancements, and its content was vastly superior to that of *Encarta*, but at $995 the CD-ROM *Britannica* sold only to corporations and libraries. Printed encyclopedias had never really sold on the basis of content; their sales had been based on intangible prestige and authority—values not highly appreciated in the world of multimedia CD-ROMs. *Forbes* reflected poignantly: "How long does it take a new computer technology to wreck a 200-year-old publishing company with sales of $650 million and a brand name recognized all over the world? Not long at all. There is no clearer, or sadder, example of this than how the august Encyclopaedia Britannica Inc. has been brought low by CD-ROM technology."[37] In 1996, Encyclopaedia Britannica Inc. was acquired by a Swiss-based financier. A new management team was installed, and the price of the CD-ROM *Britannica* was cut to less than $100.

Personal Finance Software: Quicken vs. Money

Until recently, the market for personal finance software was dominated by Intuit's Quicken. Indeed, by the early 1990s Quicken had emerged as the best-selling consumer software product of all time, far exceeding the sales of any other software package, computer game, or multimedia product.[38]

Intuit was established in 1983 by Scott Cook, a former sales executive. Cook's undergraduate education was in economics and mathematics. After receiving an MBA degree from Harvard in 1976, Cook got his formative professional experience at Procter & Gamble, where he spent 4 years as a marketer for Crisco cooking fat—an experience to which he later attributed his strong customer focus.[39] After a 3-year spell in the Menlo Park office of the management consultancy Bain and Company, he resigned to form a software company. Cook credits his decision to go into personal finance software to the experience of seeing his wife struggle to pay household bills.

In 1983, home accounting software was already a well-established software category with dozens of packages on the market, including the market leader, Continental Software's Home Accountant. However, these early packages were fairly described by one analyst as a "residual" category of "make-work" programs for technophiles and financial obsessives.[40]

Scott Cook's main strategy for entering this crowded market was to develop an easy-to-use program with a particularly "intuitive" interface. To complement his marketing know-how, Cook recruited as a founding partner Tom Proulx, a computer science student about to graduate from Stanford University, and gave him the responsibility for writing the program. This was perhaps the last moment when it was still possible for a talented programmer to single-handedly develop a major software product. Cook made overtures to more than twenty venture capitalists for funding, without success. This outcome was no doubt due to Cook's seemingly naive plan to enter a crowded market, in which he had no track record, with an untried product developed by a single programmer. Undaunted, Cook scraped together startup capital of $350,000 from friends and family members and by means of a home-equity loan.

As a latecomer to the personal finance software market, Cook was unable to induce retail computer stores to favor Quicken above Home Accountant and a dozen or so other personal finance packages—including Managing Your Money, a new market leader developed by the finance guru Andrew Tobias and published by MECA. Instead, he cajoled a number of retail banks into placing display stands in their lobbies. Quicken was shipped in December 1984 at a price of $100 a copy. The price was quickly reduced to $50 to stimulate sales. It took about 2 years for Intuit's cash flow to turn positive, and even then sales amounted to only a few thousand units.

Quicken version 2, with new features for managing credit cards and loans, was released in 1987. By then, Quicken had risen above some of the competition to be stocked in retail computer outlets. Cook drew on his background in consumer marketing to get Quicken sold through regular retail outlets such as Wal-Mart and Price/Costco, which had not previously stocked software for personal computers. He advertised Quicken in airlines' in-flight magazines, and he pioneered "that dubious contribution to advertising's Hall of Fame, the infomercial shown to the captive audience of airline passengers."[41] Version 3 arrived in 1989, with additional financial capabilities and ever-improving ease of use.

By the end of 1989, Intuit had 50 employees and sales of $19 million. Quicken's sales continued to accelerate, partly as a result of good marketing but partly on the strength of its ease of use, word of which had been spread by mouth and by carefully orchestrated press reviews. Intuit was probably the first software company to focus on ensuring that a product would be suitable for immediate use, without recourse to training, manuals or telephone help lines. In Intuit's "Follow Me Home" program,

Table 9.6
Intuit financial statistics.

	Revenues	Annual growth	Employees
1989	$18,600,000		50
1990	$33,100,000	74%	110
1991	$44,500,000	34%	175
1992	$83,800,000	88%	484
1993	$121,400,000	45%	589
1994	$194,100,000	60%	1,228
1995	$419,200,000	116%	2,732
1996	$567,200,000	35%	3,474
1997	$649,700,000	15%	3,000
1998	$689,300,000	6%	2,860
1999	$940,400,000	36%	4,025
2000	$1,093,800,000	16%	4,850

Data from *Hoover's Guide,* 1996 and Intuit annual reports.

a staffer accompanied a randomly selected novice purchaser home to study his or her "out-of-the-box experience."

In 1989, Quicken had had only a 15 percent market share; Making Your Money had had nearly 60 percent. In 1990, Quicken was the market leader, at 50 percent; Managing Your Money's share was down to 10 percent and falling.[42]

In the early 1990s, the number of Quicken users (which, being cumulative, was always much greater than the number of new sales) soared to several million. Though most users bought the full product just once in a lifetime, Intuit successfully managed the problem of turning each sale into a permanent revenue stream through annual upgrade fees and stationery supplies. New products were also introduced. The first was QuickBooks (1992), a complete accounting package for very small businesses. Intuit made an IPO in 1993 and used the proceeds to acquire TurboTax (a tax-preparation program for the self-employed, the retired, and the wealthy) from ChipSoft Inc.

By the early 1990s, personal finance software had matured into a mainstream software category that was perceived as strategically important to the future of home banking and electronic commerce. At that time no one could have foreseen or articulated how the future of on-line financial services would unfold; however, with the increasing penetration of home on-line services such as AOL and CompuServe, the personal finance software package had the potential to become a "toll booth" between consumers and financial institutions.[43] As this realization dawned on major players in the software industry, Intuit experienced heavyweight competition.

The most significant competition came from Microsoft's Consumer Division, then known as its Entry Business Unit. With the success of Windows version 3, Microsoft took the opportunity in 1991 to introduce a Windows-compatible personal finance program called Money. Though this tactic had seen Microsoft Excel and Word eclipse Lotus 1-2-3 and WordPerfect, Quicken was little affected by Money—probably because Money was a relatively immature product, only 2 years old. (Word and Excel were 7 or 8 years old.) Even though Money was bundled with Microsoft's other home products, such as Works and *Encarta*, in many consumer machines, few Quicken users switched to the Microsoft product. The reluctance of Quicken users to switch brands was probably due to a combination of loyalty and the lock-in effect of a user's having created a database of personal financial information over a period of years. However, as with its other forays into new software categories, Microsoft accepted short-term losses while continuing to invest in and perfect Money.

In 1993, Computer Associates, then the largest software company in the world, entered the personal finance software market, launching a product called Simply Money. In an attempt to build market share rapidly, Computer Associates distributed a million trial copies of Simply Money free of charge (apart from a $6.95 handling fee) through an 800 order line. However, as with Microsoft Money, there were few switchers. In 1994, the accountancy firm H&R Block entered the personal finance software market by acquiring MECA Software, the publisher of Managing Your Money. Under the new ownership, MECA's staff grew from 80 to 360, but Managing Your Money remained a distant third behind Quicken and Money. Within a few years, Simply Money and Managing Your Money were abandoned by their new owners. Simply Money, a victim of Computer Associates' inability to market personal or consumer software, was sold to 4Home Productions and relaunched as Kiplinger's Simply Money. In 1997, H&R Block sold Managing Your Money to a consortium of banks headed by Bank of America for $35 million. Managing Your

Money was never reintroduced as a product, but the software was used to develop home banking services.[44]

Continual investment in new features and in ease of use kept Quicken ahead of its competitors. Among the new features incorporated in the mid 1990s was electronic bill payment, for which Intuit established a relationship with Checkfree, an Ohio-based clearing house. This was the first of hundreds of relationships with financial institutions that Intuit established during the 1990s. Quicken's ease of use continued to attract critical acclaim. It was by now a complex piece of software, and first-time users were coached by means of innovative multimedia tutorials on CD-ROM. In addition, a minor publishing industry of the "Quicken for dummies" genre had emerged, with more than 50 titles published by 1995.

As in other software categories in which it was a second-tier player, Microsoft played a waiting game, continuing to invest in Money and publish new versions. Like Intuit, Microsoft established its own web of relationships with financial institutions; for example, it secured an exclusive arrangement with MasterCard in 1994. The "features gap" between Money and Quicken was gradually closed. However, Money remained a distant second. As Computer Associates and H&R Block were discovering at the same time, it was almost impossible to dislodge the incumbent in personal finance software.

In October 1994, Microsoft attempted to acquire Intuit, offering more than $2 billion. The offer was made in Microsoft stock—a handsome premium on Intuit's pre-bid market capitalization of $813 million. Intuit was willing, though less than enthusiastic. In Intuit's internal strategy meetings, it was judged possible—even likely—that Microsoft would eventually dominate the market for personal finance software, much as it had come to dominate every other category it had set its sights on.[45]

On April 27, 1995, the Antitrust Division of the US Department of Justice filed suit to block the Microsoft-Intuit deal on the grounds that it would lead to higher prices and less innovation. In a press release issued the same day, the Department of Justice argued that sales of personal finance software were already highly concentrated, with Intuit and Microsoft sharing more than 90 percent of a $90 million market—Intuit's Quicken with a 69 percent market share and 7 million users, Microsoft's Money with a 22 percent market share and about a million users (table 9.7).[46] Microsoft, aware that the Department of Justice might object, had prepared carefully. It had arranged to sell Money to Novell, under whose ownership it would continue to compete with Quicken. "This so-called fix

Table 9.7
Market shares of personal finance software, 1994.

	Percentage of unit sales	Percentage of revenues
Intuit (Quicken)	69%	85%
Microsoft (Money)	22%	7%
H & R Block (Managing Your Money)	<5%	5%
Computer Associates (Simply Money)	<5%	2%

Source: Antitrust Division, US Department of Justice, *U.S. vs. Microsoft and Intuit*, 1995.

just won't work," the Department of Justice argued. "Novell simply can't replace Microsoft—with its leading position in personal computer software—in competing against an entrenched, dominant product like Quicken."[47] Indeed, Microsoft's ruse of selling Money to Novell was disingenuous. Microsoft, with all its resources, had failed to generate a positive return with Money after 4 years of competing with Quicken. The Department of Justice concluded that, under Novell, Money would not be able to compete in any reasonable time.

The Department of Justice's main objection, however, was not to concentration in the market. Recognizing that personal finance software was likely to become a major conduit for home banking, the Department of Justice argued as follows:

If consummated, the proposed transaction, even accounting for the asset sale to Novell, likely would add to the dominance of the number one product (Quicken), would weaken greatly the number two product (Money), and would substantially increase concentration and substantially reduce competition in the PF/Checkbook Software Market. Because these products are a crucial springboard into other important, but emerging, areas of commerce, the effect on consumers would likely be higher prices and lessened innovation not only in PF/Checkbook software products but in other related products and services, such as PC-based home banking.[48]

The case was so strong that Bill Gates pragmatically withdrew Microsoft's bid to take over Intuit.

It is often supposed that a software entrepreneur's dream is to create a business and grow it to such a size that Microsoft can no longer resist making a takeover bid; the entrepreneur then walks into the sunset with the proceeds of the sale. For Intuit, there was no love affair with

Microsoft; however, that company was a formidable competitor that would likely one day dominate its software niche, and hence falling into its arms had some attractions. When the deal fell through, Intuit shrugged and went on with its business.

Intuit continues to dominate the personal finance software category, claiming a market share of 80 percent in 1998.[49] Though it remains a David to Microsoft's Goliath, Intuit is not insubstantial. It has 4,000 employees, a turnover approaching $1 billion, and relationships with more than 1,000 financial institutions. By an accident of history, Intuit is classified as a producer of consumer software, and so it does not feature in most software-industry rankings. If it did, it would be no lower than number 20.

Summary

Recreational software is a significant sector of the software industry, though its conflation with home entertainment and publishing of various kinds makes its extent difficult to quantify in a useful way. The three most important forms of recreational software discussed in this chapter—videogames, CD-ROM publishing, and personal finance software—do not share evolutionary roots or business models. The videogame industry, for example, has much more in common with the recorded-music and movie industries. That videogames are based on software is a secondary and somewhat parochial issue relative to the creative processes by which hit games are generated. The primacy of cultural inputs and the absence of linguistic barriers have allowed videogames to cross international borders with ease, and the videogame sector is the most internationalized sector of the software industry. As in recorded music and in movies, the United States has a dominant share; however, there are powerful players elsewhere, particularly in Japan and in Britain.

Like videogames, CD-ROM encyclopedias raise this question: Exactly what constitutes a software good? Program code is clearly an important ingredient in the package, but it is secondary to the intellectual content that constitutes the encyclopedia. In this context, software is simply an enabling technology.

Even with personal finance software, an apparently classic case of shrink-wrapped software, the boundaries of software and service are becoming blurry. In the early 1980s personal finance software was a stand-alone product for an isolated personal computer; however, by the end of the 1980s, with the arrival of consumer networks such as

CompuServe and Prodigy, personal finance software had begun to include a bundle of services such as electronic payments, on-line home banking, and portfolio valuation. This trend was accelerated by the popularization of the Internet.

In each of the examples of consumer software discussed in this chapter, there was a historical trend for software to become subordinate to the intellectual content or the complementary services offered. This suggests that the concept of a "software industry" may one day be too fuzzy to be meaningful. Programming—broadly interpreted to include development of video games, creation of templates for spreadsheets, visual programming, and database design—is becoming a widely diffused skill. Indeed, it may be as widely diffused today as writing was 150 years ago. One does not speak of a "writing industry"; one speaks of a multi-sector publishing industry whose only point in common is putting ink on paper. In 20 years, we will likely think of the software industry in the same way.

A stack of 60,000 punched cards for the SAGE master program, c. 1956. Such defense projects gave the United States an early-start advantage in software technologies.

10
Reflections on the Success of the US Software Industry

People often exaggerate the value of "the lessons of history," as if the successes and failures of the past were a reliable guide to the future. Unfortunately it is not so; otherwise, historians would be sure-fire stock-market players, and they most assuredly are not. Nonetheless, if one wants to emulate the success of another institution, history is usually all one has to go on. Though history cannot provide prescriptions, it can certainly provide insights.

In the last 20 or 30 years, in Europe and many other parts of the world, planners have attempted to reproduce the success of Silicon Valley in their local economies. Great Britain, for example, has a Silicon Fen (in the lowlands around Cambridge), a Silicon Ditch (in the Thames Valley corridor to the west of London), and a Silicon Glen (in Scotland). When planners attempt to clone Silicon Valley, they generally do so by trying to establish an economic and physical environment similar to that which enabled the original Silicon Valley to flourish. Such attempts are always deeply informed by history. For example, it is widely understood that Silicon Valley did not happen overnight. Though it was first called "Silicon Valley" in the mid 1970s, it was the culmination of a process that had begun at least 40 years earlier with the appointment of Frederick Terman as a professor at Stanford University. Its symbolic birth was the formation of Hewlett-Packard in a garage in 1939.[1]

Usually, a Silicon Valley clone has a technological university as a source of innovation, venture capital is available, and an attempt is made to foster certain cultural factors, such as ease of transfer between academe and commerce and forgiveness of entrepreneurial failure. The cultural factors are usually the hardest to reproduce. Not all Silicon Valley clones have been successful, but the results are encouraging enough to have spawned dozens of attempts, some of which have been successful indeed. Silicon Valley clones have mainly fostered firms in the computer hardware, networking, and telecommunications industries. They do not appear to be especially good Petri dishes for software firms.

In this closing chapter, I will identify some of the historical factors that have led to US dominance of the software industry and consider why these factors have been so difficult to replicate elsewhere.

Early Start and Market Size

Throughout the history of computing, the United States has been several years ahead of other developed nations in the diffusion of computers—perhaps 5 years ahead in the 1960s, perhaps 2 years by the 1990s.

Commercially produced software always has developed in response to the evolving technological capabilities of computers and their diffusion.[2] In the 1950s, software contractors evolved in response to the growing capabilities of mainframes. In the 1960s and the early 1970s, corporate software products were a response to the proliferation of standardized System/360 mainframes and small-business systems. In the late 1970s and the 1980s, shrink-wrapped software was a response to the arrival of the personal computer. Because these technological developments always happened first in the United States, the American software industry has always been ahead.

An early example, described in chapter 2, was the development of the FACT data processing compiler by the Computer Sciences Corporation in the years 1959–1962. In 1962, Dick Clippinger, the project manager, presented a lecture on the FACT system to a meeting of the British Computer Society. In another lecture, a Briton described the only comparable domestic effort at that time: a COBOL compiler for the ICT 1301, a small computer produced by what was then Britain's leading computer firm. Christopher Strachey (later a professor of computation at Oxford University) rose to his feet, incredulous, and remarked: "The lecture from Dr. Clippinger and [ICT's] contribution seems to illustrate very clearly something which must considerably confuse anybody who does not know a great deal about the construction of compilers. Dr. Clippinger has described a compiler for FACT which, he says, uses 200,000 instructions and which is enormously large and costly. . . . On the other hand, [ICT] had implemented COBOL for a machine . . . which in America would not even be regarded as an off-line printer."[3] In short, the American computer for which the FACT compiler was being developed was so far ahead of the British machine that it opened up a new vista of programming possibilities. Although the FACT compiler was not particularly successful, it enabled the Computer Sciences Corporation to develop a capability in systems software that led to its winning many contracts with European computer manufacturers and users in the late 1960s. When indigenous software contractors did get established in Europe, they operated mainly in fields (such as defense, government, and finance) where they were preferred suppliers or had specialized local knowledge.

The greater the diffusion of computers, the greater the potential market for software products. Consider the case of Informatics' Mark IV file-management system, launched in 1967. John Postley, the product's champion inside Informatics, estimated there was a market for a few hundred sales of Mark IV. Though this was perhaps better characterized as a "guesstimate," it was based on Postley's deep knowledge of the US computer market, of the probable impact of the emerging System/360 standard, and of the propensity of data processing managers to buy software. As was described in chapter 4, Mark IV was a highly successful product, with annual corporate sales of several hundred units. At that time, however, there were 35,600 computers in the United States, 2,252 in Britain, 2,008 in France, and 2,963 in Germany.[4] Moreover, the European computer stock was much older, less capable, and less homogeneous than the American. Thus, no matter how one interprets the numbers, the market for software products in the United States was at least 10 times as large as that in any European country. No sane person would have considered developing a major software product in Europe at that time. Once Mark IV had established itself in the United States, Informatics was able to generate additional sales in Europe, and these revenues were icing on what was already a satisfactory cake. In the early 1970s, when Europe had finally caught up with the United States in the diffusion of mainframe computers, major vendors such as Informatics, ADR, Pansophic, and MSA had mature products at reasonable prices against which national suppliers could not compete.

The pattern persisted into the personal computer era. Thus, in 1983, the population of desktop computers in the United States was estimated at 5.8 million—about 60 percent of worldwide shipments to that date. All the major application programs came from US suppliers—VisiCorp, MicroPro, Ashton-Tate, Sorcim, Lotus, and others. The average price of products from these vendors was about $500. The process by which that average price was arrived at has never been articulated, but the most significant factor was market size. In no other country could the market have supported software of comparable quality for $500. Statistics on the sales of individual software products at this time are difficult to find; however, it is known that by the end of 1983 VisiCalc and SuperCalc, the market leaders in spreadsheet programs, had sold 1.5 million copies, and that Lotus 1-2-3, launched in January 1983, had sold a million copies by mid 1985 and 3 million by late 1987. But only 10 percent of Lotus's sales were to Europe. In 1985, the year Lotus made its millionth sale, the number-one spreadsheet in France, Microsoft Multiplan, sold just 28,000 copies. France was not a land of opportunity for the developer of a new spreadsheet program.[5]

While would-be producers of shrink-wrapped software outside the United States were waiting for their domestic markets to mature, an entire business and publishing culture sprang up around American productivity programs. For example, by 1985 there were more than 200 books on the use of spreadsheets in business, nearly every one of them tied to one of the top US products. Europe has never since been able to wrest the market for desktop productivity software from US firms.

Information Asymmetry and Clustering Effects

The clustering of firms was a major factor in the success of Silicon Valley. A cluster is a group of companies (usually small ones) that produce the subassemblies and components of a complex product (say, a personal computer or a network switch). The firms are coordinated entirely through market mechanisms. In a ruthlessly Darwinian process, the fittest firms survive. It was this business environment that most distinguished Silicon Valley from the other nexus of electronics and telecommunications in the United States: Route 128 in Massachusetts, where large vertically integrated firms such as DEC and RCA predominated.[6] It is now accepted that a cluster of firms is much more dynamic and responsive than a giant corporation. In the world of the personal computer, the IBM-compatible PC and the Macintosh exemplify the two approaches, the former being sourced from dozens of different suppliers and the latter being a product of a single vertically integrated firm. Clusters ultimately outperformed the "command economy" of vertically integrated corporations.

Geographical proximity is crucial to the operation of a cluster. A complex network of social relationships, consolidated in professional meetings, bars, and health clubs, enables individuals working in different firms to know in what direction the market is moving and to develop the right product at the right time. This "information asymmetry" is what makes it so difficult for firms outside the network to compete. One has to be there to know what the market wants, and one has to be there to know what the hot new developments are. Another factor is the supply of labor. The firm cluster allows technical specialists to move from one firm to another with better prospects almost overnight without breaking up their social networks or uprooting their families.

At first, there was no active clustering in the software industry; indeed, the opposite held true. In software contracting it was an advantage to be close to one's customers, and this resulted in an almost random geographical distribution of firms (apart from some concentration in metropolitan areas). This pattern continued into the early software products era, when the top 10 firms included Computer Associates (in Ithaca), Cincom (in Cincinnati), MSA (in Atlanta), and Pansophic (in Chicago). During this period, ADAPSO played a large role in fostering networks in which firms could cooperate technically and could share market information and knowledge of sources of venture finance.

Perhaps the only well-studied example of clustering in the software industry is that of the relational database industry of Northern California, home to Oracle, Sybase, and Informix. Clustering allowed database-software specialists to exchange technical information, to find new career opportunities, and to market their unique skills without having to leave the region. As a corollary of this clustering, it was difficult to establish relational database technology beyond Northern California. A second example concerns the creation of complementary products in the PC software industry.[7] The Lotus 1-2-3 spreadsheet was launched in January 1983, and by the end of 1985 there were about 450 complementary products (known as "add-ons"). How did a firm become a vendor of a complementary product? How did a firm gain the technical knowledge needed to interface with the 1-2-3 spreadsheet? Through social networks and geographical proximity. The process started with a few firms located close to Lotus's Cambridge headquarters, including Funk Software (Cambridge), Personics (Wilmington), Intex (Wellesley), and the Cambridge Software Collaborative. In some cases, a firm hired a former Lotus developer; in others, a firm had close informal relationships with Lotus insiders. Other nearby firms produced training materials—for example, Arthur Young Business Systems (in Belmont) produced videos, and Addison-Wesley (in Reading) co-published Lotus's textbooks. At first, Lotus perceived complementary products as parasitic and did not actively encourage their development. However, by 1986 Lotus was beginning to see that add-ons enabled the base product to be extended in new and unpredictable directions and that they increased user lock-in. In the spring of 1986, at the Hyatt Regency Hotel in Cambridge, Lotus held its first annual developers' conference, allowing out-of-town developers to share in the creation of complementary products. In January 1987, Lotus launched ADK (Add-in Development Kit), a product that made the interfaces of 1-2-3 publicly available and provided a range of programming tools to simplify development without the need for insider knowledge. From this point on, geographical proximity was irrelevant.[8] The number of "add-ins" exploded, most of them coming from US suppliers but a few from overseas. Nonetheless, firms such as Funk Software and Personics retained their early-start advantage.

In the 1980s, as software firms proliferated, regional software industry trade associations came into being. The first of these, founded in 1984, was the Washington Software Alliance; today it has more than 1,600 corporate members.[9] In the early 1990s, particularly with the popularization of the Internet, several more trade associations came into being, including the Software Council of Southern California and the New York Software and Internet Association. There were associations in less obvious regions, too—for example, the Arizona Software and Internet Association and the Colorado Software and Internet Association.

Government Support for Research and Development

The importance of R&D in the success of the software industry is little understood. Software companies typically spend 10–15 percent of the revenues on R&D, well above the percentage in most other industries. Proportionately, the R&D spending of software companies is comparable to that of the pharmaceutical industry; however, R&D plays markedly different roles in these two industries. In the pharmaceutical industry, much of the R&D takes place in university and industrial research laboratories, and field trials are directed by PhD-qualified scientists. In the software industry, most of the R&D is done by youthful programmers, usually not trained past the bachelor's degree level, who crank out code in an intuitive but effective fashion—R&D with a small r and a large D.

One of the policy levers used by governments is the sponsorship of basic software research that would be too long term, too expensive, or too risky for the individual firm. The aim of such an intervention is to improve the international competitiveness of the national industry.

By far the most important publicly funded software R&D program was the US government's sponsorship of the Systems Development

Corporation in connection with the SAGE defense system. As much as $1 billion was spent on developing real-time software technologies at a time when such expenditures would have been unthinkable for any firm, even one as large as IBM. However, the process by which this investment was made is instructive. The US government did not create SAGE to establish real-time technology—indeed, it is unlikely that anyone in the government could have articulated such a concept. Rather, the government paid a group of systems integrators and software developers to solve a major technological problem in the way they judged best. Through SAGE, not only did the United States gain a world lead in real-time systems that lasted at least a decade; in addition, the great increase in trained programming manpower served as a springboard for the US software industry. Neither of these outcomes was predicted at the time; indeed, neither was predictable.

The shelves of university libraries groan with the weight of reports of government-funded research inquiries. Few of them are related to software. An exception is *Software: A Vital Key to UK Competitiveness*, which reports the results of an inquiry into the use and production of software in 1985 by the UK's Advisory Council on Applied Research and Development (ACARD). The committee of inquiry was composed of the great and good—scientific academics, executive officers of UK systems integrators and large-scale users, and heads of government laboratories. While sales of personal computers were increasing exponentially, and while Bill Gates was making his second billion, the ACARD inquiry ignored the personal computer. And one can almost sense the disdain with which the committee would have greeted any suggestion that it pay attention to videogames. The tone of the British committee's "analysis of technological development" was set by the opening sentence: "Software engineering is the application of sound scientific, mathematical, management, and engineering principles to the production of programs, within estimated costs and at a competitive level of performance and price."[10] Try telling that to the developers of Sonic the Hedgehog. This was a statement addressing yesterday's problems. It could have been written in the 1960s.

The report of the ACARD inquiry concludes with 15 pages of recommendations that make tedious reading, even for an enthusiast. Let us skip past those that are parochial and largely irrelevant to those that concern R&D. Predictably, reflecting the vested interests of academe, old-style systems integrators, and government departments, there were calls for improved technology transfer from the public to the private sector, for research into software engineering and formal programming methods, for the development of improved software tools, and for the Ministry of Defence to take "strong and immediate steps to improve current UK Ada technology." Prime Minister Margaret Thatcher was not much in favor of spending public money, so the ACARD report was quietly shelved to gather dust, and little money was wasted on the cul-de-sac of Ada (a programming language that for 20 years had been irrelevant to the software industry).

Within 5 years, nine of the ten leading British software companies examined by the ACARD committee had been acquired by overseas interests or gone bankrupt. Nothing in the report, even had it been implemented, would have had the slightest bearing on that outcome.

Summary

There is no magic formula for creating a successful national software industry, but history suggests some avenues that are worth exploring.

The evidence of clustering effects is ambiguous but promising. There is some evidence from the Northern California relational database industry that a cluster of firms can act as a permanent magnet for specialized programming talent. However, the example of Lotus Development's complementary producers suggests that where the programming competence is of a lower level the benefit of clustering is short-lived and is quickly overcome by knowledge transferred through development kits and training materials. However, the existence of regional trade associations suggests that, though a region may not be able to create specific programming technologies and keep them from leaking out, a region can sustain a permanent concentration of programming activity. There are numerous examples of such regions around the world, from the Thames Valley corridor in the United Kingdom to India's software cities, Bangalore and Hyderabad.

R&D subventions can be effective, but they are difficult to target successfully. They seem to have been most effective in the 1950s and the 1960s, when the US government simply acted as a customer, paying the price necessary to pull the technology through. Private firms were then able to reap the spinoffs. This process was unpredictable. Much the same can be said of the US government's role in the development of the Internet. In contrast, European attempts to direct R&D into formal methods for software development may have been harmful. It has been argued that those attempts misdirected the industrial focus and distorted the programming culture from good-enough competence to an exclusive "only PhDs need apply" ethos.[11]

Among the things a government can do to foster a software industry, manpower training may be the most beneficial and may create the least market distortion. Manpower training has been a major strategy of such emerging software nations as India and Ireland. However, the evidence so far suggests that it is likely to result in an industry performing "offshore" software development for the benefit of US-based multinational corporations. This is a new variant of the old business of "body shopping," not the rise of a leading-edge software products industry.

Of all the factors discussed in this chapter, the most important in creating a successful software products enterprise is market size. The bigger the market, the better the prospect of getting a return on investment. Sadly, there is little an individual country can do about this. Promoting the use of computers to enlarge the domestic market in a simple-minded way (through tax incentives, for example) simply has the effect of increasing US imports, making the situation worse, before it can get better. However, the overwhelming significance of market size suggests that policy makers should address strategies of international marketing as well as the more tractable issues of manpower training, technology, and industrial structure.

Notes

Chapter 1

1. "Office Robots," *Fortune*, January 1952: 82–87, 112, 114, 116, 118.

2. "Software Gap—A Growing Crisis for Computer," *Business Week*, November 5, 1966: 127–134.

3. "Missing Computer Software," *Business Week*, September 1, 1980: 46–49, 52–56.

4. "Software: The New Driving Force," *Business Week*, February 27, 1984: 54–63; reprinted in *The Information Technology Revolution*, ed. Tom Forrester (Blackwell, 1985).

5. "Missing Computer Software," p. 46.

6. "Software: The New Driving Force," *Business Week*, February 27, 1984: 54–61.

7. Detlev J. Hoch, Cyriac R. Roeding, Gert Purkert, and Sandro K. Lindner, *Secrets of Software Success: Management Insights from 100 Software Firms Around the World* (Harvard Business School Press, 1999), pp. 5–7.

8. PriceWaterhouseCoopers, *Forecasting a Robust Future: An Economic Study of the US Software Industry* (Business Software Alliance, 1999), pp. 6–7. The authors predicted: "Given current trends, beginning in 2000 the industry's contribution to the US economy will exceed that of all other manufacturing industry groups."

9. F. P. Brooks Jr., *The Mythical Man-Month: Essays on Software Engineering* (Addison-Wesley, 1975), p. 7.

10. For an excellent modern discussion of classification, see Geoffrey C. Bowker and Susan Leigh Star, *Sorting Things Out: Classification and Its Consequences* (MIT Press, 1999).

11. There is a fair consensus among the recent historical overviews of the software industry on this periodization and sectorization: Martin Campbell-Kelly, "Development and Structure of the International Software Industry, 1950–1990," *Business and Economic History* 24, no. 2 (1995): 73–110; W. Edward Steinmueller, "The US Software Industry: An Analysis and Interpretive History," in *The*

International Computer Software Industry, ed. David C. Mowery (Oxford University Press, 1996); Hoch et al., *Secrets of Software Success*.

12. In this respect, software firms fit very well the economic models discussed in Richard Nelson and Sidney G. Winter, *An Evolutionary Theory of Economic Change* (Harvard University Press, 1982).

13. Richard Heeks, *India's Software Industry* (Sage, 1996).

14. Yasunori Baba, Shinji Takai, and Yuji Mizuta, "The User-Driven Evolution of the Japanese Software Industry: The Case of Customized Software for Mainframes," in *The International Computer Software Industry*, ed. Mowery; Michael A. Cusumano, *Japan's Software Factories: A Challenge to US Management* (Oxford University Press, 1991).

15. Hoch et al., *Secrets of Software Success*, pp. 269–270.

16. Ibid., pp. 38, 276.

17. A better measure still would be the industry's value added to the economy. That is the total receipts of the industry minus its expenditures on products and services consumed.

18. Private communications with Peter Cunningham, founder and president of INPUT.

19. Table 1.2 has been reconstructed from ADAPSO Annual Reports and internal documents. Where there are gaps, it is usually because there are gaps in the historical record—though the statistics published were not entirely consistent from year to year.

20. Jerome L. Dreyer, "The ADAPSO Story," *Datamation*, March 1970: 55–58. ADAPSO, *25 Years* (ADAPSO, March 1986).

21. Letter from Luanne Johnson (former president of ADAPSO, now with the Software History Center), May 18, 2000.

22. US Department of Commerce, *A Competitive Assessment of the United States Software Industry* (1984); OECD, *Software: An Emerging Industry* (1985).

23. The Software Publishers Association was renamed the Software and Information Industry Association (SIIA) in January 1999 as a result of a merger with the Information Industry Association.

24. See, e.g., Howard Marks, "It's Time to Disband Software's Secret Police," *Data Communications*, September 21, 1992: 27–28.

25. US Department of Commerce, *A Competitive Assessment of the United States Software Industry*, p. 36.

26. Richard Brandt et al., "Can the US Stay Ahead in Software?" *Business Week*, March 11, 1991: 62–67. Data attributed to IDC.

27. Cusumano, *Japan's Software Factories*, p. 49.

28. Peter C. Grindley, "The Future of the Software Industry in the United Kingdom: The Limits of Independent Production," in *The International Computer Software Industry*, ed. Mowery.

29. Richard L. Forman, *Fulfilling the Computer's Promise: The History of Informatics 1962–1982* (Informatics General, 1985); Claude Baum, *The System Builders: The Story of SDC* (SDC, 1981).

30. Sandra L. Kurtzig, *CEO: Building a $400 Million Company from the Ground Up* (Norton, 1991); W. E. (Pete) Petersen, *AlmostPerfect: How a Bunch of Regular Guys Built WordPerfect Corporation* (Prima, 1994); John Walker, ed., *The Autodesk File: Bits of History, Words of Experience*, third edition (New Riders, 1989); Douglas G. Carlston, *Software People: An Insider's Look at the Personal Computer Software Industry* (Simon & Schuster, 1985); John P. Imlay with Dennis Hamilton, *Jungle Rules: How to Be a Tiger in Business* (Dutton, 1994); Ben Voth, *A Piece of the Computer Pie* (Gulf, 1974).

31. Mike Wilson, *The Difference between God and Larry Ellison: Inside Oracle Corporation* (Morrow, 1997); Gerd Meissner, *SAP: Inside the Secret Software Power* (McGraw-Hill, 2000); Tristan Gaston-Breton, *La Saga Cap Gemini* (Point de Mire, 1999).

32. Elmer C. Kubie, "Recollection of the First Software Company," *Annals of the History of Computing* 16 (1994): 65–71; J. Lesourne and R. Armand, "A Brief History of the First Decade of SEMA," *Annals of the History of Computing* 13 (1991): 341–349.

33. Richard H. MacNeal, *The MacNeal-Schwendler Corporation: The First Twenty Years* (MacNeal-Schwendler, 1988).

34. US Department of Commerce, *A Competitive Assessment of the United States Software Industry* (1984); OECD, *Software, an Emerging Industry* (1985); Advisory Council for Applied Research and Development, *Software: A Vital Key to UK Competitiveness* (HMSO, 1986).

35. Mowery, ed., *The International Computer Software Industry*; Salvatore Torrisi, *Industrial Organisation and Innovation: An International Study of the Software Industry* (Elgar, 1998); Stephen E. Siwek and Harold W. Furchtgott-Roth, *International Trade in Computer Software* (Quorum Books, 1993); Hoch et al., *Secrets of Software Success*.

36. The URL of *Software Magazine* is http://www.softwaremag.com.

37. The publishing history of *Business Software Review* is uncertain, and I know of no complete holding.

38. At the time of writing, Computer Associates, the world's number-3 software company, is reported to be organizing its archive in preparation for its 25th anniversary. In view of the hundreds of firm acquisitions that Computer Associates embodies, the archive promises to be spectacular. No policy on public access has been disclosed.

39. Some of these have ceased trading or have been taken over.

40. A few examples: Frost & Sullivan., *The Computer Software and Services Market* (Frost & Sullivan, 1971); Business Communications Corp., *Software Packages: An Emerging Market* (1980); Efrem Sigel and the staff of Communications Trends, *Business/Professional Microcomputer Software Market, 1984–86* (Knowledge Industry Publications, 1984).

41. The URL of the Software History Center is http://www.softwarehistory.org.

Chapter 2

1. Paul Armer, "SHARE—A Eulogy to Cooperative Effort" (1956) (reprint: *Annals of the History of Computing* 2, 1980: 122–129).

2. C. L. Baker, "The 701 at Douglas, Santa Monica," *Annals of the History of Computing* 5 (1983): 187–188. This and other articles cited below appeared in a special issue on the IBM 701 computer published in April 1983.

3. Emerson W. Pugh, Lyle R. Johnson, and John H. Palmer, *IBM's 360 and Early 370 Systems* (MIT Press, 1991), p. 293.

4. Emerson W. Pugh, *Building IBM: Shaping an Industry and Its Technology* (MIT Press, 1995), p. 190.

5. Ibid., p. 183.

6. Ibid., 63.

7. The Technical Computing Bureau had been established in 1949 as a contract service for performing scientific and engineering calculations. See Charles J. Bashe, Lyle R. Johnson, John H. Palmer, and Emerson W. Pugh, *IBM's Early Computers* (MIT Press, 1986), p. 84.

8. John Greenstadt, "Recollections of the Technical Computing Bureau," *Annals of the History of Computing* 5 (1983): 148–154.

9. Elliot C. Nohr, "FORTRAN Activities at SHARE Meeting," *Annals of the History of Computing* 6 (1984): 65–69.

10. W. F. McClelland and D. W. Pendery, "701 Installation in the West," *Annals of the History of Computing* 5 (1983): 167–170.

11. The best historical account of SHARE is Atsushi Akera, "Voluntarism and the Fruits of Collaboration: The IBM User Group, Share," *Technology and Culture* 42 (2001): 710–736.

12. R. Blair Smith, "The IBM 701—Marketing and Customer Relations," *Annals of the History of Computing* 5 (1983): 170–172.

13. Ibid.

14. Armer, "SHARE—A Eulogy to Cooperative Effort," p. 122.

15. Akera, "The IBM User Group, Share," p. 725.

16. The best general account of early programming systems is Paul E. Ceruzzi, *A History of Modern Computing* (MIT Press), especially chapter 3.

17. Donald E. Knuth and Luis Trabb Pardo, "The Early Development of Programming Languages," in *A History of Computing in the Twentieth Century*, ed. N. Metropolis, J. Howlett, and G.-C. Rota (Academic Press, 1980).

18. The best historical account of FORTRAN is *Annals of the History of Computing* 6 (1984), no. 1 (25th-anniversary special issue).

19. John Backus, "The History of FORTRAN I, II, and III," *Annals of the History of Computing* 1 (1979): 21–37.

20. Bashe et al., *IBM's Early Computers*, p. 357.

21. Ceruzzi, *History of Modern Computing*, pp. 85–86.

22. The best historical account of COBOL is *Annals of the History of Computing* 7 (1985), no. 4 (25th-anniversary special issue).

23. Ibid., p. 308.

24. Ceruzzi, *History of Modern Computing*, p. 94; OECD, *Software, An Emerging Industry* (1985), p. 31.

25. The best source for the general history of SAGE is *Annals of the History of Computing* 5 (1983), no. 4 (special issue). See also Paul N. Edwards, *The Closed World: Computer and the Politics of Discourse in Cold War America* (MIT Press, 1997); Claude Baum, *The System Builders: The Story of SDC* (SDC, 1981).

26. Thomas J. Watson Jr. and Peter Petre, *Father and Son & Co: My Life at IBM and Beyond* (Bantam, 1990), p. 230.

27. H. S. Tropp et al., "A Perspective on SAGE: Discussion," *Annals of the History of Computing* 5 (1983): 375–398, esp. p. 386.

28. Ibid., p. 387.

29. Baum, *The System Builders*, p. 23.

30. Tropp et al., "A Perspective on SAGE," p. 386.

31. Thomas C. Rowan, *Selection and Training of Computer Programmers*, March 15, 1958 (Charles Babbage Institute Archives, SDC Records 1946–1982, item 1–21). Such aptitude tests became a standard recruiting technique and remained so until the late 1960s, when the job market began to mature and colleges were producing computer science graduates in significant numbers.

32. Baum, *The System Builders*, p. 42.

33. Ibid., p. 51.

34. John F. Jacobs, *The SAGE Air Defense System: A Personal History* (MITRE, 1981), p. 108.

35. As quoted in Baum, *The System Builders*, p. 31.

36. For an account of SDC's factory style of software production, see Cusumano, *Japan's Software Factories*, pp. 119–160.

37. Baum, *The System Builders*, p. 47.

38. Ibid., p. 51.

39. Herbert D. Benington, "Production of Large Computer Programs," *Annals of the History of Computing* 5 (1983): 350–361, esp. p. 351.

40. As quoted in James L. McKenney, *Waves of Change: Business Evolution through Information Technology* (Harvard Business School Press, 1995), p. 97.

41. Jon Ecklund, "The Reservisor Automated Airline Reservation System: Combining Communications and Computing," *Annals of the History of Computing* 16 (1994): 62–69; William D. Bell, *A Management Guide to Electronic Computers* (McGraw-Hill, 1957), esp. pp. 259–273; James D. Gallagher, *Management Information Systems and the Computer* (American Management Association, 1961), esp. pp. 150–176.

42. McKenney, *Waves of Change*, p. 105.

43. Bashe et al., *IBM's Early Computers*, 518.

44. McKenney, *Waves of Change*, p. 111.

45. Gilbert Burck, *The Computer Age* (Harper & Row, 1965), p. 31.

46. Bashe et al., *IBM's Early Computers*, p. 521.

47. This appellation appears in Burck, *The Computer Age*, p. 34.

48. R. W. Parker, "The SABRE System," *Datamation*, September 1965: 49–52.

49. "A Survey of Airline Reservation Systems," *Datamation*, June 1962: 53–55.

50. The original SABRE software was very long-lived. The same a code base was still being used more than two decades later, in 1987, when the system had expanded to process over 1,000 messages per second on a system that used eight 3090 mainframes, to support 12,000 agent terminals. In 1996 SABRE became an independent entity, the SABRE Group, with its own NYSE listing. At the end of the millennium it had revenues of over $2 billion, 6,000 employees, and its 31 mainframe computers had a capacity of 7,300 messages per second. See www.sabre.com/about for a short corporate history.

51. Davis Dyer, *TRW: Pioneering Technology and Innovation Since 1900* (Harvard Business School Press, 1998), p. xi.

52. MITRE Corp., *MITRE: The First Twenty Years* (MITRE, 1979), p. 9. "MITRE" is sometimes said to be a contraction of "MIT Research," but no etymology is offered in the official history.

53. Karl L. Wildes and Nilo A. Lindgren, *A Century of Electrical Engineering and Computer Science at MIT, 1882–1982* (MIT Press, 1986), p. 300.

54. Edwards, *Closed World*, p. 107.

55. Baum, *The System Builders*, p. 53.

56. Ibid., pp. 53–57.

57. Dyer, *TRW*, pp. 306–307.

58. Kenneth Flamm, *Targeting the Computer: Government Support and International Competition* (Brookings Institution, 1987). Kenneth Flamm, *Creating the Computer: Government, Industry, and High Technology* (Brookings Institution, 1988).

59. By mid 1962 it was reported that there were 11 reservation systems under development, with one already "on the air." See "Survey of Airlines Reservation Systems," p. 53.

60. M. Campbell-Kelly, *ICL: A Business and Technical History* (Clarendon, 1989), pp. 180–186.

61. The surprising choice of BTM was due to a personal connection between Chase Manhattan and R. Holland-Martin, a director of Martin's Bank, London, and of BTM. The name Diana—the God of the Chase—was coined by a BTM sales executive.

62. McKenney, *Waves of Change*, esp. chapter 3; Homer R. Oldfield, *King of the Seven Dwarfs: General Electric's Ambiguous Challenge to the Computer Industry* (IEEE Computer Society Press, 1996).

63. McKenney, *Waves of Change*, p. 57.

64. Ibid., p. 64.

65. Frost and Sullivan, *The Computer Software and Service Market* (1971), pp. 107–108.

66. Little has been published on the early startups. A useful source is a set of profiles of the three major software firms (CSC, CUC, and CEIR) produced in 1963 by the editor of *Datamation*, Harold Bergstein: "Computerized Reflections at CSC," *Datamation*, March 1963: 39–42; "Its Elmer's Turn," April 1963: 48–51; "The CEIR View," May 1963: 55–59.

67. Franklin M. Fisher, James W. McKie, and Richard B. Mancke, *IBM and the US Data Processing Industry: An Economic History* (Praeger, 1983), p. 322.

68. "Software Gap—A Growing Crisis for Computers" *Business Week*, November 5, 1966: 127–134, esp. p. 134.

69. Elmer C. Kubie, "Recollection of the First Software Company," *Annals of the History of Computing* 16 (1994): 65–71.

70. Bashe et al., *IBM's Early Computers*, p. 338.

71. Kubie, "Recollection of the First Software Company," p. 67.

72. Bergstein, "Its Elmer's Turn," p. 48.

73. Two fairly detailed company histories were published in the April 1979 and April 1984 issues of *CSC News*, to celebrate the company's 20th and 25th anniversaries, respectively. Available in Product Literature Collection, Item 37.1.7.A, Charles Babbage Institute Archives, University of Minnesota.

74. "One Compiler, Coming Up! Jones, Nutt, Patrick form Computer Sciences Corporation," *Datamation*, May-June, 1959: 15.

75. William D. Smith, "Texan Guides Software Unit to Big Board," *New York Times*, December 1, 1968.

76. Edith Myers, "CSC: A Hectic 25 Years," *Datamation*, March 1984: 96ff.

77. Jack A. Strong and Richard F. Clippinger, "Recollections of the Intermediate Range [CODASYL] Committee," *Annals of the History of Computing* 7 (1985): 326–328, esp. p. 328.

78. Richard F. Clippinger, "FACT," *Computer Journal* 5 (1962): 112–125, esp. p. 112.

79. Strong and Clippinger, "Recollections," p. 328.

80. Ibid.

Chapter 3

1. "Software Suffers Unprogrammed Woes," *Business Week*, June 6, 1970: 68, 70–71.

2. For a history of ACT, see Katherine D. Fishman, *The Computer Establishment* (Harper & Row, 1981), pp. 268–280.

3. Ibid., p. 276. "Multinational and Multilingual [Profile of Charles P. Lecht]," *Datamation*, April 1976: 13.

4. Charles Philip Lecht, *The Waves of Change: A Techno-Economic Analysis of the Data Processing Industry* (Advanced Computer Techniques, 1977).

5. For a capsule histories of ADR see Vin McLellan, "ADR: Well Enough to Lease Again," *Datamation*, April 1977: 152–154; Don Leavitt, "A Silver Anniversary in a 15-Year-Old World," *Software News*, July 1984: 38.

6. For a history of Informatics, see Richard L. Forman, *Fulfilling the Computer's Promise: The History of Informatics, 1962–1982* (Informatics General, 1985).

7. International Resource Development Inc., *Computer Services and Software Markets* (International Resource Development, 1975), p. 145.

8. For a history of LEASCO, see John Brooks, *The Go-Go Years* (Wiley, 1973; reprinted 1999), pp. 227–259.

9. In 1973 CDC acquired IBM's Service Bureau Corporation as a result of the settlement of its lawsuit with IBM.

10. For a history of ADP see Fishman, *Computer Establishment*, pp. 280–288.

11. Ibid., p. 281.

12. Franklin M. Fisher, James W. McKie, and Richard B. Mancke, *IBM and the US Data Processing Industry: An Economic History* (Praeger, 1983), p. 321.

13. Doron P. Levin, *Irreconcilable Differences: Ross Perot versus General Motors* (Little, Brown, 1989).

14. Montgomery Phister Jr., *Data Processing: Technology and Economics* (Digital Press and Santa Monica Publishing Company, 1979), Table II.1.26, p. 277. Phister's statistics, used in figure 3.1, are not wholly consistent with the more authoritative statistics used in table 1.2, but they cover an earlier time period.

15. Jerome L. Dreyer, "The ADAPSO Story," *Datamation*, March 1970: 55–58.

16. Fisher, McKie, and Mancke, *IBM and the US Data Processing Industry*, pp. 322–323.

17. For the early history of UCC, see Ben Voth, *A Piece of the Computer Pie* (Gulf, 1974).

18. Forman, *Fulfilling the Computer's Promise*, p. 5/8.

19. Ibid., p. 1/20.

20. Frost & Sullivan, *The Computer Software and Services Market* (1971), p. 129.

21. CSC advertisement, *Datamation*, January 1964: 1.

22. Frost & Sullivan, *The Computer Software and Services Market*, p. 129.

23. These were major themes of the first conference on software engineering, held in 1968. See chapter 4.

24. Philip Kraft, "The Industrialization of Computer Programming," in *Case Studies on the Labor Process*, ed. Andrew Zimbalist (Monthly Review Press, 1979), pp. 1–17.

25. Jacobs, *The SAGE Air Defense System*, pp. 111, 175–263. See also Herbert D. Benington, "Production of Large Computer Programs" (1956) (reprint: *Annals of the History of Computing* 5, 1983: 350–361).

26. Charles Philip Lecht, *The Management of Computer Programming Projects* (American Management Association, 1967).

27. Fisher, McKie, and Mancke, *IBM and the US Data Processing Industry*, p. 325.

28. Ibid.

29. Forman, *Fulfilling the Computer's Promise*, p. 1/15.

30. Nancy Foy, "Hard Recession in Software," *Management Today*, April 1971: 95–97, 136, esp. p. 97.

31. Edith Myers, "CSC Seeks the Midas Touch," *Datamation*, July 1972: 45; Edith Myers, "CSC: 25 Hectic Years," *Datamation*, March 1984: 96–97, 100, 104.

32. "The Manager CSC Wanted," *Datamation*, November 1976: 13.

33. Elmer C. Kubie, "Recollection of the First Software Company," *Annals of the History of Computing* 16 (1994): 65–71, esp. p. 68.

34. Baum, *The System Builders*, p. 112.

35. Ibid., p. 130.

36. Ibid., p. 141.

37. Fishman, *The Computer Establishment*, pp. 276–279.

38. Ibid.

39. OECD, *Gaps in Technology: Electronic Computers* (OECD, 1969). See also Paul Gannon, *Trojan Horses and National Champions: The Crisis in Europe's Computing and Telecommunications Industry* (Apt-Amatic Books, 1997), esp. pp. 247–274.

40. The only European countries to develop significant computer industries were Britain and France. They had begun well, with 80 percent and 50 percent indigenous manufacture in 1962, respectively, but this did not last. By 1967, American imports accounted for well over half the computer sales in both countries. M. Campbell-Kelly, *ICL: A Business and Technical History* (Clarendon, 1989), p. 250.

41. Edward K. Yasaki, "European Software Market," *Datamation*, December 1967: 27–31.

42. Some background to the UK's software industry was given in a one-day conference held by the Computer Conservation Society at London's Science Museum on November 6, 1997.

43. See Campbell-Kelly, *ICL*, p. 206. The most influential polemic on the American invasion was J. J. Servan-Schreiber, *The American Challenge* (Hamilton, 1968).

44. Campbell-Kelly, *ICL*, pp. 246–248. The National Computer Centre eventually became a major software house. It was privatized in 1984.

45. Brian Harris, *BABS, BEACON and BOADICEA: A History of Computing in British Airways and Its Predecessor Airlines* (Speedwing, 1993), pp. 175–176.

46. As cited in Yasaki, "European Software Market."

47. J. Lesourne and R. Armand, "A Brief History of the First Decade of SEMA," *Annals of the History of Computing* 13 (1991): 341–349.

48. For a good contemporary survey of the UK's software industry, see Nancy Foy, "Software's Hard Currency," *Management Today*, December 1975: 68–72, 74, 76, 78. See also Peter C. Grindley, *The UK Software Industry: A Survey of the Industry and Evaluation of Policy* (London Business School, 1988).

49. Tristan Gaston-Breton, *La Saga Cap Gemini* (Editions Point de Mire, 1999).

50. "Computer Management Group Limited" in John M. Ryan, *It Can Be Done!* (Scope Books, 1979), pp. 77–82, esp. p. 78.

51. Ibid., p. 78.

52. Debra Issac, "The Logic in Logica," *Management Today*, August 1985), pp. 47–51, 81–82.

53. Foy, "Software's Hard Currency," p. 68.

54. "Vaughan Systems & Programming," in Ryan, *It Can Be Done!*

55. Tom Lloyd, *Dinosaur & Co: Studies in Corporate Evolution* (Routledge & Kegan Paul, 1984), pp. 61–79.

56. Teresa Poole, "The IT Woman," *Independent*, January 5, 2000). The FI Group was renamed Xansa in 2001.

57. Brooks, *The Go-Go Years*, pp. 127–128.

58. Gilbert Burck, "The Computer Industry's Great Expectations," *Fortune*, August 1968: 92–97, 142, 145–146, esp. p. 93.

59. Voth, *Computer Pie*, p. 28.

60. Burck, "The Computer Industry's Great Expectations," p. 96.

61. Voth, *A Piece of the Computer Pie*, p. 83.

62. Forman, *Fulfilling the Computer's Promise*, pp. 3/48–3/50.

63. "Computer Software Companies: How Many Are Houses of Cards?" *Forbes*, February 15, 1970: 40–42.

64. Brooks, *The Go-Go Years*, p. 1.

65. Williams D. Smith, "Tight Money Hits Software Field," *New York Times*, May 16, 1970.

66. Williams D. Smith, "Software Maker Files Bankruptcy," *New York Times*, October 21, 1970.

67. Robert Metz, "Market Place," *New York Times*, July 30, 1970.

68. D. F. Parkhill, *The Challenge of the Computer Utility* (Addison-Wesley, 1966); C. C. Barnett Jr. et al., *The Future of the Computer Utility* (American Management Association, 1967).

69. Burck, "The Computer Industry's Great Expectations," pp. 142, 145.

70. Foy, "Hard Recession in Software," p. 96.

71. "CSC Seeks the Midas Touch," *Business Week*, July 29, 1972: 46.

72. The CSC-Western Union merger was never consummated.

73. Todd Mason, "Sam Wyly: Will the Hunter Become the Hunted?" *Business Week*, July 13, 1987: 62–63.

74. "A Fresh Start for Wyly Corp.," *Datamation*, April 1978: 187.

75. Uccel was acquired by Computer Associates in 1987, giving Haefner a 23 percent stake in that company. See chapter 6.

76. Forman, *Fulfilling the Computer's Promise*, pp. 4/17–4/19.

77. "Software Suffers Unprogrammed Woes," *Business Week*, June 20, 1970: 68, 70–71.

78. Fred Gruenberger, ed., *Expanding Use of Computers in the 70's: Markets—Needs—Technology* (Prentice-Hall, 1971), p. 40. This conference was funded by Informatics.

79. "Software Giant Goes Commercial," *Business Week*, August 23, 1969: 84, 86.

80. Baum, *The System Builders*, p. 141.

81. Frost & Sullivan, *The Computer Software and Services Market*, pp. 108–109.

82. Foy, "Hard Recession in Software," p. 136.

83. "Software Suffers Unprogrammed Woes," p. 70.

Chapter 4

1. "Software Gap—A Growing Crisis for Computers," *Business Week*, November 5, 1965: 126–134. One should temper reports of early programmer shortages with the fact that the scarcity of programming talent has been a constant feature of the software landscape. Some solution to the current crisis has always emerged: in the 1950s high-level programming languages eased the programming problem, while in the 1990s it was off-shore software development that came to the rescue. See Peter Freeman and William Aspray, *The Supply of Information Technology Workers in the United States* (Computing Research Association, 1999).

2. Montgomery Phister Jr., *Data Processing Technology and Economics* (Digital Press and Santa Monica Publishing Company, 2nd ed., 1979), p. 26.

3. Werner L. Frank, "Software for Terminal Oriented Systems," *Datamation*, June 1968: 30–34.

4. Werner L. Frank, "The History of Myth no. 1," *Datamation*, May 1983: 252–256, esp. p. 252.

5. Barry W. Boehm, "Software and Its Impact: A Quantitative Assessment," *Datamation*, May 1973: 48–59.

6. Peter Naur and Brian Randell, *Software Engineering* (Scientific Affairs Division, NATO, 1969).

7. Cited in Boehm, "Software and Its Impact."

8. Pugh et al., *IBM's 360 and Early 370 Systems*, pp. 331–345.

9. Boehm, "Software and Its Impact," p. 57.

10. Naur and Randell, *Software Engineering*, pp. 68–69.

11. Ibid., p. 69.

12. Frederick P. Brooks Jr., *The Mythical Man-Month: Essays on Software Engineering* (Addison-Wesley, 1975), p. 14

13. Thomas J. Watson Jr., *Father, Son & Co.: My Life at IBM and Beyond* (Bantam, 1990), p. 353.

14. "Final Report of the SPREAD Task Group, December 28, 1961," *Annals of the History of Computing* 5, no. 1 (1983): 6–26.

15. Ibid., p. 6.

16. Walter Bauer, interview conducted by A.L. Norberg, May 16, 1983, Charles Babbage Institute, University of Minnesota, ref. OH61.

17. L. Welke, "The Origins of Software," *Datamation*, December 1980: 127–130, esp. p. 127.

18. JoAnne Yates, "Application Software for Insurance in the 1960s and Early 1970s," *Business and Economic History* 24, no. 1 (1995): 123–134.

19. Ibid. (this and the following quotations).

20. The NCR advertisement appears in *Datamation* (July, 1965, p. 52).

21. Gilbert Burck, The Assault on Fortress I.B.M.," *Fortune*, June 1964: 112–116, 196, 198, 200, 202, 207. "The Series 200 From Honeywell," *Datamation*, April 1965: 54.

22. The Honeywell advertisement appears in *Datamation* (April 1965, p. 96).

23. The catalytic role of ICP was referred to frequently at the "One for the History Books" Workshop (Software History Center, Benicia, California, September 24, 2000).

24. Larry Welke, the founder of ICP, subsequently became a president of ADAPSO's software section and a principal government witness in the IBM antitrust lawsuit as a leading authority on the software industry.

25. R. V. Head and E. F. Linick, "Software Package Acquisition," *Datamation*, October 1968: 22–27

26. Luanne James Johnson, "A View from the 1960s: How the Software Industry Began," *Annals of the History of Computing* 20, no. 1 (1998): 36–42. Martin Goetz, "ADR's Autoflow," *Unbundling History: The Emergence of the Software Product*, Palo Alto, September 22–23, 2000. (Transcripts in the archives of the Charles Babbage Institute, University of Minnesota.)

27. A. Pantages, "IBM's Vigorous Defense Spreads Thin, as ADR Files # Four," *Datamation*, June 1969: 121–123.

28. Johnson, "One for the History Books" Workshop.

29. The SyncSort advertisement appears in the *1975 ICP Software Directory*. SyncSort has been one of the longest-lived software products. Today SyncSort Inc. claims that it is in use by 95 of the *Fortune* 100, with sales in 50 countries.

30. Johnson, "One for the History Books" Workshop.

31. R. Forman, *Fulfilling the Computer's Promise: The History of Informatics 1962–1982* (Informatics General Corp., 1985).

32. Ibid., p. 3/21.

33. W. F. Bauer, interview.

34. J. A. Postley, "Mark IV: Evolution of a Software Product," *Annals of the History of Computing* 20, no. 1 (1998): 43–50; J. A. Postley, "The Mark IV System," *Datamation*, January 1968: 28–30.

35. Bauer, interview; Head and Linick, "Software Package Acquisition," p. 24.

36. W. F. Bauer, "Informatics: An Early Software Company," *Annals of the History of Computing* 18, no. 2 (1996): 70–76.

37. See, e.g., R. P. Biglow, "Legal Aspects of Proprietary Software," *Datamation*, October 1968: 40–47.

38. Johnson, "A View from the 1960s: How the Software Industry Began," p. 40.

39. F. M. Fisher, J. W. McKie, and R. B. Mancke, *IBM and the US Data Processing Industry: An Economic History* (Praeger, 1983), p. 172.

40. Ibid., p. 176.

41. Ibid., p. 177.

42. Burton Grad Associates Inc., *Evolution of the U. S. Packaged Software Industry* (Burton Grad Associates Inc., October 14, 1992), p. II-1.

43. Ibid., p. II-3.

44. Ibid., p. II-6

45. M. Goetz, "IBM's Operating System Monopoly," *Datamation*, July 1974: 168–169.

46. Frost & Sullivan, *The Computer Software and Services Market* (1971), p. 58.

47. Pantages, "IBM's Vigorous Defense."

48. Watts S. Humphrey, "Reflections on a Software Life," in *In the Beginning: Personal Recollections of Software Pioneers*, ed. R. L. Glass (IEEE Computer Society Press, 1997), p. 42.

49. Ibid.

50. See M. E. Conway, "On the Economics of the Software Market," *Datamation*, October 1968: 28–31.

51. As late as 1980, IBM's software revenues were estimated to be only about 3 percent of its sales ($800 million of $21 billion in total revenues).

52. Yates, "Application Software for Insurance," p. 123.

53. "Software Firms Unwrap Packages and Hope for Presents from IBM," *Datamation*, May 1969: 128–129.

54. D. B. Steig, "File Management Systems," *Datamation*, October 1972: 48–51; L. Welke, "A Review of File Management Systems," *Datamation*, October 1972: 52–54.

Chapter 5

1. Vin McLellan, "ADR: Well Enough to Lease Again," *Datamation*, April 1977: 152–154, esp. 152.

2. Richard A. McLaughlin, "Software Packages for System/3," *Datamation*, June 1973: 66–71.

3. McLellan, "ADR: Well Enough to Lease Again," p. 152.

4. Frost & Sullivan, *Data Base Management Services Software Market* (1979), p. 154.

5. McLellan, "ADR: Well Enough to Lease Again," p. 152.

6. Pugh, *Building IBM*, pp. 307–313.

7. Yankee Group, *IBM's Future Software Strategy* (Boston: Yankee Group, 1989), p. 3.

8. International Resource Development Inc., *Computer Services and Software Markets* (1975), p. 123.

9. Business Communications Corp., *Software Packages: An Emerging Market* (Business Communications Corp., 1980), p. 88.

10. Richard H. MacNeal, *The MacNeal-Schwendler Corporation: The First Twenty Years* (MacNeal-Schwendler, 1987).

11. In fairness to MSC (now MSC Software), it should be noted that its steady growth was maintained in the 1980s and 1990s, so that by 1999 it had 740 employees, sales exceeding $100 million, and occupied a place in the lower part of the top 100 software firms.

12. Stephen T. McLellan, *The Coming Computer Industry Shakeout* (Wiley, 1984), p. 259. Triad Systems now operates as CCI/Triad Inc. (www.cci-triad.com).

13. Private communication. Atlantic Software was acquired by AGS Computers in the mid 1970s.

14. Tom McCusker, "COSMIC Moves to Plug Gaps in Software Distribution Project," *Datamation*, September 1, 1970: 41.

15. M. Campbell-Kelly, "From National Champions to Little Ventures: The NEB and the Second Wave of IT in Britain 1975–1985," forthcoming.

16. Ralph Emmett, "Insac: Are There Any Survivors?" *Datamation*, March 1980: 226HH–226LL.

17. Martin Campbell-Kelly, "The Rise and Rise of the Spreadsheet," in *From Sumer to Spreadsheets: The Curious History of Mathematical Tables*, ed. M. Campbell-Kelly, M. G. Croarken, R. G. Flood, and E. Robson (Oxford University Press, forthcoming).

18. All these examples appeared in *Datamation* in the early 1970s.

19. Dennis Hamilton, "The ICP Million Dollar Awards," ICP *Business Software Review*, June 1985: 32–40. Syncsort, a privately owned company, has never been the target of a takeover. At the time of writing it remains one of the longest-lived medium-size software product companies, with 100 employees and 1999 sales of $7.8 million. It ranks about 375 in *Software Magazine*'s "Software 500."

20. International Resource Development Inc., *Computer Services and Software Markets*, pp. 63–64.

21. Ibid., p. 68.

22. Business Communications Inc., *Software Packages: An Emerging Market*, p. 118.

23. These managed a fixed number of tasks and a variable number of tasks, respectively; the latter variant came to dominate.

24. M. S. Auslander et al., "The Evolution of the MVS Operating System," *IBM Journal of Research and Development* 25 (September 1981): 471–482; R. J. Creasy, "The Origin of the VM/370 Time-Sharing System," *IBM Journal of Research and*

Development 25 (September 1981): 483–490; Jim Hoskins, *IBM ES/9000: A Business Perspective* (Wiley, 1994).

25. ESA stood for "enterprise system architecture."

26. For example: Martin A. Goetz, "IBM's Operating System Monopoly," *Datamation*, July 1974: 168–169; C. Donald Berteau, "IBM and the Structure of the Industry: The Software Monopoly," *Datamation*, October 1975: 111; Tom McCusker, "Will IBM Unbundle Its Operating Systems?" *Datamation*, August 1976: 102–103.

27. Franklin M. Fisher, James. W. McKie, and Richard B. Mancke, *IBM and the US Data Processing Industry: An Economic History* (Praeger, 1983), pp. 415–417.

28. Cusumano, *Japan's Software Factories*.

29. Peter H. Salus, *A Quarter Century of Unix* (Addison-Wesley, 1994); D. M. Ritchie, "The Evolution of the Unix Time-Sharing System," *AT&T Bell Laboratories Technical Journal* 63 (October 1984): 1577–1593.

30. Elliot I. Organic, *The MULTICS System: An Examination of Its Structure* (MIT Press, 1980).

31. Mark Hall and John Barry, *Sunburst: The Ascent of Sun Microsystems* (Contemprary Books, 1990).

32. Ed Dunphy, *The Unix Industry*, second edition (Wiley, 1994).

33. W. C. McGee, "Database Technology," *IBM Journal of Research and Development* 25 (September 1981): 505–519.

34. David K. Alison, *Transcript of a Video History Interview with Mr. Thomas M. Nies, Founder and Chief Executive Officer of Cincom Systems, Inc.* (National Museum of American History, 1995).

35. Frost & Sullivan, *Data Base Management Services Software Market*.

36. [David Mindell], "The Rise of Relational Databases," in National Research Council, *Funding a Revolution: Government Support for Computing Research* (National Academy Press, 1999).

37. Pugh et al., *IBM's 360 and Early 370 Systems*, pp. 587–593.

38. Ian Palmer, *Database Systems: A Practical Reference* (CACI, 1975).

39. Jim Geraghty, *CICS Concepts and Uses: A Management Guide* (McGraw-Hill, 1994), p. 3.

40. The exception was American Airlines, whose SABRE reservation system pre-dated CICS.

41. Curt Monash, "Software Strategies," *Datamation*, February 1984: 171–172, 176, 181–182.

42. David C. Mounce, *CICS: A Light Hearted Chronicle* (IBM UK Laboratories, 1994).

43. Jeff Moad, "IBM Puts New Muscle into CICS," *Datamation*, November 1, 1991: 62, 64, 67.

44. Steve Homer, "Battling on with Veteran Computers," *New Scientist*, November 14, 1992: 32–35

45. David Kirkpatrick, "Why the Internet Is Boosting IBM's Mainframe Sales," *Fortune*, January 11, 1999: 76–77.

46. Charles Newman, "Software: Build or Buy," *Banking Technology*, May 1984: 43–46.

47. "1989 Directory of Computer Software for Banking," *Bank Administration* 65 (October 1989): 56–96.

48. See, e.g., James L. McKenney, *Waves of Change: Business Evolution Through Information Technology* (Harvard Business School Press, 1995), pp. 41–95.

49. Tom Lawton, "A Banking Software Story [re. Hogan Systems]," *Datamation*, November 1, 1985: 98, 100, 102.

50. "Banking on Big Blue," *Datamation*, July 15, 1986: 34, 38.

51. In 1995, Hogan Systems was acquired by Continuum, a medium-size developer of insurance software.

52. Richard Bourke, "A Compilation of Available MRP Systems and Their Vendors," *Datamation*, October 1980: 101–106.

53. For a detailed history of ASK up to 1990, see Kurtzig, *CEO*.

54. John Imlay Jr. with Dennis Hamilton, *Jungle Rules: How to Be a Tiger in Business* (Dutton, 1994); Frost & Sullivan, *MSA: The Software Company* (New York: Frost & Sullivan, 1982).

55. Amy D. Wohl, "What's Happening in Word Processing," *Datamation*, April 1977: 65–74.

56. Frost & Sullivan, *The Text Processing Software Market in the US* (1982). A typical word processing software firm was Satellite Software Inc., later the WordPerfect Corporation.

57. An Wang, *Lessons: An Autobiography* (Addison-Wesley, 1986).

58. Wohl, "What's Happening in Word Processing."

59. Charles C. Kenney, *Riding the Runaway Horse: The Rise and Decline of Wang Laboratories* (Little, Brown, 1992), pp. 94–97.

60. There is no monograph or single article on the history of Computer Vision. The details below have been taken from disparate sources, including *Datamation* and *Business Week*. A good general source on the CAD/CAM industry is Carl

Machover, ed., *The CAD/CAM Handbook* (McGraw-Hill, 1989; second edition 1996).

61. A similar use of specialized processors was made by videogame console makers in the 1980s.

62. Barnaby J. Feder, "Bolts and Brackets by (Computer) Design," *New York Times*, January 18, 1981.

Chapter 6

1. "Software: The New Driving Force," *Business Week*, February 27, 1984: 54–63, esp. p. 63. The estimates of 8,000 packages and 3,000 vendors are consistent with other contemporary guestimates—e.g., Larry Welke, "The Origins of Software," *Datamation*, December 1980: 127–130; Frederic G. Withington, "The Golden Age of Packaged Software," *Datamation*, December 1980: 131–134.

2. US Department of Commerce, *A Competitive Assessment of the United States Software Industry* (1984).

3. Ware Myers, "An Assessment of the Competitiveness of the United States Software Industry," *Computer*, March 1985: 81–92. See also Grindley, *The UK Software Industry*; Grindley, "The Future of the Software Industry in the United Kingdom."

4. "Best Selling Software Packages in Japan," in OECD, *Software: An Emerging Industry* (Paris, 1985), pp. 201–202.

5. OECD, *Software: An Emerging Industry*; ACARD, *Software: A Vital Key to UK Competitiveness*.

6. Broadview Associates' statistics were published as follows: years 1970–80 in "Pain and Pleasure in Going Public," *Datamation*, August 25, 1981: 60–66; years 1980–84 in OECD, *Software: An Emerging Industry*, p. 78.

7. "Software: The New Driving Force," p. 60.

8. Monash, "Software Strategies," *Datamation*, February 1984: 171–172, 176, 181–182.

9. Cusumano, *Japan's Software Factories*.

10. Bob Djurdjevic, "Up the Software Curve," *Datamation*, May 15, 1985: 96, 98, 102, 104–105.

11. David W. Barron, *Computer Operating Systems* (Chapman and Hall, 1971), p. 1.

12. Monash, "Software Strategies," p. 172.

13. IBM software rentals cited in Djurdjevic, "Up the Software Curve."

14. Johanna Ambrosio, "[IBM's] Value-based Pricing Arrives," *Software News*, December 1986: 20–21, 24.

15. Yankee Group, *IBM's Future Software Strategy*, p. 7.

16. The *New York Times* article is cited in Michael Killen's *IBM: The Making of the Common View* (Harcourt Brace Jovanovich, 1988), p. xv.

17. The story was told in Michael Killen's *IBM: The Making of the Common View*. However, while that account was strong on incident, the overarching theme was too abstract to reach the common reader (or sometimes apparently the author). In contrast, James Chposky and Ted Leonsis's *Blue Magic: The People, Power and Politics Behind the IBM Personal Computer* (Facts on File, 1988), published in the same year, was widely read inside and outside the industry.

18. Deidre A. Depke et al., "Suddenly, Software Houses Have a Big Blue Buddy," *Business Week*, August 7, 1989: 51–52.

19. Amy Cortese, "Sexy? No. Profitable? You Bet," *Business Week*, November 11, 1996: 70–72.

20. Hesh Kestin, *Twenty-First-Century Management: The Revolutionary Strategies That Have Made Computer Associates a Multibillion Dollar Software Giant* (Atlantic Monthly Press, 1992).

21. Kenny MacIver, "CA: The Hidden Dimension," *Computer Business Review*, July 1988: 14–16, 18, 20.

22. Ibid.

23. Charles B. Wang, *Techno Vision* (McGraw-Hill, 1994). A second edition, *Techno Vision II*, was published in 1997.

24. The business history of Computer Associates is mainly taken from Computer Associates, The First Twenty Years (http://www.ca.com) and from Jonathan Flatt, Computer Associates—Strategic Mergers and Acquisitions in the Software Industry (unpublished report, Department of Computer Science, University of Warwick, 1999).

25. Damian Rinaldi, "CA-Uccel: Now They Are One," *Software News*, July 1987: 22–23.

26. Kestin, *Twenty-First-Century Management*, p. 177.

27. Katherine M. Hafner, "How Computer Associates Climbed to No. 1 in Software," *Business Week*, July 11, 1988: 50–51.

28. Evan I. Schwartz, "Computer Associates Gets User Friendly," *Business Week*, January 21, 1991: 86–87.

29. Ibid.

30. Steve Hamm et al., "Why Oracle Is Cool Again," *Business Week*, May 8, 2000: 42–47.

31. Wilson, *The Difference between God and Larry Ellison**. (The asterisk in the title led to a footnote: "God doesn't think He's Larry Ellison." This was apparently a popular joke inside and outside Oracle while Wilson was researching his book.)

32. AnnaLee Saxenian, *Regional Advantage: Culture and Competition in Silicon Valley and Route 128* (Harvard University Press, 1994).

33. Richard S. Rosenbloom and Clayton M. Christensen, "Technological Discontinuities, Organizational Capabilities, and Strategic Commitments," *Industrial and Corporate Change* 3 (1994): 655–685. For a more accessible account, see Joseph L. Bower and Clayton M. Christensen, "Disruptive Technologies: Catching the Wave," *Harvard Business Review*, January-February 1995: 43–53.

34. [David Mindell], "The Rise of Relational Databases," in National Research Council, *Funding a Revolution: Government Support for Computing Research* (National Academy Press, 1999); W. C. McGee, "Database Technology," *IBM Journal of Research and Development* 25 (September 1981): 505–519.

35. For a history of Oracle, see Wilson, *The Difference between God and Larry Ellison*. See also Stuart Read, *The Oracle Edge* (Adams Media, 1999).

36. International Resource Development Inc., *Database Management Systems* (International Resource Development Inc., 1987), pp. 129–130.

37. Ibid., p. 122.

38. David K. Alison, Transcript of a Video History Interview with Mr. Thomas M. Nies, Founder and Chief Executive Officer of Cincom Systems, Inc. (National Museum of American History, 1995); Kenny MacIver, "Re-Inventing the Corporation [re. Cincom]," *Computer Business Review*, September 1995: 24–26.

39. AmeriTech was one of the seven regional companies created by the breakup of AT&T in 1984.

40. Stephen T. McClellan, *The Coming Computer Industry Shakeout* (Wiley, 1984), p. 243.

41. Mike Bucken, "CAI Picks Up Cullinet," *Software Magazine*, August 1989: 24–27.

42. Gary McWilliams, "Oracle's Olympian Challenge," *Datamation*, November 15, 1988: 31, 34–35, 38.

43. John Desmond, "Here Comes DB2: DBMS, Application Vendors Respond," *Software News*, July 1986: 32–33, 36, 38–39, 44–45, 48, 50, 52.

44. Evan I. Schwartz, "No More Funny Money for Software Makers," *Business Week*, February 18, 1991: 122B, 122G.

45. Janice Maloney, "Larry Ellison Is Captain Ahab and Bill Gates Is Moby Dick," *Fortune*, October 28, 1996: 75–78; Richard Brandt, "Can Larry Beat Bill?" *Business Week*, May 15, 1995: 38–43, 46.

46. Richard Brandt, "Oracle's Prognostication: Tomorrow Looks Terrific," *Business Week*, September 20, 1993: 57–58.

47. Gerd Meissner, *SAP: Inside the Secret Software Power* (McGraw-Hill, 2000); Grant Norris et al., *SAP: An Executive's Comprehensive Guide* (Wiley, 1998).

48. Andrew Lawrence, "Accidental Empire [SAP]," *Computer Business Review*, August 1996: 9–12.

49. In German: Systeme, Anwendungen, Produkte in der Datenverarbeitung.

50. Gail Edmondson et al., "Silicon Valley on the Rhine [re. SAP]," *Business Week*, November 3, 1997: 40–47

51. Meissner, *SAP*, p. 71.

52. Michael Hammer and J. Champy, *Re-engineering the Corporation: A Manifesto for Business Revolution* (Harper Business, 1993).

53. Meissner, *SAP*, pp. 76–77, 80.

54. "The Great Pretenders [re. Baan, SSA, Oracle, and PeopleSoft]," *Computer Business Review*, September 1996: 17–20, 22, 24. "Competitors to SAP," in Norris et al., *SAP*.

55. Edmondson et al., "Silicon Valley on the Rhine," p. 42.

56. For an extensive discussion on system building, see A. D. Chandler, *The Visible Hand: The Managerial Revolution in American Business* (Harvard University Press, 1977).

57. D. Edgerton, *Science, Technology and the British Industrial "Decline," 1870–1970* (Cambridge University Press, 1996).

Chapter 7

1. The best account of the early development of the personal computer industry is Paul Frieberger and Michael Swaine, *Fire in the Valley: The Making of the Personal Computer* (Osborne/McGraw-Hill, 1984; second edition McGraw-Hill, 1999). The best source for the later development of the industry, and for statistics, is John Steffens, *Newgames: Strategic Competition in the PC Revolution* (Pergamon, 1994).

2. "Lift Off Time for Microcomputers," *Business Week*, September 22, 1986: 34.

3. Quoted in M. Campbell-Kelly and W. Aspray, *Computer: A History of the Information Machine* (Basic Books, 1996), p. 240.

4. The details of Microsoft here are mainly from the following: James Wallace and Jim Erickson, *Hard Drive: Bill Gates and the Making of the Microsoft Empire* (Wiley, 1992); Stephen Manes and Paul Andrews, *Gates: How Microsoft's Mogul Reinvented an Industry—and Made Himself the Richest Man in America* (Simon and

Schuster, 1994); Daniel Ichbiah and Susan L. Knepper, *The Making of Microsoft* (Prima, 1991).

5. In 1995 I met Gates's former boss at TRW at a conference. He joked the Gates had been quite a talented programmer, and that had he stayed with TRW he would surely have become a department head.

6. "It's Going to Happen" is a chapter title in Wallace and Erickson, *Hard Drive*.

7. The term OEM, meaning original equipment manufacturer, had been used in the mainframe and minicomputer markets for many years for suppliers of software or hardware subsystems bought in and integrated into finished products.

8. There are no detailed accounts of the early years of Digital Research. The best sources are interviews with Kildall published in the following books: Robert Slater, *Portraits in Silicon* (MIT Press, 1987), pp. 251–262; Susan Lammers, *Programmers at Work: Interviews with 19 Programmers Who Shaped the Computer Industry* (Tempus-Microsoft, 1986), pp. 56–69.

9. Chposky and Leonsis, *Blue Magic*.

10. The operating system was developed by Seattle Computer Products (SCP), a nearby microcomputer manufacturer that was developing a computer with the same Intel chip that IBM was planning to use. The charge that Microsoft exploited SCP does not stand up. Microsoft, fully capable of developing an operating system, was primarily buying product lead time.

11. Sigel et al., *Business/Professional Microcomputer Software Market, 1984–86*, pp. 32–33.

12. Although IBM subsequently allowed users choice among MS-DOS and two other operating system (one of which was a new version of Digital Research's CP/M), the later operating systems entered the market too late to compete with MS-DOS.

13. Statistic cited in Sigel et al., *Business/Professional Microcomputer Software Market*, p. 7.

14. Efrem Sigel and Louis Giglio, *Guide to Software Publishing: An Industry Emerges* (Knowledge Industry Publications, 1984).

15. Statistic cited in Sigel et al., *Business/Professional Microcomputer Software Market*, p. 54.

16. Robert Levering, Michael Katz, and Milton Moskowitz, *The Computer Entrepreneurs: Who's Making It Big and How in America's Upstart Industry* (New American Library, 1984), pp. 326–333.

17. Sigel et al., *Business/Professional Microcomputer Software Market*, pp. 67–69.

18. Efrem Sigel, "The Selling of Software," *Datamation*, April 15, 1984: 125–128.

19. Peter Petre, "The Man Who Keeps the Bloom on Lotus," *Fortune*, June 10, 1985: 92–94, 96, 98, 100.

20. The retail value of PC software was approximately twice that of publishers' receipts. Wholesalers and retailers typically applied a markup of 100 percent on software. For example, the 1983 retail value of shrink-wrapped software was estimated to be $860 million, compared with publishers receipts of $468 million.

21. Sigel et al., *Business/Professional Microcomputer Software Market*, p. 73.

22. Ibid., p. 52.

23. There is no detailed account of the early development of VisiCalc. The most useful sources are interviews of Bricklin published in Slater, *Portraits in Silicon* (pp. 282–294), in Levering et al., *Computer Entrepreneurs* (pp. 128–133), and in Lammers, *Programmers at Work* (pp. 130–151).

24. Slater, *Portraits in Silicon*, pp. 285–286.

25. See, e.g., "Killer Application," in *Encyclopedia of Computers and Computer History*, ed. R. Rojas (Fitzroy Dearborn, 2001).

26. Robert T. Fertig, *The Software Revolution: Trends, Players, Market Dynamics in Personal Computer Software* (North-Holland, 1985), p. 178.

27. The best sources for the early development of Lotus are interviews with Kapor published in Levering et al., *Computer Entrepreneurs* (pp. 188–195) and in William Aspray, ed., *Engineers as Executives: An International Perspective* (IEEE Press, 1995) (pp. 219–241).

28. Frost & Sullivan, *The Text Processing Software Market in the US*.

29. The best sources on the early development of MicroPro are Frieberger and Swaine, *Fire in the Valley* (pp. 152–153) and an interview with Rubenstein published in Levering et al., *Computer Entrepreneurs* (pp. 214–221).

30. International Resource Development Inc., *Microcomputer Software Packages* (1983), p. 67.

31. Ibid., p. 68.

32. Exact "street prices" of shrink-wrapped software cannot be stated with certainty, since packages were often discounted below the recommended selling price. Here, $140 was an average retail store price.

33. Interview with Ratcliffe in Lammers, *Programmers at Work*, pp. 110–129.

34. Ibid., p. 115.

35. Profile of George Tate in Levering et al., *Computer Entrepreneurs*, pp. 228–235.

36. Profile of Gibbons in Levering et al., *Computer Entrepreneurs*, pp. 164–169; interview with Page in Lammers, *Programmers at Work*, pp. 92–109.

37. Frieberger and Swaine, *Fire in the Valley*, p. 150.

38. Frost & Sullivan, *MSA: The Software Company*.

39. Formerly Magic Wand and MagicCalc, published by Small Business Applications, Houston.

40. Kurtzig, *CEO*, pp. 252–265.

41. Michael Moritz, *The Little Kingdom* (Morrow 1984), p. 224.

42. Yankee Group, *Software Strategies: The Home Computer and Video Game Marketplace* (Boston: Yankee Group, 1984). See also Mary-Beth Santarelli, "The Home-Software Challenges: Knowing What You Want and Where to Get It," *Software News*, November 1984: 43–44.

43. Ralph Watkins, *A Competitive Assessment of the US Video Game Industry* (US International Trade Commission, 1984).

44. Creative Strategies International, *Computer Home Software* (San Jose: Creative Strategies International, 1983), pp. 14–15.

45. Although domestic productivity applications had lower functionality than their professional counterparts, this was largely a means of segmenting the market. The cost of developing software was such a small fraction of the overall cost that it did not materially affect the retail price.

46. Like textbook authors, program developers typically earned royalties of 10 percent. Few titles sold more than a few thousand copies.

47. Although the IBM PC arrived in August 1981, there was an installed base of millions of non-IBM-compatible machines. Not until the mid 1980s would the IBM PC standard be irrevocably established.

Chapter 8

1. The chapter has been published in a slightly extended form as M. Campbell-Kelly, "Not Only Microsoft: The Maturing of the Personal Computer Software Industry, 1982–1995," *Business History Review* 75 (spring 2001): 103–145.

2. The literature on Microsoft is not rewarding to study in its entirety. The best semi-technical account of the development of Microsoft and its products is Ichbiah and Knepper, *The Making of Microsoft*. For good economic perspectives, see Randall E. Stross, *The Microsoft Way* (Addison-Wesley, 1996) and Stanley J. Liebowitz and Stephen E. Margolis, *Winners, Losers and Microsoft: Competition and Antitrust in High Technology* (Independent Institute, 1999). The two best journalistic accounts are James Wallace and J. Erickson, *Hard Drive: Bill Gates and the Making of the Microsoft Empire* (Wiley, 1992) and Stephen Manes and Paul Andrews, *Gates: How Microsoft's Mogul Reinvented an Industry—and Made Himself the Richest Man in America* (Simon and Schuster, 1994). A book in a different genre that

contains many historical insights is Michael A. Cusumano and Richard W. Selby, *Microsoft Secrets* (Free Press, 1995).

3. "The Shakeout in Software: It's Already Here," *Business Week*, August 23, 1984: 96–98.

4. Ibid.

5. On the history of the corporate software industry, see M. Campbell-Kelly, "Development and Structure of the International Software Industry, 1950–1990," *Business and Economic History* 24 (winter 1995): 73–110; W. Edward Steinmueller, "The US Software Industry: An Analysis and Interpretive History," in *The International Computer Software Industry*, ed. Mowery; US Department of Commerce, *A Competitive Assessment of the United States Software Industry* (1984).

6. James Aley, "The Theory That Made Microsoft," *Fortune*, April 29, 1996: 23–24.

7. Brian W. Arthur, "Competing Technologies, Increasing Returns, and Lock-In by Historical Events," *Economic Journal* 99 (1989): 116–131; "Positive Feedbacks in the Economy," *Scientific American*, February 1990: 92–99; Arthur, "Increasing Returns and the New World of Business," *Harvard Business Review*, July-August, 1996: 100–109; Jeffrey Church and Neil Gandal, "Network Effects, Software Provision, and Standardization," *Journal of Industrial Economics* 60 (1992): 85–103.

8. John Steffens, *Newgames: Strategic Competition in the PC Revolution* (Pergamon, 1994), pp. 211–216.

9. Tim Paterson, "An Insider Look at MS-DOS," *Byte*, June 1983: 230ff.

10. Richard J. Gilbert "Networks, Standards, and the Use of Market Dominance: Microsoft (1995)," in *The Antitrust Revolution*, ed. John E. Kwoka Jr. and Lawrence J. White (Oxford University Press 1999).

11. Chposky and Leonsis, *Blue Magic*.

12. Tim Paterson, "Insider Look at MS-DOS."

13. The advertisement quoted appeared in the May 1983 issue of *Byte*.

14. Fertig, *The Software Revolution*, p. 118.

15. Ibid., p. 132.

16. Sigel et al., *Business/Professional Microcomputer Software Market, 1984–86*, p. 32.

17. This and the foregoing statistics appear in Ichbiah and Knepper, *The Making of Microsoft* (p. 86ff.).

18. Arthur, "Increasing Returns and the New World of Business."

19. Bill Gates, *The Road Ahead* (Viking, 1995).

20. Jim Carlton, *Apple: The Inside Story of Intrigue, Egomania, and Business Blunders* (Random House, 1997), pp. 40–43.

21. *The Autodesk File*, published as a 600-page book in 1989 (Thousand Oaks, Calif.: New Riders), was updated and published on the World Wide Web in 1994 (http://www.fourmilab.ch/autofile/). See also Jonathan Richardson, "A Decade of CAD," *CAD User*, March 1998: 20–22, 26, 28.

22. Walker, *The Autodesk File*, p. 24

23. Ibid., p. 299.

24. Ibid., pp. 296–304.

25. The concepts of paradigm shift, technological closure, and critical problems are informed by the works of Thomas Kuhn, Thomas Parke Hughes, and Nathan Rosenberg, among others. See Thomas Kuhn, *The Structure of Scientific Revolutions* (University of Chicago Press, 1962); Thomas Parke Hughes, *Networks of Power: Electrification in Western Society, 1880–1930* (Johns Hopkins University Press, 1983); Nathan Rosenberg, *Inside the Black Box: Technology and Economics* (Cambridge University Press, 1982).

26. Douglas K. Smith and R. C. Alexander, *Fumbling the Future: How Xerox Invented, Then Ignored, the First Personal Computer* (Morrow, 1988); Michael A. Hiltzik, *Dealers of Lightning: Xerox PARC and the Dawn of the Computer Age* (HarperBusiness, 1999).

27. Steven Levy, *Insanely Great: The Life and Times of Macintosh, the Computer That Changed Everything* (Penguin, 1994).

28. See, e.g., "A Fierce Battle Brews Over the Simplest Software Yet," *Business Week*, November 21, 1983: 61–63.

29. Phil Lemmons, "A Guided Tour of VisiOn," *Byte*, June 1983: 256ff.

30. Irene Fuerst, "Broken Windows," *Datamation*, March 1, 1985: 46, 51–52.

31. John Markoff, "Five Window Managers for the IBM PC," *Byte Guide to the IBM PC*, fall 1984: 65–66, 71–76, 78, 82, 84, 87.

32. Efrem Sigel, "Alas Poor VisiCorp," *Datamation*, January 15, 1985: 93–94, 96.

33. Lawrence D. Graham, *Legal Battles That Shaped the Computer Industry* (Quorum Books, 1999), pp. 35–41.

34. The best account of the complicated history of OS/2 appears in Paul Carroll, *Big Blues: The Unmaking of IBM* (Crown, 1993).

35. As quoted in Wallace and Erickson, *Hard Drive*, p. 362.

36. Cusumano and Selby, *Microsoft Secrets*.

37. Petre, "The Man Who Keeps the Bloom on Lotus."

38. Kelly R. Conatser, "1-2-3 Through the Years," *Lotus*, June 1992: 38–45.

39. Keith H. Hammonds, "The Spreadsheet That Nearly Wore Lotus Out," *Business Week*, July 3, 1989: 50–51.

40. For spreadsheet sales statistics for 1988–97, see Liebowitz and Margolis, *Winners, Losers and Microsoft*, pp. 175–176.

41. W. E. Pete Petersen, *AlmostPerfect: How a Bunch of Regular Guys Built WordPerfect Corporation* (Prima, 1994).

42. Frost & Sullivan, *The Text Processing Market in the US* (1982).

43. For word processing market shares by units, see Ichbiah and Knepper, *The Making of Microsoft*, p. 132; Petersen, *AlmostPerfect*, passim. For 1988–1997 market share by revenues, see Liebowitz and Margolis, *Winners, Losers and Microsoft*, p. 181. Sales revenues are difficult to interpret in the word processing market because some high-function packages were priced much higher than the average, while others were priced low to capture market share. Unit sales, used here, was the usual measure.

44. Sandra D. Atchison, "A Perfectly Good Word for WordPerfect: Gutsy," *Business Week*, October 2, 1989: 79–80.

45. Kathy Rebello, "The Glitch at WordPerfect," *Business Week*, May 17, 1993: 56–57.

46. Robert A. Sehr, "Beefing Up Software," *Datamation*, February 15, 1985: 148.3–148.6; David W. Carroll, "The dbase Phenomenon: Nurtured by dBase II, Another Aftermarket has Developed," *Software News*, August 1985: 62–64.

47. Market data, attributed to IDC, in Patrick E. Cole, "dBase IV Is a Godsend—To the Competition," *Business Week*, November 13, 1989: 79.

48. Patrick E. Cole, "dBugs in dBase IV Spread to the Bottom Line," *Business Week*, July 17, 1989: 78–79.

49. Cole, "dBase IV Is a Godsend—to the Competition," p. 79.

50. Petre, "The Man Who Keeps the Bloom on Lotus."

51. "A Toe-to-Toe Duel in Personal Software," *Business Week*, April 9, 1984: 52–53.

52. Irene Feurst, "So Where Is the Market?" *Datamation*, April 1, 1985: 45, 48

53. Richard Brandt, "Software Will Play Hardball Again," *Business Week*, January 10, 1994: 48.

54. Amy Cortese, "Once Again, Software Is Seething," *Business Week*, January 9, 1995: 46.

55. Stross, *The Microsoft Way*, p. 56.

56. Bill Lawrence, "Three Suite Deals," *Byte*, March 1994.

57. Amy Cortese and Ira Sager, "Gerstner at the Gates," *Business Week*, June 19, 1995: 30–32.

58. Patrick E. Cole, "Lost a Computer File? Call on Dr. Norton," *Businesss Week*, May 23, 1986: 116.

59. Evan Schwartz, "The Industry Needs an Alternative—But Will It Be Novell?" *Business Week*, February 1, 1993: 48–49.

60. Interview with Gordon Eubanks in Rama Dev Jager and Rafael Ortiz, *In the Company of Giants: Candid Conversations with Visionaries of the Digital World* (McGraw-Hill, 1997), pp. 45–59, esp. p. 55.

61. Interview with John Warnock in Susan Lammers, *Programmers at Work: Interviews with 19 Programmers Who Shaped the Computer Industry* (Tempus, 1986), pp. 40–55. For an interview with John Warnock and Charles Geschke, see Jager and Ortiz, *In the Company of Giants*, pp. 99–113.

62. Paul M. Horvitz, "Efficiency and Antitrust Considerations in Home Banking: The Propsoed Microsoft-Intuit Merger," *Antitrust Bulletin*, summer 1996: 427–446.

63. Richard Brandt, "It's Grab-Your-Partner Time for Software Makers," *Business Week*, February 8, 1988: 52–53.

64. Katherine M. Hafner, "How Two Pioneers Brought Publishing to the Desktop," *Business Week*, October 15, 1987: 61–62; Richard Brandt, "Does Adobe Have a Paper Cutter?" *Business Week*, November 6, 1992: 98–99; Amy Cortese, "This Acrobat Has Really Limbered Up," *Business Week*, September 26, 1994: 73–74.

65. Michael A. Cusumano and David B. Yoffie, *Competing on Internet Time: Lessons from Netscape and Its Battle with Microsoft* (Free Press, 1998).

66. Cusumano and Selby, *Microsoft Secrets*, p. 141.

67. Kathy Rebello et al., "Novell: End of Era?" *Business Week*, November 22, 1993: 43–44.

68. Richard Brandt, "A Tricky Tack for Borland," *Business Week*, August 2, 1993: 44–45.

69. Interview with Gordon Eubanks in Jager and Ortiz, *In the Company of Giants*.

70. Ibid., p. 55.

71. Steve Lohr, *Go To: The Story of The Programmers Who Created the Software Revolution* (Basic Books, 2001).

72. Ibid., pp. 96–97.

73. Robert X. Cringely, *Accidental Empires: How the Boys of Silicon Valley Made Their Millions, Battle Foreign Competition, and Still Can't Get a Date* (Addison-Wesley, 1992).

Chapter 9

1. Ralph Watkins, *A Competitive Assessment of the US Video Game Industry* (US International Trade Commission, 1984); Yankee Group, *Software Strategies*; Creative Strategies International, *Computer Home Software* (1983); Creative Strategies International, *Cartridge-Based Software: Further Developments* (1984).

2. Enthusiasts have not done such a good job of recording the corporate and intellectual history of videogames, although there are some important exceptions. The best and most systematic historical account of the industry is Leonard Herman's *Phoenix: The Fall and Rise of Videogames* (second edition: Rolenta, 1997).

3. Steven Levy, *Hackers: Heroes of the Computer Revolution* (Penguin, 1994).

4. Interview with Nolan Bushnell in Slater, *Portraits in Silicon*, pp. 296–307.

5. Scott Cohen, *Zap! The Rise and Fall of Atari* (McGraw-Hill, 1984), p. 30.

6. Watkins, *Competitive Assessment of the US Video Game Industry*, p. 7.

7. Cohen, *Zap!*, p. 77.

8. Watkins, *Competitive Assessment*, p. 42.

9. Cohen, *Zap!*, p. 94.

10. Yankee Group, *Software Strategies*, p. 179.

11. Douglas G. Carlston, *Software People* (Simon & Schuster, 1985).

12. J. C. Herz, *Joystick Nation: How Videogames Gobbled Our Money, Won Our Hearts, and Rewired Our Minds* (Abacus, 1997), p. 49.

13. Herman, *Phoenix*, p. 112.

14. Rama Dev Jager and Rafael Ortiz, *In the Company of Giants: Candid Conversations with Visionaries of the Digital World* (McGraw-Hill, 1997), p. 186.

15. Watkins, *Competitive Assessment of the US Video Game Industry*, p. xi.

16. The concept of creative inputs is discussed in Stephen E. Siwek and Harold W. Furchtgott-Roth, *International Trade in Computer Software* (Quorum Books, 1993), pp. 119–134.

17. Electronic Arts was formed in 1982, on the eve of the 1983 crash, and had a number of cyclical downturns and layoffs before achieving stability at the end of the 1980s. It is currently the largest videogame producer with year 2000 annual revenues of $1.4 billion and 2500 employees.

18. Interview with Trip Hawkins, in Jager and Ortiz, *In the Company of Giants*, pp. 175–189, esp. p. 178.

19. Electronic Arts' PGA Tour Golf remains a hugely popular game. In its latest incarnation, Tiger Woods' PGA Tour Golf 2001, it is one of the few games with significant sales to people over the age of 30.

20. David Sheff, *Game Over: How Nintendo Zapped an American Industry, Captured Your Dollars, and Enslaved Your Children* (Random House, 1993).

21. Herman, *Phoenix*, p. 113.

22. The price was later reduced to $199.

23. Sheff, *Game Over*, pp. 195, 203.

24. Reiji Asakura, *Revolutionaries at Sony: The Making of the Sony PlayStation and the Visionaries Who Conquered the World of Video Games* (McGraw-Hill, 2000).

25. Paul du Gay et al., *Doing Cultural Studies: The Story of the Sony Walkman* (Sage, 1997).

26. Michael A. Cusumano, Yiorgos Mylonadis, and Richard S. Rosenbloom, "Strategic Manoeuvring and Mass-Market Dynamics: The Triumph of VHS over Beta," *Business History Review*, spring 1992: 51–94.

27. Asakura, *Revolutionaries at Sony*, p. ix.

28. Herz, *Joystick Nation*; Steven Poole, *Trigger Happy: The Inner Life of Videogames* (Fourth Estate, 2000).

29. Interview with Gary Kildall in Susan Lammers, *Programmers at Work: Interviews with 19 Programmers Who Shaped the Computer Industry* (Tempus-Microsoft, 1986), pp. 56–69, esp. p. 69.

30. Gary Samuels, "CD-ROM's First Big Victim," *Forbes*, February 28, 1994: 42ff. Richard A. Melchar, "Dusting Off Britannica," *Business Week*, October 20, 1997: 143ff.

31. For the best account of Microsoft's multimedia developments, see Randall E. Stross *The Microsoft Way: The Real Story of How the Company Outsmarts Its Competition* (Addison-Wesley, 1996), pp. 78–94. See also Cheryl Tsang: *Microsoft First Generation* (Wiley, 2000), pp. 149–177.

32. Stross, *The Microsoft Way*, p. 67.

33. Gates, *The Road Ahead*, p. 116.

34. "The CD-ROM Hit Parade," *Fortune*, September 10, 1994.

35. Stross *The Microsoft Way*, p. 91.

36. Samuels, "CD-ROM's First Big Victim."

37. Ibid.

38. For an excellent history of Intuit, see Stross, *The Microsoft Way*, pp. 110–125. See also Paul M. Horvitz, "Efficiency and Antitrust Considerations in Home Banking: The Proposed Microsoft-Intuit Merger," *Antitrust Bulletin*, summer 1996: 427–446; Stanley J. Liebowitz and Stephen E. Margolis, *Winners, Losers and Microsoft: Competition and Antitrust in High Technology* (Independent Institute, 1999), pp. 201–206.

39. Interview with Scott Cook in Jager and Ortiz, *In the Company of Giants*, pp. 73–85.

40. Yankee Group, *Software Strategies*, pp. 41–48.

41. Stross, *The Microsoft Way*, p. 113.

42. Liebowitz and Margolis, *Winners, Losers and Microsoft*, p. 204.

43. Michael Oneal, "Scott Cook Wants to Control Your Checkbook," *Business Week*, September 26, 1994: 50–52.

44. "Meca Software's Turnaround," *ABA Banking Journal Online*, September 1997 (http://www.banking.com, accessed December 2000).

45. William Harris [Intuit executive vice president], "Summary of Testimony in Microsoft Antitrust Trial" (http://www.usdoj.gov, accessed December 2000).

46. US Department of Justice, Justice Department Files Antitrust Suit to Challenge Microsoft's Purchase of Intuit, April 27, 1995 (http://www.usdoj.gov, accessed December 2000).

47. Ibid.

48. US Department of Justice, Complaint Case 0184, July 27, 2000 (http://www.usdoj.gov, accessed December 2000).

49. Intuit press releases (http://www.intuit.com, accessed December 2000). A more recent source (*US Business Reporter*, http://www.activemedia-guide.com, accessed March 2001) suggests that Microsoft Money's market share is 40 percent and rising.

Chapter 10

1. The history of Silicon Valley has been well documented. See Martin Kenney, ed., *Understanding Silicon Valley: The Anatomy of an Entrepreneurial Region* (Stanford University Press, 2000).

2. A. D. Chandler and James W. Cortada, "The Information Age: Continuities and Differences," in *A Nation Transformed by Information*, ed. A. D. Chandler and James W. Cortada (Oxford University Press, 2000), esp. pp. 290–299.

3. British Computer Society, "Conference Proceedings: Automatic Programming Languages for Business and Science," *Computer Journal* 5 (1962): 107–139, esp. p. 124.

4. OECD, *Gaps in Technology: Electronic Computers*, passim.

5. Ichbiah and Knepper, *The Making of Microsoft*, p. 118.

6. Saxenian, *Regional Advantage*.

7. Campbell-Kelly, "The Rise and Rise of the Spreadsheet."

8. An "add-in" was more tightly integrated with the base product than an "add-on."

9. An index to US software industry trade associations is maintained by the Council of Regional Information Technology Trade Associations (http://www.crita.org).

10. ACARD, *Software: A Vital Key to UK Competitiveness*, p. 68.

11. Grindley, *The UK Software Industry*.

Sources of Chapter Frontispieces

Chapter 1: *Fortune*, April 1965, p. 213.

Chapter 2: ICP Software Directory, volume 1: Data Processing Management, January 1977, p. 97. Courtesy of Charles Babbage Institute, University of Minnesota, Minneapolis.

Chapter 3: *Datamation*, November 1965, p. 121. Reproduced with permission of Computer Associates and Walter F. Bauer.

Chapter 4: From personal papers of Martin Goetz. Reproduced with permission of Computer Associates and Martin Goetz.

Chapter 5: System Development Corporation monthly magazine, April 1959, p. 11. Courtesy of Charles Babbage Institute, University of Minnesota, Minneapolis.

Chapter 6: *Datamation* special report, September 1981, p. 64. Reproduced with permission of *Software Magazine* and Wiesner Publishing.

Chapter 7: *Byte*, January 1980, p. 81. Reproduced with permission of International Business Machines.

Chapter 8: *Lotus*, October 1985, p. 113. Reproduced with permission of International Business Machines.

Chapter 9: *The Official Sonic the Hedgehog Yearbook* (London: Grandream, 1992).

Chapter 10: Lincoln Laboratory photo P547-10. Reproduced with permission of Lincoln Laboratory and MITRE Corporation.

Bibliography

This bibliography includes the principal works cited in the text—primarily published books, government reports, industry-analyst publications, and substantial academic or business press articles. Non-print sources, oral histories, web sites, and ephemeral press articles are cited in the notes and are not separately listed here.

Advisory Council for Applied Research and Development (ACARD). *Software: A Vital Key to UK Competitiveness.* HMSO, 1986.

Akera, Atsushi. Voluntarism and the Fruits of Collaboration: The IBM User Group, Share. *Technology and Culture* 42 (2001): 710–736.

Annals of the History of Computing 5 (1983), no. 2. Special issue on the IBM 701 computer.

Annals of the History of Computing 5 (1983), no. 4. Special issue on SAGE.

Annals of the History of Computing 6 (1984), no. 1. Special issue on FORTRAN.

Annals of the History of Computing 7 (1985), no. 4. Special issue on COBOL.

Armer, Paul. SHARE—A Eulogy to Cooperative Effort. 1956. Reprinted in *Annals of the History of Computing* 2 (1980): 122–129.

Arthur, Brian W. Competing Technologies, Increasing Returns, and Lock-in by Historical Events. *Economic Journal* 99 (1989): 116–131.

Arthur, Brian W. Increasing Returns and the New World of Business. *Harvard Business Review*, July-August, 1996: 100–109.

Arthur, Brian W. Positive Feedbacks in the Economy. *Scientific American*, February 1990: 92–99.

Asakura, Reiji. *Revolutionaries at Sony: The Making of the Sony PlayStation and the Visionaries Who Conquered the World of Video Games.* McGraw-Hill, 2000.

Aspray, William, ed. *Engineers as Executives: An International Perspective.* IEEE Press, 1995.

Association of Data Processing Service Organizations (ADAPSO), *25 Years.* Washington: ADAPSO, 1986.

Auslander, M. S., et al. The Evolution of the MVS Operating System. *IBM Journal of Research and Development* 25 (1981), September: 471–482.

Barnett, C. C., et al. *The Future of the Computer Utility.* New York: American Management Association, 1967.

Barron, David W. *Computer Operating Systems.* Chapman and Hall, 1971.

Bashe, Charles J., Lyle R. Johnson, John H. Palmer, and Emerson W. Pugh. *IBM's Early Computers.* MIT Press, 1986.

Bauer, W. F. Informatics: An Early Software Company. *Annals of the History of Computing* 18 (1996), no. 2: 70–76.

Baum, Claude. *The System Builders: The Story of SDC.* Santa Monica: SDC, 1981.

Bell, William D. *A Management Guide to Electronic Computers.* McGraw-Hill, 1957.

Biglow, R. P. Legal Aspects of Proprietary Software. *Datamation,* October 1968: 40–47.

Boehm, Barry W. Software and Its Impact: A Quantitative Assessment. *Datamation,* May 1973: 48–59.

Bower, Joseph L., and Clayton M. Christensen. Disruptive Technologies: Catching the Wave. *Harvard Business Review,* January-February 1995: 43–53.

Bowker, Geoffrey C., and Susan Leigh Star. *Sorting Things Out: Classification and Its Consequences.* MIT Press, 1999.

Brandt, Richard, et al. Can the US Stay Ahead in Software. *Business Week,* March 11, 1991: 62–67.

Brooks, Frederick P. Jr. *The Mythical Man-Month: Essays on Software Engineering.* Addison-Wesley, 1975.

Brooks, John. *The Go-Go Years: The Drama and Crashing Finale of Wall Street's Bullish 60s.* Wiley, 1973, 1999.

Burck, Gilbert. The Assault on Fortress I.B.M.. *Fortune,* June 1964: 112–116, 196, 198, 200, 202, 207.

Burck, Gilbert, *The Computer Age.* Harper & Row, 1965.

Burck, Gilbert. The Computer Industry's Great Expectations. *Fortune,* August 1968: 92–97, 142, 145–146.

Burton Grad Associates Inc. *Evolution of the US Packaged Software Industry.* Tarrytown, N.Y.: Burton Grad Associates Inc., 1992.

Business Communications Corp. *Software Packages: An Emerging Market.* Stamford, Conn.: Business Communications Corp., 1980.

Campbell-Kelly, Martin. *ICL: A Business and Technical History.* Clarendon, 1989.

Campbell-Kelly, Martin. Development and Structure of the International Software Industry, 1950–1990. *Business and Economic History* 24 (1995), winter: 73–110.

Campbell-Kelly, Martin, and William Aspray. *Computer: A History of the Information Machine.* Basic Books, 1996.

Campbell-Kelly, Martin. The Rise and Rise of the Spreadsheet. In M. Campbell-Kelly, M. G. Croarken, R. Flood, and E. Robson, eds., *From Sumer to Spreadsheet.* Oxford University Press, forthcoming.

Carlston, Douglas G. *Software People: An Insider's Look at the Personal Computer Software Industry.* Simon & Schuster, 1985.

Carlton, Jim. *Apple: The Inside Story of Intrigue, Egomania, and Business Blunders.* Random House, 1997.

Carroll, Paul. *Big Blues: The Unmaking of IBM.* Crown, 1993.

Ceruzzi, Paul E. *A History of Modern Computing.* MIT Press.

Chandler, A. D. *The Visible Hand: The Managerial Revolution in American Business.* Harvard University Press, 1977.

Chandler, A. D., and James W. Cortada. The Information Age: Continuities and Differences. In A. D. Chandler Jr. and James W. Cortada, eds., *A Nation Transformed by Information.* Oxford University Press, 2000.

Chposky, James, and Ted Leonsis. *Blue Magic: The People, Power and Politics Behind the IBM Personal Computer.* Facts on File, 1988.

Church, Jeffrey, and Neil Gandal. Network Effects, Software Provision, and Standardization. *Journal of Industrial Economics* 60 (1992): 85–103.

Clippinger, Richard F. FACT. *Computer Journal* 5 (1962): 112–125.

Cohen, Scott, *Zap! The Rise and Fall of Atari.* McGraw-Hill, 1984.

Conatser, Kelly R. 1-2-3 Through the Years. *Lotus,* June 1992: 38–45.

Cortese, Amy, and Ira Sager. Gerstner at the Gates. *Business Week,* June 19, 1995: 30–32.

Creasy, R. J. The Origin of the VM/370 Time-Sharing System. *IBM Journal of Research and Development* 25 (1981), September: 483–490.

Creative Strategies International. *Computer Home Software.* San Jose: Creative Strategies International, 1983.

Creative Strategies International. *Cartridge-Based Software: Further Developments.* San Jose: Creative Strategies International, 1984.

Cringely, Robert X. *Accidental Empires: How the Boys of Silicon Valley Made Their Millions, Battle Foreign Competition, and Still Can't Get a Date.* Addison-Wesley, 1992.

Cusumano, Michael A. *Japan's Software Factories: A Challenge to US Management.* Oxford University Press, 1991.

Cusumano, Michael A., and Richard W. Selby. *Microsoft Secrets.* Free Press, 1995.

Cusumano, Michael A., and David B. Yoffie. *Competing on Internet Time: Lessons from Netscape and Its Battle with Microsoft.* Free Press, 1998.

Cusumano, Michael A., Yiorgos Mylonadis, and Richard S. Rosenbloom. Strategic Manoeuvring and Mass-Market Dynamics: The Triumph of VHS over Beta. *Business History Review,* spring 1992: 51–94.

Dreyer, Jerome L. The ADAPSO Story. *Datamation,* March 1970: 55–58.

du Gay, Paul, et al. *Doing Cultural Studies: The Story of the Sony Walkman.* Sage, 1997.

Dunphy, Ed. *The Unix Industry,* second edition. Wiley, 1994.

Dyer, Davis. *TRW: Pioneering Technology and Innovation Since 1900.* Harvard Business School Press, 1998.

Ecklund, Jon. The Reservisor Automated Airline Reservation System: Combining Communications and Computing. *Annals of the History of Computing* 16 (1994): 62–69.

Edgerton, D. *Science, Technology and the British Industrial Decline, 1870–1970.* Cambridge University Press, 1996.

Edmondson, Gail, et al. Silicon Valley on the Rhine [re. SAP]. *Business Week,* November 3, 1997: 40–47.

Edwards, Paul N. *The Closed World: Computer and the Politics of Discourse in Cold War America.* MIT Press, 1997.

Fertig, Robert T. *The Software Revolution: Trends, Players, Market Dynamics in Personal Computer Software.* North-Holland, 1985.

Fisher, Franklin M., James W. McKie, and Richard B. Mancke. *IBM and the US Data Processing Industry: An Economic History.* Praeger, 1983.

Fishman, Katherine D. *The Computer Establishment.* Harper & Row, 1981.

Flamm, Kenneth. *Targeting the Computer: Government Support and International Competition.* Brookings Institution, 1987.

Flamm, Kenneth. *Creating the Computer: Government, Industry, and High Technology.* Brookings Institution, 1988.

Forman, Richard L. *Fulfilling the Computer's Promise: The History of Informatics 1962–1982.* Woodland Hills, Calif.: Informatics General Corp., 1985.

Frank, Werner L. The History of Myth No. 1. *Datamation*, May 1983: 252–256.

Frank, Werner L. *Critical Issues in Software: A Guide to Software Economics, Strategy, and Profitability.* Wiley-Interscience, 1983.

Freeman, Peter, and William Aspray. *The Supply of Information Technology Workers in the United States.* Washington: Computing Research Association, 1999.

Frieberger, Paul, and Michael Swaine. *Fire in the Valley: The Making of the Personal Computer.* Osborne/McGraw-Hill, 1984; McGraw-Hill, 1999.

Frost & Sullivan. *The Computer Software and Services Market.* New York: Frost & Sullivan, 1971.

Frost & Sullivan. *Data Base Management Services Software Market.* New York: Frost & Sullivan 1979.

Frost & Sullivan. *MSA, The Software Company.* New York: Frost & Sullivan, 1982.

Frost & Sullivan. *The Text Processing Software Market in the US.* New York: Frost & Sullivan, 1982.

Gallagher, James D. *Management Information Systems and the Computer.* New York: American Management Association, 1961.

Gannon, Paul. *Trojan Horses and National Champions: The Crisis in Europe's Computing and Telecommunications Industry.* London: Apt-Amatic Books, 1997.

Gaston-Breton, Tristan. *La Saga Cap Gemini: 30 Milliards en 30 Ans.* Paris: Point de Mire, 1999.

Gates, Bill. *The Road Ahead.* Viking, 1995.

Geraghty, Jim. *CICS Concepts and Uses: A Management Guide.* McGraw-Hill, 1994.

Gilbert, Richard J. Networks, Standards, and the Use of Market Dominance: Microsoft (1995). In J. Kwoka Jr. and L. White, eds., *The Antitrust Revolution.* Oxford University Press, 1999.

Glass, Robert L., ed. *In the Beginning: Personal Recollections of Software Pioneers.* IEEE Computer Society Press, 1997.

Graham, Lawrence D. *Legal Battles that Shaped the Computer Industry.* Quorum Books, 1999.

Grindley, Peter C. *The UK Software Industry: A Survey of the Industry and Evaluation of Policy.* London Business School, 1988.

Grindley, Peter C. The Future of the Software Industry in the United Kingdom: The Limits of Independent Production. In David C. Mowery, ed., *The International Computer Software Industry.* Oxford University Press, 1996.

Gruenberger, Fred, ed. *Expanding Use of Computers in the 70's: Markets—Needs—Technology.* Prentice-Hall, 1971.

Hafner, Katherine M. How Computer Associates Climbed to No. 1 in Software. *Business Week,* July 11, 1988: 50–51.

Hall, Mark, and John Barry. *Sunburst: The Ascent of Sun Microsystems.* Contemporary Books, 1990.

Hammer, Michael, and J. Champy. *Re-engineering the Corporation: A Manifesto for Business Revolution.* Harper Business, 1993.

Harris, Brian. *BABS, BEACON and BOADICEA: A History of Computing in British Airways and Its Predecessor Airlines.* Speedwing, 1993.

Head, Robert V. *A Guide to Packaged Systems.* Wiley-Interscience, 1971.

Heeks, Richard. *India's Software Industry.* Sage, 1996.

Herman, Leonard. *Phoenix: The Fall and Rise of Videogames,* second edition. Rolenta, 1997.

Herz, J. C. *Joystick Nation: How Videogames Gobbled Our Money, Won Our Hearts, and Rewired Our Minds.* London: Abacus, 1997.

Hiltzik, Michael A. *Dealers of Lightning: Xerox PARC and the Dawn of the Computer Age.* Harper Business, 1999.

Hoch, Detlev J., Cyriac R. Roeding, Gert Purkert, and Sandro K. Lindner. *Secrets of Software Success: Management Insights from 100 Software Firms Around the World.* Harvard Business School Press, 1999.

Horvitz, Paul M. Efficiency and Antitrust Considerations in Home Banking: The Propsoed Microsoft-Intuit Merger. *Antitrust Bulletin,* summer 1996: 427–446.

Hoskins, Jim. *IBM ES/9000: A Business Perspective.* Wiley, 1994.

Hughes, Thomas Parke. *Networks of Power: Electrification in Western Society, 1880–1930.* Johns Hopkins University Press, 1983.

IBM. Final Report of the SPREAD Task Group, December 28, 1961. *Annals of the History of Computing* 5 (1983), no. 1: 6–26.

Ichbiah, Daniel, and Susan L. Knepper. *The Making of Microsoft.* Rocklin, Calif.: Prima, 1991.

IEEE Annals of the History of Computing 24 (2002), no. 2. Special issue on the early software industry.

Imlay, John. *Jungle Rules: How to Be a Tiger in Business.* Dutton, 1994.

International Resource Development Inc. *Computer Services and Software Markets.* New Canaan: International Resource Development Inc., 1975.

International Resource Development Inc. *Microcomputer Software Packages.* Norwalk: International Resource Development Inc., 1983.

International Resource Development Inc. *Database Management Systems.* Norwalk: International Resource Development Inc., 1987.

Jacobs, John F. *The SAGE Air Defense System: A Personal History.* Bedford: MITRE Corp., 1981.

Jager, Rama Dev, and Rafael Ortiz. *In the Company of Giants: Candid Conversations with Visionaries of the Digital World.* McGraw-Hill, 1997.

Johnson, Luanne James. A View from the 1960s: How the Software Industry Began. *Annals of the History of Computing* 20 (1998), no. 1: 36–42.

Kenney, Charles C. *Riding the Runaway Horse: The Rise and Decline of Wang Laboratories.* Little, Brown, 1992.

Kenney, Martin, ed. *Understanding Silicon Valley: The Anatomy of an Entrepreneurial Region.* Stanford University Press, 2000.

Kestin, Hesh. *Twenty-First-Century Management: The Revolutionary Strategies that Have Made Computer Associates a Multibillion Dollar Software Giant.* Atlantic Monthly Press, 1992.

Killen, Michael. *IBM: The Making of the Common View.* Harcourt Brace Jovanovich, 1988.

Kraft, Philip. The Industrialization of Computer Programming in Andrew Zimbalist (ed.) *Case Studies on the Labor Process.* Monthly Review Press, 1979), 1–17.

Kubie, Elmer C. Recollection of the First Software Company. *Annals of the History of Computing* 16 (1994): 65–71.

Kuhn, Thomas. *The Structure of Scientific Revolutions.* University of Chicago Press, 1962.

Kurtzig, Sandra L. *CEO: Building a $400 Million Company From the Ground Up.* Norton, 1991.

Lammers, Susan. *Programmers at Work: Interviews with 19 Programmers Who Shaped the Computer Industry.* Redmond, Washington: Tempus-Microsoft Press, 1986.

Lawrence, Bill. Three Suite Deals. *Byte,* March 1994: 120–124, 6.

Lecht, Charles Philip. *The Management of Computer Programming Projects.* American Management Association, 1967.

Lecht, Charles Philip. *The Waves of Change: A Techno-Economic Analysis of the Data Processing Industry.* New York: Advanced Computer Techniques Corp., 1977.

Lesourne, J., and R. Armand. A Brief History of the First Decade of SEMA. *Annals of the History of Computing* 13 (1991): 341–349.

Levering, Robert, Michael Katz, and Milton Moskowitz. *The Computer Entrepreneurs: Who's Making it Big and How in America's Upstart Industry.* New American Library, 1984.

Levin, Doron P. *Irreconcilable Differences: Ross Perot versus General Motors.* Little, Brown, 1989.

Levy, Steven. *Hackers: Heroes of the Computer Revolution.* Penguin, 1994.

Levy, Steven. *Insanely Great: The Life and Times of Macintosh, the Computer that Changed Everything.* Penguin, 1994.

Liebowitz, Stanley J., and Stephen E. Margolis. *Winners, Losers and Microsoft: Competition and Antitrust in High Technology.* Oakland: Independent Institute, 1999.

Lloyd, Tom. *Dinosaur & Co: Studies in Corporate Evolution.* Routledge & Kegan Paul, 1984.

Lohr, Steve. *Go To: The Story of The Programmers Who Created the Software Revolution.* Basic Books, 2001.

Machover, Carl, ed. *The CAD/CAM Handbook.* McGraw-Hill, 1989, 1996.

MacNeal, Richard H. *The MacNeal-Schwendler Corporation: The First Twenty Years.* Los Angeles: MacNeal-Schwendler Corp., 1987.

Manes, Stephen, and Paul Andrews. *Gates: How Microsoft's Mogul Reinvented an Industry—and Made Himself the Richest Man in America.* Simon and Schuster, 1994.

McClellan, Stephen T. *The Coming Computer Industry Shakeout.* Wiley, 1984), 243.

McGee, W. C. Database Technology. *IBM Journal of Research and Development* 25 (1981), September: 505–519.

McKenney, James L. *Waves of Change: Business Evolution through Information Technology.* Harvard Business School Press, 1995.

McLaughlin, Richard A. Software Packages for System/3. *Datamation,* June 1973: 66–71.

McLellan, Stephen T. *The Coming Computer Industry Shakeout: Winners, Losers, and Survivors.* Wiley, 1984.

Meissner, Gerd. *SAP: Inside the Secret Software Power.* McGraw-Hill, 2000.

Metropolis, N., J. Howlett, and G.-C. Rota, eds. *A History of Computing in the Twentieth Century.* Academic Press, 1980.

Mindell, David. The Rise of Relational Databases. in National Research Council, *Funding a Revolution: Government Support for Computing Research.* Washington: National Academy Press, 1999.

MITRE Corp. *MITRE: The First Twenty Years.* Bedford: MITRE Corp., 1979.

Moritz, Michael. *The Little Kingdom.* Morrow, 1984.

Mounce, David C. *CICS: A Light Hearted Chronicle.* Winchester: IBM UK Laboratories, 1994.

Mowery, David C., ed. *The International Computer Software Industry.* Oxford University Press, 1996.

Myers, Edith. CSC: A Hectic 25 Years. *Datamation*, March 1984: 96 ff.

Myers, Ware. An Assessment of the Competitiveness of the United States Software Industry. *Computer*, March 1985: 81–92.

Naur, Peter, and Brian Randell, eds. *Software Engineering.* Brussels: Scientific Affairs Division, NATO, 1969.

Nelson, Richard, and Sidney G. Winter. *An Evolutionary Theory of Economic Change.* Harvard University Press, 1982.

Norris, Grant, et al. *SAP: An Executive's Comprehensive Guide.* Wiley, 1998.

OECD. *Gaps in Technology: Electronic Computers.* Paris: OECD, 1969.

OECD. *Software, an Emerging Industry.* Paris: OECD, 1985.

Oldfield, Homer R. *King of the Seven Dwarfs: General Electric's Ambiguous Challenge to the Computer Industry.* IEEE Computer Society Press, 1996.

Organic, Elliot I. *The MULTICS System: An Examination of Its Structure.* MIT Press, 1980.

Palmer, Ian. *Database Systems: A Practical Reference.* London: CACI, 1975.

Parker, R. W. The SABRE System. *Datamation*, September 1965: 49–52.

Parkhill, D. F. *The Challenge of the Computer Utility.* Addison-Wesley, 1966.

Petersen, W. E. *AlmostPerfect: How a Bunch of Regular Guys Built WordPerfect Corporation.* Prima, 1994.

Petre, Peter. The Man Who Keeps the Bloom on Lotus. *Fortune*, June 10, 1985: 92–94, 96, 98, 100.

Phister, Montgomery Jr. *Data Processing: Technology and Economics*, second edition. Digital Press, 1979.

Poole, Steven. *Trigger Happy: The Inner Life of Videogames.* London: Fourth Estate, 2000.

Postley, J. A. Mark IV: Evolution of a Software Product. *Annals of the History of Computing* 20 (1998), no. 1: 43–50.

Postley, J. A. The Mark IV System. *Datamation*, January 1968: 28–30.

PriceWaterhouseCoopers. *Forecasting a Robust Future: An Economic Study of the US Software Industry.* Washington DC: Business Software Alliance, 1999.

Pugh, Emerson W. *Building IBM: Shaping an Industry and Its Technology.* MIT Press, 1995), 190.

Pugh, Emerson W., Lyle R. Johnson, and John H. Palmer. *IBM's 360 and Early 370 Systems.* MIT Press, 1991), 293.

Read, Stuart. *The Oracle Edge.* Adams Media, 1999.

Ritchie, D. M. The Evolution of the Unix Time-Sharing System. *AT&T Bell Laboratories Technical Journal* 63 (1984), October: 1577–1593.

Rosenberg, Nathan. *Inside the Black Box: Technology and Economics.* Cambridge University Press, 1982.

Rosenbloom, Richard S., and Clayton M. Christensen. Technological Discontinuities, Organizational Capabilities, and Strategic Commitments. *Industrial and Corporate Change* 3 (1994): 655–685.

Ryan, John M. *It Can Be Done!* Newbury, Berkshire: Scope Books, 1979.

Salus, Peter H. *A Quarter Century of Unix.* Addison-Wesley, 1994.

Saxenian, AnnaLee. *Regional Advantage: Culture and Competition in Silicon Valley and Route 128.* Harvard University Press, 1994.

Servan-Schreiber, J. J. *The American Challenge.* Hamish Hamilton, 1968.

Sheff, David. *Game Over: How Nintendo Zapped an American Industry, Captured Your Dollars, and Enslaved Your Children.* Random House, 1993.

Sigel, Efrem, and Louis Giglio. *Guide to Software Publishing: An Industry Emerges.* Knowledge Industry Publications, 1984.

Sigel, Efrem, et al. *Business/Professional Microcomputer Software Market, 1984–86.* Knowledge Industry Publications, 1984.

Siwek, Stephen E., and Harold W. Furchtgott-Roth. *International Trade in Computer Software.* Quorum Books, 1993.

Slater, Robert. *Portraits in Silicon.* MIT Press, 1987.

Smith, Douglas K., and R. C. Alexander. *Fumbling the Future: How Xerox Invented, Then Ignored, the First Personal Computer.* Morrow, 1988.

Steffens, John. *Newgames: Strategic Competition in the PC Revolution.* Pergamon, 1994.

Stross, Randall E. *The Microsoft Way: The Real Story of How the Company Outsmarts Its Competition.* Addison-Wesley, 1996.

Torrisi, Salvatore. *Industrial Organisation and Innovation: An International Study of the Software Industry.* Elgar, 1998.

Tsang, Cheryl. *Microsoft First Generation.* Wiley, 2000), 149–177.

Turkle, Sherry. *The Second Self: Computers and the Human Spirit.* Simon & Schuster, 1984.

US Department of Commerce. *A Competitive Assessment of the United States Software Industry.* Washington: US Department of Commerce, 1984.

Voth, Ben. *A Piece of the Computer Pie.* Houston: Gulf, 1974.

Walker, John, ed. *The Autodesk File: Bits of History, Words of Experience,* third edition. Thousand Oaks, Calif.: New Riders, 1989.

Wallace, James, and Jim Erickson. *Hard Drive: Bill Gates and the Making of the Microsoft Empire.* Wiley, 1992.

Wang, An. *Lessons: An Autobiography.* Addison-Wesley, 1986.

Wang, Charles B. *Techno Vision.* McGraw-Hill, 1994. Second edition, *Techno Vision II,* 1997.

Watkins, Ralph. *A Competitive Assessment of the US Video Game Industry.* Washington: US International Trade Commission, March 1984.

Watson, Thomas J. Jr., and Peter Petre. *Father and Son & Co: My Life at IBM and Beyond.* Bantam, 1990.

Welke, Larry. The Origins of Software. *Datamation,* December 1980: 127–130.

Wildes, Karl L., and Nilo A. Lindgren. *A Century of Electrical Engineering and Computer Science at MIT, 1882–1982.* MIT Press, 1986.

Wilson, Mike. *The Difference between God and Larry Ellison: Inside Oracle Corporation.* Morrow, 1996.

Withington, Frederic G. The Golden Age of Packaged Software. *Datamation,* December 1980: 131–134.

Wohl, Amy D. What's Happening in Word Processing. *Datamation,* April 1977: 65–74.

Yankee Group. *Software Strategies: The Home Computer and Video Game Marketplace.* Boston: Yankee Group, 1984.

Yankee Group. *IBM's Future Software Strategy.* Boston: Yankee Group, 1989.

Yasaki, Edward K. European Software Market. *Datamation,* December 1967: 27–31.

Yates, JoAnne. Application Software for Insurance in the 1960s and early 1970s. *Business and Economic History* 24 (1995), no. 1: 123–134.

Index

Abacus, 99
Accounting software, 139, 156–161, 183, 221, 223
Activenture, 289
Activision, 7, 275, 284
Ada, 310
Adabas, 146–148, 151
Addison-Wesley, 307
Adobe Systems, 236, 261, 262
  Acrobat, 262
  Illustrator, 255, 263
Advanced Computer Techniques, 57, 58, 66–69, 74
Advanced Information Systems, 104
Advanced Micro Devices, Inc., 239
Adventure International, 277
Advisory Council for Applied Research and Development, 24, 309
Airline reservation systems, 65, 112, 115, 149, 152, 153. *See also* SABRE
Aldus, 236, 255, 261, 263
Allen-Bradley Inc., 106
Allen, Paul, 202–207, 253
Allied Leisure, 273
Allways, 253
Altair, 202, 204, 221
Amdahl Computer Corporation, 143
Amdahl, Gene, 143
American Airlines, 41–45
American Banking Association, 99
American Management Association, 66–69
American Management Systems, 165, 172, 178
America Online, 297
AmeriTech, 184, 189
Ampex Corporation, 49, 188
Anacomp, 154
Ansa, 256, 263
Apple Computer, 202, 243, 247, 250, 260
  Apple [I], 202
  Apple II, 202–206, 213, 217, 220, 221, 224, 227, 253
  Lisa, 247
  Macintosh, 247, 253, 256, 258, 261, 262, 306
  vs. Microsoft, 250
Application Junction, 256
Application Service Providers, 162
Application Software, 117
Applicon, 160, 244
Applied Data Research, 6, 36, 57–59, 66, 67, 73, 74, 79, 100, 101, 105, 107, 111–116, 122, 123, 148, 151, 168, 169, 184, 187, 189, 305
  Autoflow, 6, 58, 69, 100–107, 113–116, 135
  Librarian, 116
  MetaCOBOL, 116, 135
  Roscoe, 116
Armand, R., 24
Arthur Andersen and Co., 190, 195
Arthur, Brian, 242
Arthur Young Business Systems, 307
Artzt, Russell, 182
Ashton, Alan, 254
Ashton-Tate, 203, 210–212, 219–220, 227, 236, 251, 256–258, 263, 305

ASK Computer Systems, 123, 131, 154–156, 168, 223
Asset Management Company, 102
Association of Computing Machinery, 66, 146
Association of Data Processing and Services, 1, 13–17, 20–22, 62, 63, 77, 307
Association of Independent Software Companies, 74, 85, 130
AT&T, 144, 207, 240, 255
Atari, 224, 249, 272–275, 280, 284, 288
  Atari 800, 276, 280
  VCS 2600, 275, 279, 280
  VCS 5200, 280
Atlantic Software, 129
Attachmate, 172
Auerbach, Issac, 75
Auerbach Publications, 27, 75, 132
Autodesk, 236, 242–246, 261, 265
  AutoCAD, 244, 245
  AutoDesk, 244
  AutoScreen, 244
Autoflow, *See* Applied Data Research
Automatic Data Processing, 62, 73
Automatic Payrolls Inc., 62
Auto-trol, 160
AVI Institut, 148

Baan, 172, 197
Bachman, Charles, 146
Backus, John, 31, 34
Bally Midway, 273, 284
Bank of America, 48, 49, 298
Bank Street Writer, 226
Barclays Bank, 78
BARIC, 78
Barnaby, Rob, 217
BASIC, 204, 205, 260
Bastian, Bruce, 254
Bauer, Walter F., 58, 65, 66, 71, 75, 80, 96, 104–107
Bedford Software, 183
Bedke, Janelle, 220
Bell Labs, 37, 38, 144
Bendix Aviation Group, 37
Benjamin, Alan, 77

BMC Software, 102, 172
Boehm, Barry, 92–94
Boeing, 110, 115, 197
Boole and Babbage, 101, 102
Borland, 236, 253, 257, 259, 260, 263
BPI Systems, 183, 221, 223
Bricklin, Dan, 213, 214, 252
Britain, 10, 22, 75–80, 130, 138, 150, 166, 175, 223, 300, 303, 309, 310
British Tabulating Machine Company, 48, 49
Broderbund, 7, 210, 226, 277–279, 292
Brooks, Fred, 95
Brown, Walter, 129
Bundling, 177, 258, 285
Burroughs Corporation, 26, 37, 48, 65, 70, 82, 85, 98, 174
Bushnell, Nolan, 272–275
Business Communications, 26, 132
Business Software Alliance, 20

C, 260
CADCON, 244
CalcStar, 258
California Research Association, 51
Calman, 160, 244
Cambridge Software Collaborative, 307
Cap Gemini, 24, 77
Capex, 182
Carlston, Douglas, 24, 277–279
CBS Entertainment, 284
CCA, 117
CD-ROMs, 289, 290
C-E-I-R, 50, 53, 75, 76
CGE, 175
Charles Babbage Institute, 26, 27
Chase Manhattan Bank, 48, 49
ChipSoft Inc., 296
Choplifter, 226
Chu, J. Chuan, 98
Cincom, 126, 145, 146, 151, 155, 168, 187, 189, 198, 307
Client-server computing, 195, 196
Clippinger, Dick, 304
Clustering of software companies, 185, 186, 306–308

Index 363

COBOL, 35, 36, 47, 53, 54, 100, 113, 135, 177, 205, 304
Codd, Edgar F., 31, 187
Cogent II, 115, 117
Coleco, 224, 275, 276, 284, 288
COMDEX, 216, 244, 248, 255
Committee on Data Systems and Languages, 35, 148
Commodore Business Machines, 203, 217, 224, 280
 PET, 203
 VIC-20, 224, 276
Communications Orientated Processing Equipment, 84
Communications Trends, 27
Compilers, 34–36, 70, 100, 135, 177
Complementary software products, 307, 310
Com-plete, 151
Compton's Encyclopedia, 292, 294
Compuflight II, 115
CompuServe, 297, 301
CompuShop, 209
Computax, 73
Computer-aided design, 128, 139, 160–162, 236, 243–245, 261, 265
Computer Analysts and Programmers, 77
Computer Applications Inc., 50, 59, 71, 73, 82
Computer Associates, 9, 10, 27, 59, 103, 126, 136, 148, 156, 165, 168, 169, 173, 178–185, 189, 190, 198, 221, 223, 231, 234, 264, 266, 297, 298, 307
 CA-Datacom/DB, 184
 CA90s architecture, 184
 CA-SORT, 136, 176, 182
Computer Craft, 209
Computer Factory, 209
Computer Industries Inc., 80, 84
Computer Information Management, 148
Computer Leasing Company, 80
Computer Machinery Group, 77
Computer Sciences Corporation, 5, 12, 50–54, 57, 70–73, 76, 83, 84, 115, 117, 195, 205, 304, 305

Computer Sciences International, 76
Computer Services and Software Industry, 57, 59
Computer Services Association, 77
Computer Space, 272
Computer Usage Company, 5, 24, 31, 50–52, 57, 59, 71, 73
Computer Vision, 128, 129, 160–162, 244, 245
ComputerLand, 209, 255
Computers, numbers of, 89, 90
Computers and Software Inc., 62
Computicket, 73
Compuware, 168, 172
Comshare, 63
Configuration Utilization Evaluator, 102
Connecticut Mutual Insurance Company, 219
Consolidated Edison, 149
Continental Software, 227, 294
Continuous Integrated System, 154
Control Data Corporation, 62, 82, 102, 109, 250
 CDC 1604 computer, 80
Cook, Scott, 261, 294, 295
Copyright, *See* Intellectual property issues
Corel, 261, 263
Cornfeld, Bernie, 83
COSMIC, 130
CP/M, 206, 215–218, 239–241, 249
Creative Strategies International, 26
Cricket Software, 183
Cross-platform software, 262
Cullinane Corporation, 86, 122, 126, 148, 165
Cullinane, John, 148, 169, 189, 190
Cullinet, 168, 184, 187–190
Cyrix, 239
Cytation Inc., 291

Data Decisions, 132
Data General, 159, 160, 254
Data Pro, 132
Data Products Corporation, 80
Data Systems Analysts, 58, 59

Data Transmission Company, 84
Databases, 6, 101, 113, 116, 123, 133, 134, 145–149, 156, 176, 184–191, 203, 210, 213–216, 219–221, 234, 244, 251, 256–259, 263, 301, 307
relational, 31, 149, 168, 169, 185–191, 198, 307, 310
DATACOM/DB, 116, 146–148, 151, 184
Dataskil, 78
Datasolv, 78
Datran, 84, 85
dBase II–IV, 7, 203, 210–212, 215, 219, 220, 254–257, 259, 263
Defense Advanced Research Projects Agency, 41
Desktop publishing software, 236, 261, 263
DESQ, 249
DeVries, John, 82
Diana project, 48, 49
Digital Computer Association, 32, 33
Digital Equipment Corporation, 143–145, 159, 160, 173, 188, 195, 213, 306
Digital Research, 203, 206, 207, 217, 234, 239, 241, 248, 249, 264, 289
Disruptive technologies, 186, 187
DistribuPro, 183
Dorling Kindersley, 292
Dow Chemical, 195
Drake, Dan, 243
Draw!, 261
DrawPerfect, 258
Dun & Bradstreet, 168, 172, 191, 195
DuPont, 195

EasyWriter, 183, 209, 215, 217
Economics of increasing returns, 236, 237, 242, 243
Educational software, 225, 227
Edu-Ware, 227
Electric Pencil, 209, 217
Electrologica, 75
Electronic Arts, 283, 284
Electronic Data Services, 12, 62, 63, 73, 77, 82, 83, 154, 195
Electronic Trend Publications, 27

Ellison, Larry, 180, 185, 188, 190
Encyclopedia, CD-ROM multimedia, 269, 288–294, 300
Encyclopaedia Britannica, 288–294
Enterprise Resource Planning software, 12, 168, 169, 172, 191–198
ENVIRON/1, 151
Equitable Life Insurance, 84
Erdwinn, Joel, 66
ERMA, 48, 49
Eubanks, Gordon, 263, 264
EXEC-8, 143
Exodus, 115

Facilities management, 59, 62, 63, 72, 73, 87
FACT, 35, 36, 53, 54, 70, 304, 305
Fairchild, 275
Famicom, 284–286
File management systems, 103–105, 113–117, 132, 133, 145, 305
Financial sector software, 98, 131, 136, 138, 154, 157
Flowcharts, 69, 100, 101, 113, 115. *See also* Autoflow, Flowcharter
FORTRAN, 31, 34–36, 47, 53, 54, 70, 72, 113, 130, 205
Fox Software, 256, 259
Foxbase, 256, 259, 263
Framework, 258
France, 22, 24, 75–77, 175, 306
Frank, Werner L., 65, 91–94, 123
Frankston, Bob, 213, 214, 252
Freelance Programmers Limited, 78
Frogger, 226
Frost & Sullivan, 26, 132
Fujitsu, 143, 174
Funk Software, 253, 307, 308
Fylstra, Dan, 214

Galactic, 277
Game Boy, 286
Garmisch, software engineering conference at, 94, 95
Gartner Group, 27
Gates, Bill, 23, 180, 185, 202–207, 231, 242, 253, 264, 265, 289, 291, 309

Gatt, Lou, 66
GEM, 251
General Electric, 30, 37, 45, 49, 51, 62, 73, 82, 144–148, 155, 197
  Time-Sharing System, 203
General Instrument, 274
General Motors, 35, 70
Genesis, 286
Gerber, 160
Germany, 22, 76, 130, 148, 166, 175, 191–194
Geschke, Charles, 261
Getty Oil, 106
Gibbons, Fred, 210, 220
Goetz, Martin, 58, 67, 100, 101, 111, 113, 123
Go-Go years, 59, 74, 79–86, 121, 165
Gorf, 284
Government funding of R&D, 308–310
Grad, Burton, 110
Grandma and Me, 292
Graphical user interface, 241, 242, 246–253, 256, 258
Graphics software, 183, 214, 216, 247
Grolier Encyclopedia, 290
GUIDE, 33
Gulf Insurance Company, 84
Gupta Technologies, 185

H&R Block, 297, 298
Haefner, Walter, 85
Hawkins, Tripp, 283
Hewlett-Packard, 145, 155, 159, 195, 207, 220, 303
  HP 3000, 220
Hills, Richard, 65
Hitachi, 143, 174
Hogan Systems, 126, 154, 172, 178
Hogan, Bertie, 154
Home Accountant, 227, 294, 295
Home banking, 261, 299, 301
Homeword, 226
Honeywell, 36, 53, 62, 82, 98, 148, 205
  Liberator, 98
  200 series, 98, 101
Hopp, Dietmar, 193

Hopper, Grace, 34, 35
Hoskyns Group, 77
Hoskyns, John, 77
Hughes, 45, 46
Hughes Dynamics, 104. *See also* Advanced Information Systems
Humphrey, Watts, 114
Hurd, Cuthbert C., 31

IBM Applied Science Division, 31
IBM computers
  AN/FSQ-7 (Q7), 38, 39, 41, 44
  AS/400, 172, 197
  Future System, 124
  650, 33, 90, 91
  701, 29–31
  702, 33
  704, 32–34
  705, 33
  1401, 41, 62, 90, 96–101, 115
  3090, 177
  7090, 44
  9370, 177
  PC, 206, 207, 214, 216, 224–227, 237–240, 253–256, 274
  PC Jr., 224
  System/3, 123, 132, 172
  System/32, 136
  System/360, 4, 72, 89–91, 94–96, 101, 102, 105, 114–116, 123, 124, 133, 142, 148, 150, 178, 182, 237, 304, 305
  System/370, 142, 157
IBM Corporation, 6, 46, 48, 59, 62–65, 69, 70, 80, 82, 91, 95–98, 101, 108–118, 123, 125, 134, 136, 138, 143, 149–154, 167, 173–178, 182, 187–190, 195, 197, 232, 236–240, 250–252, 259
IBM Finance Program Exchange Directory, 99
IBM Hursley Laboratories, 150, 151
IBM Information Processing Services, 176
IBM *Pointers*, 30
IBM Program Applications Library, 96, 97

IBM San Jose Research Laboratory, 185–188
IBM software, 173–178
  '62 CFO, 97
  AIX, 145
  Assembler, 113
  ATMS-III, 217
  BOMP, 146
  CICS, 113, 134, 149–152, 176, 266
  CMS, 142
  COMTRAN, 35, 36
  DB2, 189
  DL/1, 146
  DOS, 142
  Flowcharter, 113, 114
  GIS, 113, 117
  IBM Assistant Series, 226
  IMS, 113, 146, 176, 177, 189
  OS/2, 250–256, 265
  OS/360, 94, 95, 111, 135, 142, 150, 176
  PARS, 112
  PC/IX, 242
  SQL/DS, 189
  System/R, 188, 189
  Systems Application Architecture, 177, 178, 184
  TopView, 249
IBM Technical Computing Bureau, 30, 31, 51, 52
IBM word processing systems, 203
IBM/Delta 9072 SABRE, 48
IBM/Pan American PANAMAC, 48
ICT 1301, 304
IDMS, 148, 184, 189
Illustra, 185, 188
Imlay, John, 24, 156
Imperial Chemical Industries Ltd., 193, 195
Information Management Sciences, 202, 209, 217
  IMSAI 8080, 202, 206, 217, 220
InaComp, 209
India, 10, 310, 311
Infocom, 277
Infodata, 117
Infonet, 73, 84

Informatics, 6–8, 23, 57, 58, 64–66, 70, 71, 74, 79, 80, 84, 91, 96, 103–109, 115–118, 123, 126, 133, 145, 155, 168, 169, 180, 214, 221, 305
  Mark IV, 6–8, 58, 80, 103–109, 115–118, 133, 145, 305
Information Builders, 256
Information Sciences, 126
Information Technology Association of America, 13, 17
Information Unlimited, 183
Informix, 169, 188, 307
Ingres, 156, 169, 188
INPUT, 13–17, 20–22, 26, 27, 132, 162
INQUIRE, 117
INSAC, 130
Institute for Numerical Analysis, 104
Insyte Corporation, 148
Integrated Software Systems, 182
Intel Corporation, 148, 201, 238–240
  microprocessors, 204, 205, 238, 246, 249
Intellectual property issues, 98, 107, 108
Intellivision, 284
Interact, 244
Intergraph, 160, 244
International Computer Programs Inc., 25–27, 63, 99, 104, 162
  ICP Business Software Review, 136
  ICP Quarterly, 25, 99, 131, 133, 136, 139
International Computers Limited, 75, 78, 175
International Data Corporation, 13, 22, 26, 132
International Federation for Information Processing, 75, 114
International Reservation Corporation, 73
International Resource Development, 27, 132, 136
Internet, 11, 143, 162, 185, 231, 262, 272, 301, 308, 310
Intex, 307
Intuit, 261, 269, 294–299
Ireland, 80, 311

Italy, 76, 175
ITT, 45, 46

Jacobs, John F., 67–69
Jacquard, 160
Japan, 10, 22, 86, 116, 166, 174, 273, 274, 280, 281, 300
Japanese videogame manufacturers, 284–288
Java, 11
J. D. Edwards, 172, 197
Jobs, Steve, 202, 216
Johnson, Franklin "Pitch," 102
Johnson Systems, 182
Joint Computer Conference, 114
Jones, Fletcher, 33, 52, 53
JOVIAL, 46, 47

Kahn, Phillipe, 263
Kapor, Mitch, 216
Katch, David, 102
Kildall, Gary, 202, 206, 207, 217, 250, 264, 289, 290
Killer applications, 7, 212, 213, 285, 288, 291, 292
Kiplinger, 297
Knowledge Industry Publications, 27
Knowledge-Set, 289
Kolence, Ken, 102
Kriya Systems, 227
Kubie, Elmer, 24, 51, 52, 73
Kurtzig, Sandra, 24, 155

L systems, 5, 46, 74, 75
Laboratory for Electronics, 48, 49
Lanier, 160, 219, 254
Laser printer software, 236, 256, 261
Lautenberg, Frank, 62
Leading Edge Software, 219
LEASCO, 59
Lecht, Charles P., 58, 67, 69, 74
Legent, 168, 180, 185
Lesourne, J., 24
Lincoln Laboratory, 37, 39, 46, 67–69
Linear programming software, 112, 131
Linux, 11
Lockheed, 110

Logica, 77
London Airport Customs Entry System, 76
Look&Link, 253
Lotus Development Corporation, 8, 173, 180, 183, 210–212, 216, 228, 234, 236, 251–259, 263–265, 289, 305, 310
  AmiPro, 259
  1-2-3, 2, 7, 210–212, 216, 247, 252, 258, 289, 297, 306, 307
  SmartSuite, 258
  Symphony, 258
Lunar Lander, 272

Macintosh. *See* Apple Computer
MacNeal, Richard, 24, 126
MacNeal-Schwendler Corporation, 24, 72, 126
Magic Wand, 218
Magnuson Systems Corporation, 143
Managing Your Money, 295–298
MANMAN, 155
Marin Systems, 243
Mario Bros., 283–286
Martin Marietta, 45, 77
Massachusetts Institute of Technology, 37, 38, 48, 144
Mathematica, 117
Mattel, 275
McCormick and Dodge, 191
McDonnell Douglas, 45, 47
McDonnell Douglas Automation, 115
MCP, 143
MECA Software, 295, 297
Micro Instrumentation Telemetry Systems, 202
MicroPro International, 7, 203, 217–219, 227, 234, 240, 254, 255, 258, 305. *See also* WordStar
Microrim, 256
Microsoft Corporation, 7–10, 23, 24, 144, 173, 174, 180, 183, 185, 197, 203–207, 212, 231–242, 248–266, 290, 291, 298
  vs. Apple, 250
  and bundling, 258, 297

Microsoft software
  Access, 263
  Bookshelf, 291
  Cinemania, 292, 293
  Encarta, 288–293, 297
  Excel, 252–258, 265, 297
  Internet Explorer, 262, 263
  Money, 261, 294, 297–299
  MS-DOS, 207, 234, 238–242, 249, 253, 258, 260, 262
  Multiplan, 253, 306
  Office, 258, 259
  PowerPoint, 258, 259
  Windows, 11, 242, 249–258, 262, 265, 266, 297
  Word, 254–258, 265, 269, 297
  Works, 297
Midland Bank, 154
Millard, William, 202, 209
Mills, Bryan, 77
MITRE Corporation, 46, 50, 58, 65–69
Mock, Owen, 66, 70
MODE IV, 143
MRI Systems, 148. *See also* Intel
Material Resource Planning software, 154–156
Management Science America, 126, 139, 156–158, 165–169, 172, 178, 191, 221–223, 234, 305, 307
MULTICS, 144
MultiMate, 219
Multitasking, 246, 247
Myst, 292

Namco, 280
NASTRAN, 128, 130
National Aeronautics and Space Administration, 41, 65, 72, 128, 130, 219
National CSS, 63
National Dairy Industries, 106
National Enterprise Board, 130
National Machine Accountants Association, 32
National Semiconductor, 143
NATO, 94
NCR, 49, 62, 82, 98
  315 computer, 98

NEC, 143, 174
Netherlands, 75, 76, 80, 172, 197
Netscape Communications, 262
NetWare, 260
Nies, Tom, 145
Nintendo, 281, 284–287, 288
Nintendo Entertainment System, 285–287
Nixdorf, 175
North American Aviation, 32
Norton Utilities, 260, 264
Novell, 144, 236, 259, 260, 263, 298, 299
Nutt, Roy, 33, 34, 52, 53, 66, 70
Nutting Associates, 272, 273

Office automation, 128, 159, 160. *See also* Word processing
Office suites, 257–259
Olivetti, 175
One Source, 289
On-Line Software, 184
Open-source software, 11, 143–145
Operating systems, 11, 94, 95, 111, 113, 125, 133–135, 141–145, 149, 175, 176, 206, 207, 234, 236, 238–242, 246–251, 261, 262, 265, 266
Oracle Systems, 9, 11, 12, 24, 123, 168, 169, 180, 184–191, 196–198, 231, 266, 307
Organization for Economic Cooperation and Development, 24

Pacific Gas & Electric, 195, 196
PACT, 32
Page, John, 220
Paladin Software, 249, 252
Pansophic, 101–103, 126, 165, 168, 184, 305, 307
PANVALET, 103
Paradox, 256, 259, 263
PARS, 112
Pascal, 260, 261
Patents. *See* Intellectual property issues
Patrick, Robert, 52, 53
Patterson, Tim, 240
Payroll software, 133, 221
PC/Focus, 256

Peachtree Software, 221–223, 234
PeopleSoft, 172, 197
Performance monitoring software, 102
Perot, H. Ross, 62, 82
Personal finance software, 226, 227, 261, 269, 294–301
Personal Software, 203, 210–216
Personics, 253, 307, 308
Peter Norton Computing, 260, 264
Peterson, W. E., 24
PFS Series software, 215, 218–221, 226, 256, 259. *See also* Software Publishing Corporation
PGA Tour Golf, 283
Phase-Four Systems, 160
Philco, 70
Phillips, 70, 75, 76, 289
Piscopo, Joe, 102, 103
Pitfall!, 275
PL/1, 113
PL/M, 206
Plan Calcul, 75
Planning Research Corporation, 45, 50, 71, 73, 76, 77
PlanPerfect, 258
Plattner, Hasso, 191, 193
Policy Management Systems, 172
Pong, 269–274
Postley, John, 104–109, 305
Postscript, 261
Precision Instrument, 188
Prime, 159, 160
Problem Program Evaluator, 102
Processing services, 59–62
Program Store, 209
Programming languages, 11, 34–36, 113, 186, 198, 204–207, 236, 244, 260, 261, 310
Programming Methods Inc., 59
Programming tools, 125, 133–135
Programs Unlimited, 209
Proulx, Tom, 295
Prudential Insurance Company, 98, 106

Quantum Services, 13
Quarterdeck, 249
Quattro, 253, 259

QuickBooks, 296
Quicken, 261, 294–299

R:base, 256
Radio Shack. *See* Tandy
RAMIS, 117
Ramo, Simon, 45
Ramo-Wooldridge Corporation, 45, 46
RAND Corporation, 5, 32, 33, 38–41, 45, 46, 50, 92, 94, 104
  System Development Division, 38
Random House, 227
Rapidata, 63
Ratcliffe, Wayne, 210, 219, 220, 257
RCA, 36, 37, 45, 46, 49, 65, 70, 82, 100, 275, 306
  501 computer, 100
  601 computer, 36, 58
  Spectra 70 computer, 101
Real-time software, 153, 154, 193, 309. *See also* SABRE; SAGE
Recreational software, 4, 7, 162, 207, 210, 225, 226, 269–301
Relational Database Systems, 185
Relational Systems Inc., 188. *See also* Oracle Systems
Relational Technology Inc., 185, 188. *See also* Ingres
Ritchie, Dennis, 144
Rome Air Defense Center, 66
Rubenstein, Seymour, 217

SABRE, 41–45, 48, 149
SAGE, 5, 37–41, 44, 45, 48, 58, 67–69, 309
SAGE plc., 223
Samna, 259
Santa Cruz Operation, 144, 236, 261, 262
SAP, 12, 24, 126, 166–169, 172, 191–198
  R/2, 193–195
  R/3, 191, 195–197
  System R, 193
SAS Institute, 148, 168
Satellite Software International, 254, 255
Scantlin, 63

Scicon, 75, 78
Science of Learning, 227
SCORE, 129
Scripsit, 217
Sega, 281, 286, 287
Service bureaus, 59–62
SHARE, 31–34, 52, 54, 66, 96, 97, 109
Sheldon, John, 31, 51, 52
Shirley, Stephanie "Steve," 78, 79
Siemens, 175
Sierra On-Line, 210, 226, 277, 292
Silicon Valley, 303, 306
SimCity, 281
Simply Money, 297
Sinclair, 224
Sirius, 226
Smith, C. R., 43, 44
Smith, R. Blair, 32, 43
Sociétié d'Economie et de Mathématiques Appliquées, 24, 76
SofTech, 239
SoftSel, 209
Softwaire Center International, 209
Software, major categories of. *See also* Databases; Operating Systems; Real-time software; Turnkey products and suppliers
   applications, 97, 111–113, 123, 125, 133, 136–141, 145, 149, 176, 183, 191, 209, 253
   banking, 48, 49, 78, 97, 98, 99, 133, 136, 152–154, 172, 178, 183, 224
   cross-industry 133, 139–141, 156–161, 169–172, 176, 221–223
   health care, 133, 157, 172, 223, 224
   industry-specific, 133, 136–139, 152–156, 169–172, 176, 223, 224
   insurance, 97, 98, 133, 136, 172
   manufacturing, 133, 154–156, 157, 172, 197
   productivity applications, 214–216, 234, 236, 251–259
   project management, 5, 67–70, 112, 133, 139
   retail industry, 98, 128, 172
   utilities, 125, 133–136, 176, 182, 183, 244

Software AG, 130, 148, 151, 166, 187
Software Arts, 7, 210, 214, 216, 252
Software brokers, 124, 129, 130, 162
Software City, 209
Software Dimensions Inc., 221, 223
Software History Center, 27, 101
Software International, 182
Software Plus, 220
Software Publishers Association, 17–20, 269–271
Software Publishing Corporation, 210, 218, 219, 221, 226, 240, 256, 259
Software, errors in, 95, 125
Sonic the Hedgehog, 284, 286, 309
Sony, 281, 287–289
Sony PlayStation, 287, 288
Sorcim, 183, 212, 215–218, 234, 305
Sorting software, 101, 102, 136, 244
Space Invaders, 273
Space War, 272
Speedata, 73, 83
SpellStar, 218
Sperry Rand, 70
Spreadsheets, 7, 131, 183, 203, 213–216, 222, 231, 234, 246, 247, 251–254, 263, 269, 281, 301, 306
Spyglass, 263
SQL, 187
Stanford Research Institute, 49
Steinberg, Saul, 59
Sterling Software, 168, 169, 178–180
Stonebraker, Michael, 188
Strachey, Christopher, 304
Strategic Air Command Control System, 46
Strong, Jack, 32
Structured programming, 69
Sun Microsystems, 144
Sun Oil, 106
SunSoft, 144
SuperCalc, 7, 183, 212, 215, 216, 306
SuperWriter, 218
SUPRA, 189
Swift, Charles, 66
Sybase, 169, 185, 262, 307
Symantec, 236, 263, 264

SyncSort, 102, 103, 136, 176, 182
System 2000, 146–148
System Development Corporation, 5, 23, 26, 36–41, 45–47, 50, 54, 65, 71–74, 84, 115, 117, 174, 308, 309
System Development Laboratories, 188. *See also* Relational Systems Inc.
Systems and Programs Limited, 77
Systems software, 97, 111–114, 125, 133–136, 175, 177, 236, 253
Systems Software Associates, 172, 197

Taito, 273
Tandy, 203, 209, 224
  TRS-80, 203, 206, 214, 217, 221 224, 276, 277
  Computer Centers, 209
Tate, George, 220
Taub, Henry, 62
Taxonomy of software industry, 2, 8, 9, 131–134, 139, 141, 208
Teleprocessing, 41, 59, 63, 65, 83, 133, 134, 149–152, 176
Teleregister Corporation, 42, 43, 48
Terman, Frederick, 303
Tetris, 281
Texas Instruments, 49, 224, 276
Thatcher, Richard, 129
Thomson, Ken, 144
Time-sharing services and systems, 124, 131, 142, 162, 204
Timex, 224
TK!Solver, 214
Tobias, Andrew, 295
Tomb Raider, 288
Top Secret Software, 182
TOTAL, 146, 151, 189
Trade associations, 1, 11–22, 308
Traf-O-Data, 204
Transaction processing, 266
Triad Systems, 128
Triton, 197
TRW, 46, 47, 65, 69, 70, 115, 117, 204
Turbo Pascal, 260, 261
TurboTax, 296
Turnkey products and suppliers, 112, 124, 128–129, 139, 155, 159–162

Tymshare, 63, 155, 203
Typing Tutor, 227

Uccel, 85, 168, 182–183
Ultima, 283
ULTRA, 189
Ultrix, 145
Unbundling of software, 6, 13, 82, 98, 109–118, 121, 125, 134, 142, 146, 149, 161, 174, 193, 198, 237
Unify, 185
Unisys, 143, 173
United Aircraft Corporation, 33, 34
Univac, 34, 35, 48, 75, 82, 148, 174. *See also* Remington Rand; Sperry Rand; Unisys
Univac A-0 compiler, 34
Univac B-0 complier (FLOW-MATIC), 35, 36
Univac computers
  1107, 70
  1108, 73
  LARC, 70
Univac [I], 1, 30
University Computing Corporation, 64, 79–85, 126, 139. *See also* Uccel
Unix, 134, 143–145, 188, 206, 236, 240, 261, 262
Unix Software Laboratories Inc., 144
US Census Bureau, 1, 12, 13, 20–22
US Department of Commerce, 22, 24, 133, 165, 167, 168
US Department of Defense, 36, 65
US Department of Justice, 109, 261, 266, 298, 299
US Financial Accounting Board, 190
USCD p-System, 239
USE, 33

Valley Committee, 37
Valley, George E., 37
Value-based pricing, 177
Vaughan Systems and Programming Limited, 78
Videogame magazines, 277, 280
Videogames, 162, 207, 269–288, 300, 301. *See also* Recreational software

372   Index

Videogames cartridges, 275, 276, 284
Virtua Fighter, 288
VisiCorp, 212, 214, 216, 227, 234, 244, 248, 249, 252, 258, 305. *See also* Personal Software
VisiCalc, 2, 7, 203, 210–216, 252, 257, 306
VisiOn, 248–252
VisiSeries, 214, 216, 258
VMS, 143
Voth, Ben, 24, 80
Vulcan, 220

Wagner, Frank, 32, 109
Walker, John, 24, 243–245, 265
Wang Laboratories, 128, 129, 159, 160, 203, 219, 254
Wang, An, 159
Wang, Anthony, 182
Wang, Charles, 180–182
Warner Communications, 275, 280
Warnock, John, 261
Waterfall model of software production, 47, 69
Watson, Thomas J. Jr., 38, 43, 110
Watson, Thomas J. Sr., 32
Welke, Larry, 63, 99
Western Electric, 37, 45
Westinghouse, 45, 115
Whirlwind, 37
Whitlow Computer Systems, 101, 102. See also SyncSort
Whitlow, Duane, 101, 102
Windowing system, 247–249
Wizard of War, 284
Wooldridge, Dean, 45
Word processing software, 7, 133, 139, 159–162, 183, 186, 203, 2136–219, 222, 226, 231, 234, 247, 251, 254–259, 263, 269, 281
Word. *See* Microsoft
WordMaster, 217
WordPerfect, 7, 11, 180, 228, 234, 236, 251, 254–259, 263, 265, 297
WordPro, 217
WordStar, 2, 7, 8, 203, 215–219, 226, 243, 254, 255, 257, 258

World Book, 292
World Wide Web, 162
Wozniak, Steve, 202
Wyly, Sam, 80

XD-1 computer. *See* Whirlwind
XENIX, 240, 242, 261
Xerox Palo Alto Research Center, 247, 261
Xerox Star, 247

Yankee Group, 26

```
HD            Campbell-Kelly,
9696.63       Martin.
.A2
C35           From airline
2003          reservations to
              Sonic the Hedgehog.
```